David Sang

Cambridge IGCSE Physics Coursebook

Completely Cambridge – Cambridge resources for Cambridge qualifications

Cambridge University Press works closely with University of Cambridge International Examinations (CIE) as parts of the University of Cambridge. We enable thousands of students to pass their CIE exams by providing comprehensive, high-quality, endorsed resources.

To find out more about University of Cambridge International Examinations visit www.cie.org.uk

To find out more about Cambridge University Press visit www.cambridge.org/cie

CAMBRIDGE UNIVERSITY PRESS
Cambridge, New York, Melbourne, Madrid, Cape Town,
Singapore, São Paulo, Delhi, Tokyo, Mexico City

Cambridge University Press
The Edinburgh Building, Cambridge CB2 8RU, UK

www.cambridge.org
Information on this title: www.cambridge.org/9780521757737

First published 2010
Reprinted 2010, 2011

Printed in Poland by Opolgraf

A catalogue record for this publication is available from the British Library

ISBN 978-0-521-75773-7 Paperback with CD-ROM for Windows ® and Mac ®

Cover image: Fingers weave through optical fibres.
© Adam Hart-Davis / Science Photo Library

Contents

Acknowledgements

The publishers would like to thank the following for permission to reproduce photographs. While every effort has been made, it has not always been possible to identify the sources of all the material used, or to trace all copyright holders. If any omissions are brought to our notice we will be happy to include the appropriate acknowledgement on reprinting.

p. vii(*l*) AJ Photo/SPL; p. vii(*r*), 1.2 Andrew Brookes, National Physical Laboratory/SPL; p. viii(*l*) Mark Garlick/SPL; p. viii(*r*) Volker Steger/SPL; p. 1, 6.11, p. 85, 15.6, p.169 NASA/SPL; 1.1, 5.8, 20.1, 21.4, 23.5 SPL; 1.12 GoGo Images Corporation/Alamy; 2.1 TRL Ltd/SPL; 2.3, 6.7, 13.22, 15.1 Nigel Luckhurst; 2.7 Brian F. Peterson/Corbis; 2.8 Alejandro Ernesto/EFE/Corbis; 3.1 Nelson Jeans/Corbis; 3.2 Scott Andrews/Science Faction/Corbis; 3.9 Birdlike Images Gregory Bajor/Alamy; 3.10 Eric Schremp/SPL; 3.12 Stockshot/Alamy; 4.1 Frans Lemmens/zefa/Corbis; 4.12 NCNA, Camera Press London; 5.2, 6.13 Gustoimages/SPL; 5.3 PhotoStock-Israel/Alamy; 5.9 Colin Cuthbert/SPL; 5.10 imagebroker/Alamy; 5.11 Alexis Rosenfeld/SPL; 6.1 Jeff Rotman/naturepl; 6.5 Visions of America LLC/Alamy; 6.6a European Space Agency/SPL; 6.10a, 9.12, 10.5, 10.8, 10.9a, 10.9b, 13.2, 13.6a, 13.8, 13.9, 13.14, 13.18, 13.20, 14.4a, 14.4b, 14.10a, 14.12a, 14.12b, 18.3a, 18.4, 18.11, 19.4a, 19.5a, 19.6a, 19.17a, 19.26, 20.12, 23.10 Andrew Lambert/SPL; 7.1 Jim Wileman/Alamy; 7.3 Liba Taylor/Corbis; 7.4 Ryan Pyle/Corbis; 7.5 Martin Land/SPL; 7.6 BNFL; 7.7 Worldwide Picture Library/Alamy; 8.7 Ace Stock Limited/Alamy; 9.1 Caro/Alamy; 9.9 81A Productions/Corbis; 10.1 Bubbles Photolibrary/Alamy; 10.2a CC Studio/SPL; 10.2b Paul Whitehall/SPL; 10.12 Matt Meadows/SPL; 11.1 Staffan Widstrand/naturepl; 11.2 Karl Ammann/naturepl; 11.7 Dr Gary Settler/SPL; 11.8, 12.8 sciencephotos/Alamy; 11.11 Edward Kinsman/SPL; 11.12 Justin Kaze zsixz/Alamy; p. 123 AFP/Getty Images; 12.1 Jill Douglas/Redferns/Getty Images; 12.2 John Eccles/Alamy; 12.3 Cardiff University; 12.4(*t*) Mode Images Limited/Alamy; 12.4(*b*) Niall McDiarmid/Alamy; 12.5 David Redfern/Redferns/Getty Images; 13.1 Royal Grenwich Observatory/SPL; 13.3 Hank Morgan/SPL; 13.4 Mark Bowler Scientific Images/www.markbowler.com; 13.16a, 23.18 TEK Image/SPL; 13.17 Dr Jeremy Burgess/SPL; 14.1 David Hosking/FLPA; 14.2 Rick Strange/Alamy; 14.11a Berenice Abbott/SPL; 14.14 John Foster/SPL; 15.2, p. 239 David Parker/SPL; 15.4 CCI Archives/SPL; 15.9 David R. Frazier/SPL; 16.1 The London Art Archive/Alamy; 16.7 Cordelia Molloy/SPL; 16.10 Jeremy Walker/SPL; 17.1 Photo Researchers/SPL; 18.1 Maximilian Stock Ltd/Alamy; 18.2 Martin Dorhn/SPL; 19.1 Rosenfeld Images Ltd/SPL; 19.3, 19.27a, 23.15a Leslie Garland Picture Library/Alamy; 19.28 Sheila Terry/SPL; 21.1a, 21.1b Adam Hart-Davis/SPL; 21.3 Alex Bartel/SPL; 21.8 Ed Michaels/SPL; 21.9 D Burke/Alamy; 22.1 David Simson; 22.2 IBM/SPL; 23.1 Radiation Protection Division/Health Protection Agency/SPL; 23.2 US Air Force/SPL; 23.3 Yoav Levy/Phototake Science/Photolibrary; 23.6 Pascal Goetgheluck/SPL; 23.16 National Radiation Protection Board; 23.19 P. Deliss/Godong/Corbis

b = bottom, *l* = left, *r* = right, *t* = top, SPL = Science Photo Library

Introduction

Studying physics

Why study physics? Some people study physics for the simple reason that they find it interesting. Physicists study matter, energy and their interactions. They might be interested in the tiniest sub-atomic particles, or the nature of the Universe itself. (Some even hope to discover whether there are more universes than just the one we live in!)

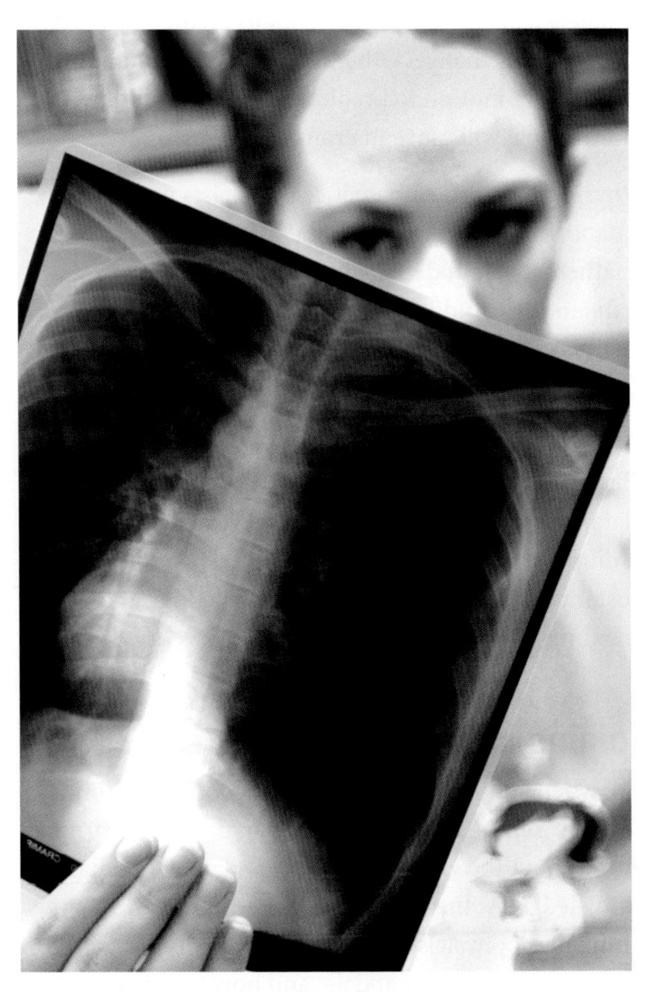

When they were first discovered, X-rays were sometimes treated as an entertaining novelty. Today, they can give detailed views of a patient's bones and organs.

On a more human scale, physicists study materials to try to predict and control their properties. They study the interactions of radiation with matter, including the biological materials we are made of.

Some people don't want to study physics simply for its own sake. They want to know how it can be used, perhaps in an engineering project, or for medical purposes. Depending on how our knowledge is applied, it can make the world a better place.

Some people study physics as part of their course because they want to become some other type of scientist – perhaps a chemist, biologist or geologist. These branches of science draw a great deal on ideas from physics, and physics may draw on them.

Thinking physics

How do physicists think? One of the characteristics of physicists is that they try to simplify problems – reduce them to their basics – and then solve them by applying

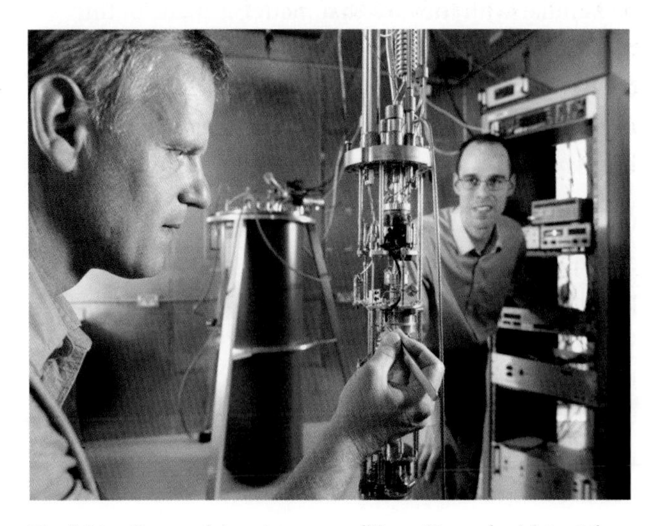

Physicists often work in extreme conditions. Here, physicists at the UK's National Physical Laboratory prepare a dilution refrigerator, capable of cooling materials down almost to absolute zero, the lowest possible temperature.

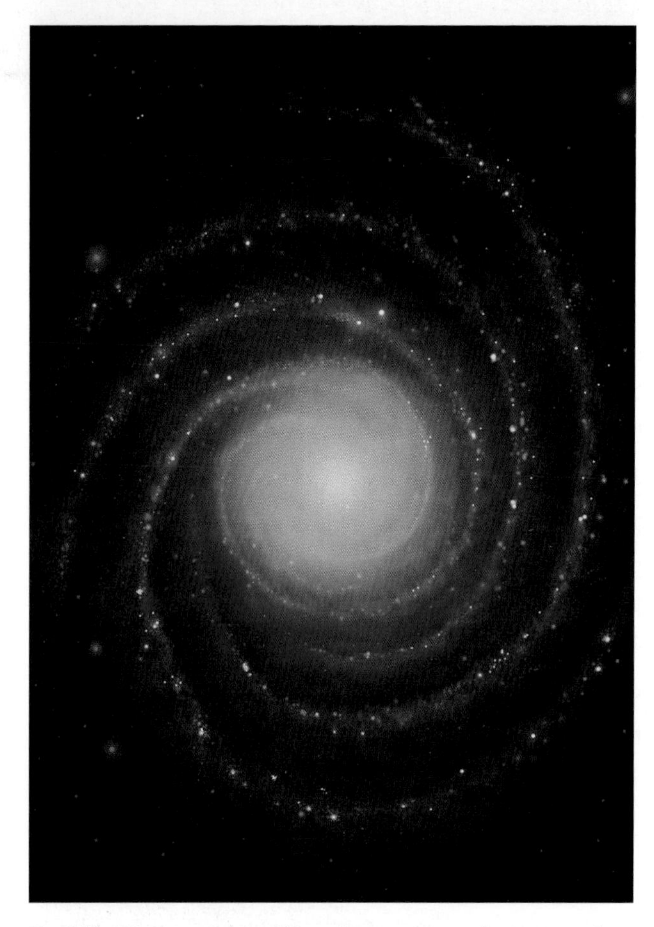

The Milky Way, our Galaxy. Although we can never hope to see it from this angle, careful measurements of the positions of millions of stars has allowed astronomers to produce this computer-generated view.

some very fundamental ideas. For example, you will be familiar with the idea that matter is made of tiny particles that attract and repel each other and move about. This is a very powerful idea, which has helped us to understand the behaviour of matter, how sound travels, how electricity flows, and so on.

Once a fundamental idea is established, physicists look around for other areas where it might help to solve problems. One of the surprises of 20th-century physics was that, once physicists had begun to understand the fundamental particles of which atoms are made, they realised that this helped to explain the earliest moments in the history of the Universe, at the time of the Big Bang.

The more you study physics, the more you will come to realise how the ideas join up. Also, physics is still expanding. Many physicists work in economics and

The Internet, used by millions around the world. Originally invented by a physicist, Tim Berners-Lee, the Internet is used by physicists to link thousands of computers in different countries to form supercomputers capable of handling vast amounts of data.

finance, using ideas from physics to predict how markets will change. Others use their understanding of particles in motion to predict how traffic will flow, or how people will move in crowded spaces.

Physics relies on mathematics. Physicists measure quantities and process their data. They invent mathematical models – equations and so on – to explain their findings. (In fact, a great deal of mathematics was invented by physicists, to help them to understand their experimental results.)

Computers have made a big difference in physics. Because a computer can 'crunch' vast quantities of data, whole new fields of physics have opened up. Computers can analyse data from telescopes, control distant spacecraft and predict the behaviour of billions of atoms in a solid material.

Joining in

So, when you study physics, you are doing two things. You are joining in with a big human project – learning more about the world around us, and applying that knowledge. At the same time, you will be learning to think like a physicist – how to apply some basic ideas, how to look critically at data, and how to recognise underlying patterns. Whatever your aim, these ideas can stay with you throughout your life.

Block 1

General physics

In your studies of science, you will already have come across many of the fundamental ideas of physics. In this block, you will develop a better understanding of two powerful ideas: (i) the idea of force and (ii) the idea of energy.

Where do ideas in physics come from? Partly, they come from observation. When Galileo looked at the planets through his telescope, he observed the changing face of Venus. He also saw that Jupiter had moons. Galileo's observations formed the basis of a new, more scientific, astronomy.

Ideas also come from thought. Newton (who was born in the year that Galileo died) is famous for his ideas about gravity. He realised that the force that pulls an apple to the ground is the same force that keeps the Moon in its orbit around the Earth. His ideas about forces are explored in this block.

You have probably studied some basic ideas about energy. However, Newton never knew about energy. This was an idea that was not developed until more than a century after his death, so you are already one step ahead of him!

In 1992, a spacecraft named Galileo was sent to photograph Jupiter and its moons. On its way, it looked back to take this photograph of the Earth and the Moon.

1 Making measurements

Core Making measurements of length, volume and time

E **Extension** Increasing the precision of measurements of length and time

Core Determining the densities of solids and liquids

How measurement improves

Galileo Galilei is often thought of as the father of modern science. He did a lot to revolutionise how we think of the world around us, and in particular how we make measurements. In 1582, Galileo was a medical student in Pisa. During a service in the cathedral there, he observed a lamp swinging (Figure **1.1**). Galileo noticed that the time it took for each swing was the same, whether the lamp was swinging through a large or a small angle. He realised that a swinging weight – a pendulum – could be used as a timing device. He went on to use it to measure a person's pulse rate, and he also designed a clock regulated by a swinging pendulum.

In Galileo's day, many measurements were based on the human body – for example, the foot and the yard (a pace). Weights were measured in units based on familiar objects such as cereal grains. These 'natural' units are inevitably variable – one person's foot is longer than another's – so efforts were made to standardise them. (It is said that the English 'yard' was defined as the distance from the tip of King Henry I's nose to the end of his outstretched arm.)

Today, we live in a globalised economy. We cannot rely on monarchs to be our standards of measurement. Instead, there are international agreements on the basic units of measurement. For example, the metre is defined as follows:

The metre is the distance travelled by light in $\frac{1}{299\,792\,458}$ second in a vacuum.

Laboratories around the world are set up to check that measuring devices match this standard.

Figure 1.1 An imaginative reconstruction of Galileo with the lamp that he saw swinging in Pisa Cathedral in 1582.

Figure **1.2** shows a new atomic clock, undergoing development at the UK's National Physical Laboratory. Clocks like this are accurate to 1 part in 10^{14}, or one-billionth of a second in a day. You might think that this is far more precise than we could ever need.

Figure 1.2 Professor Patrick Gill of the National Physical Laboratory is devising an atomic clock that will be one-thousand times more accurate than previous types.

In fact, you may already rely on ultra-precise time measurements if you use a GPS (Global Positioning Satellite) system. These systems detect satellite signals, and they work out your position to within a fraction of a metre. Light travels one metre in about $\frac{1}{300\,000\,000}$ second, or 0.000 000 003 second. So, if you are one metre further away from the satellite, the signal will arrive this tiny fraction of a second later. Hence the electronic circuits of the GPS device must measure the time at which the signal arrives to this degree of accuracy.

1.1 Measuring length and volume

In physics, we make measurements of many different lengths – for example, the length of a piece of wire, the height of liquid in a tube, the distance moved by an object, the diameter of a planet or the radius of its orbit. In the laboratory, lengths are often measured using a rule (such as a metre rule).

Measuring lengths with a rule is a familiar task. But when you use a rule, it is worth thinking about the task and just how reliable your measurements may be. Consider measuring the length of a piece of wire (Figure 1.3).

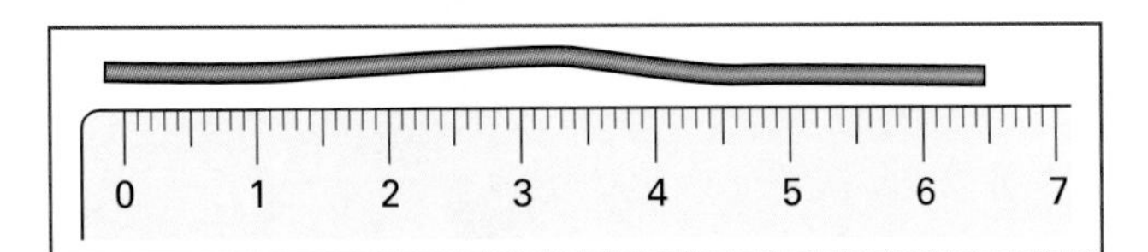

Figure 1.3 Simple measurements – for example, finding the length of a wire – still require careful technique.

- The wire must be straight, and laid closely alongside the rule. (This may be tricky with a bent piece of wire.)
- Look at the ends of the wire. Are they cut neatly, or are they ragged? Is it difficult to judge where the wire begins and ends?
- Look at the markings on the rule. They are probably 1 mm apart, but they may be quite wide. Line one end of the wire up against the zero of the scale. Because of the width of the mark, this may be awkward to judge.
- Look at the other end of the wire and read the scale. Again, this may be tricky to judge.

Now you have a measurement, with an idea of how precise it is. You can probably determine the length of the wire to within a millimetre. But there is something else to think about – the rule itself. How sure can you be that it is correctly calibrated? Are the marks at the ends of a metre rule separated by exactly one metre? Any error in this will lead to an inaccuracy (probably small) in your result.

The point here is to recognise that it is always important to think critically about the measurements you make, however straightforward they may seem. You have to consider the method you use, as well as the instrument (in this case, the rule).

More measurement techniques

If you have to measure a small length, such as the thickness of a wire, it may be better to measure several thicknesses and then calculate the average. You can use

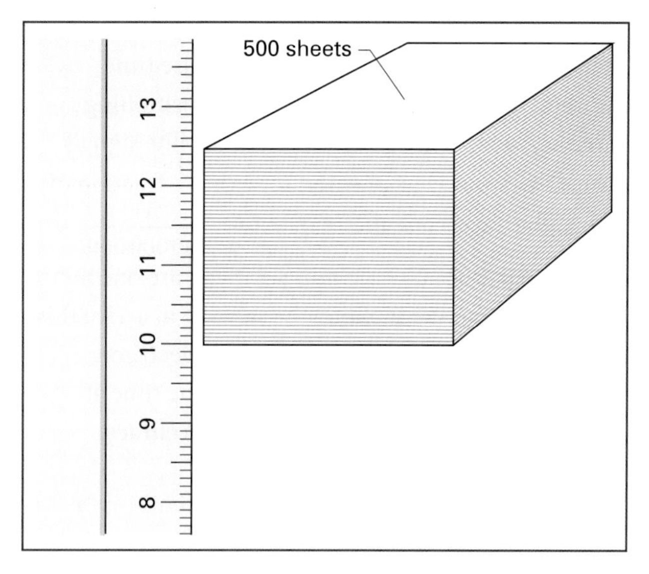

Figure 1.4 Making multiple measurements.

the same approach when measuring something very thin, such as a sheet of paper. Take a stack of 500 sheets and measure its thickness with a rule (Figure **1.4**). Then divide by 500 to find the thickness of one sheet.

For some measurements of length, such as curved lines, it can help to lay a thread along the line. Mark the thread at either end of the line and then lay it along a rule to find the length. This technique can also be used for measuring the circumference of a cylindrical object such as a wooden rod or a measuring cylinder.

Measuring volumes

There are two approaches to measuring volumes, depending on whether or not the shape is regular.

For a regularly shaped object, such as a rectangular block, measure the lengths of the three different sides and multiply them together. For objects of other regular shapes, such as spheres or cylinders, you may have to make one or two measurements and then look up the formula for the volume.

For liquids, measuring cylinders can be used. (Recall that these are designed so that you look at the scale **horizontally**, not at an oblique angle, and read the level of the **bottom** of the meniscus.) Think carefully about the choice of cylinder. A one-litre cylinder is unlikely to be suitable for measuring a small volume such as $5\,cm^3$. You will get a more accurate answer using a $10\,cm^3$ cylinder.

Units of length and volume

In physics, we generally use SI units (this is short for *Le Système International d'Unités* or The International System of Units). The SI unit of length is the metre (m). Table **1.1** shows some alternative units of length, together with some units of volume.

Quantity	Units
length	metre (m) 1 centimetre (cm) = 0.01 m 1 millimetre (mm) = 0.001 m 1 micrometre (μm) = 0.000 001 m 1 kilometre (km) = 1000 m
volume	cubic metre (m^3) 1 cubic centimetre (cm^3) = 0.000 001 m^3 1 cubic decimetre (dm^3) = 0.001 m^3 1 litre (l) = 0.001 m^3 1 litre (l) = 1 cubic decimetre (dm^3) 1 millilitre (ml) = 1 cubic centimetre (cm^3)

Table 1.1 Some units of length and volume in the SI system.

QUESTIONS

1 A rectangular block of wood has dimensions 240 mm × 20.5 cm × 0.040 m. Calculate its volume in cm^3.

2 Ten identical lengths of wire are laid closely side-by-side. Their combined width is measured and found to be 14.2 mm. Calculate:

a the radius of a single wire

b the volume in mm^3 of a single wire if its length is 10.0 cm. (Volume of a cylinder = $\pi r^2 h$, where r = radius and h = height.)

1.2 Improving precision in measurements

A rule is a simple measuring instrument, with many uses. However, there are instruments designed to give greater precision in measurements. Here we will look at how to use two of these.

Vernier callipers

The callipers have two scales, the main scale and the vernier scale. Together, these scales give a measurement of the distance between the two inner faces of the jaws (Figure **1.5**).

The method is as follows:

- Close the callipers so that the jaws touch lightly but firmly on the sides of the object being measured.
- Look at the zero on the vernier scale. Read the main scale, just to the left of the zero. This tells you the length in millimetres.
- Now look at the vernier scale. Find the point where one of its markings is **exactly** aligned with one of the markings on the main scale. Read the value on the vernier scale. This tells you the fraction of a millimetre that you must add to the main scale reading.

For the example in Figure **1.5**:

thickness of rod
= main scale reading + vernier reading
= 35 mm + 0.7 mm
= 35.7 mm

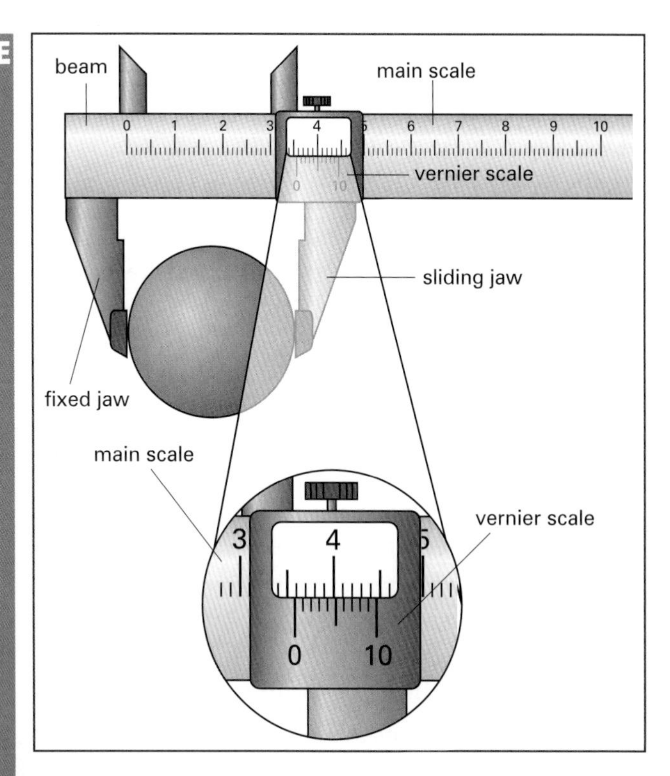

Figure 1.5 Using vernier callipers.

Micrometer screw gauge

Again, this has two scales. The main scale is on the shaft, and the fractional scale is on the rotating barrel. The fractional scale has 50 divisions, so that one complete turn represents 0.50 mm (Figure **1.6**).

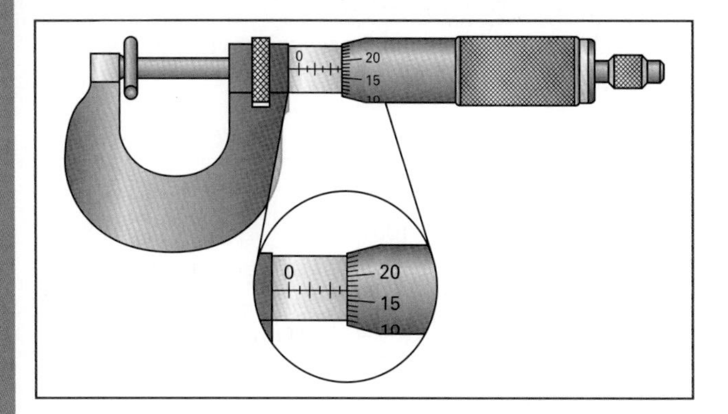

Figure 1.6 Using a micrometer screw gauge.

The method is as follows:

- Turn the barrel until the jaws just tighten on the object. Using the friction clutch ensures just the right pressure.

- Read the main scale to the nearest 0.5 mm.
- Read the additional fraction of a millimetre from the fractional scale.

For the example in Figure **1.6**:

thickness of rod
= main scale reading + fractional scale reading
= 2.5 mm + 0.17 mm
= 2.67 mm

Activity 1.1 Precise measurements

Practise reading the scales of vernier callipers and micrometer screw gauges.

Measuring volume by displacement

It is not just instruments that improve our measurements. Techniques also can be devised to help. Here is a simple example, to measure the volume of an irregularly shaped object:

- Select a measuring cylinder that is somewhat (three or four times) larger than the object. Partially fill it with water (Figure **1.7**), enough to cover the object. Note the volume of the water.
- Immerse the object in the water. The level of water in the cylinder will increase. The increase in its volume is equal to the volume of the object.

This technique is known as measuring volume by displacement.

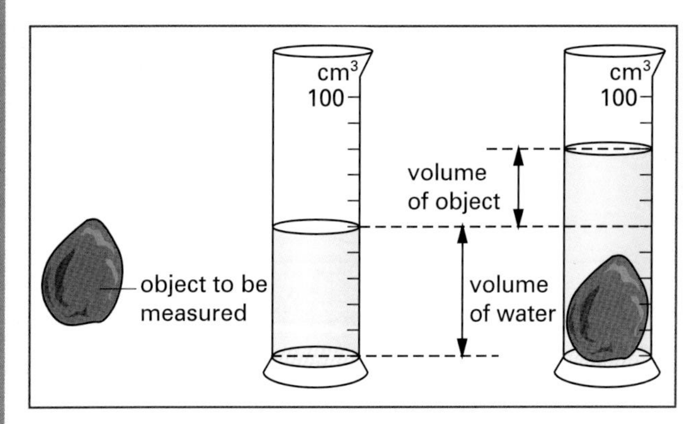

Figure 1.7 Measuring volume by displacement.

QUESTIONS

3 State the measurements shown in Figure **1.8** on the scale of

a the vernier callipers

b the micrometer screw gauge.

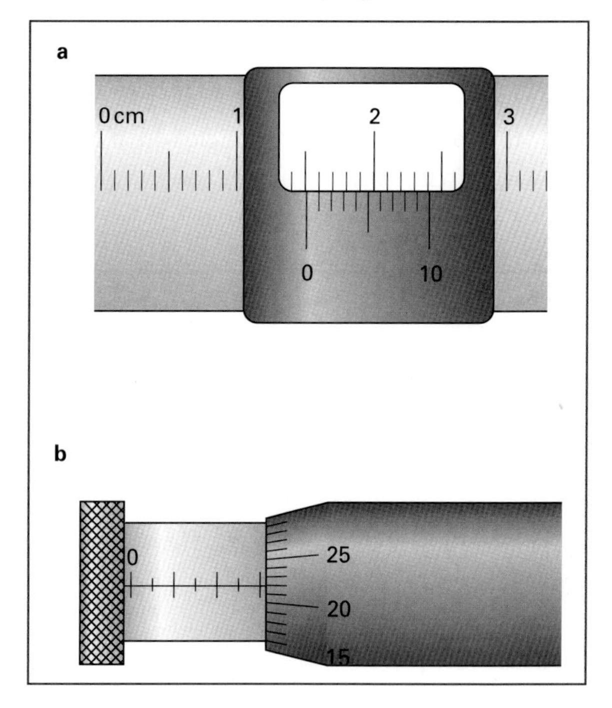

Figure 1.8 For Question **3**.

4 Figure **1.9** shows how the volume of a piece of wood (which floats in water) can be measured. Write a brief paragraph to describe the procedure. State the volume of the wood.

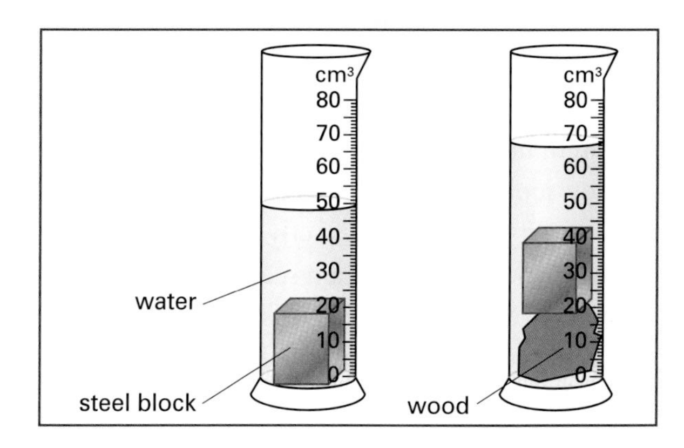

Figure 1.9 For Question **4**.

1.3 Density

Our eyes can deceive us. When we look at an object, we can judge its volume. However, we can only guess its mass. We may guess incorrectly, because we misjudge the density. You may offer to carry someone's bag, only to discover that it contains heavy books. A large box of chocolates may have a mass of only 200 g, a great disappointment!

The **mass** of an object is the amount of matter it is made of. Mass is measured in kilograms. But **density** is a property of a material. It tells us how concentrated its mass is. (There is more about the meaning of **mass** and how it differs from **weight** in Chapter **3**.)

In everyday speech, we might say that lead is heavier than wood. We mean that, given equal volumes of lead and wood, the lead is heavier. In scientific terms, the density of lead is greater than the density of wood. So we define density as follows:

$$\text{density} = \frac{\text{mass}}{\text{volume}} \qquad D = \frac{M}{V}$$

The SI unit of density is kg/m^3 (kilograms per cubic metre). You may come across other units, as shown in Table **1.2**. A useful value to remember is the density of water (Table **1.3**):

density of water = $1000\,kg/m^3$

Unit of mass	Unit of volume	Unit of density	Density of water
kilogram, kg	cubic metre, m^3	kilograms per cubic metre	$1000\,kg/m^3$
kilogram, kg	litre, l	kilograms per litre	1.0 kg/litre
kilogram, kg	cubic decimetre, dm^3	kilograms per cubic decimetre	$1.0\,kg/dm^3$
gram, g	cubic centimetre, cm^3	grams per cubic centimetre	$1.0\,g/cm^3$

Table 1.2 Units of density.

	Material	Density / kg/m^3
gases	air	1.29
	hydrogen	0.09
	helium	0.18
	carbon dioxide	1.98
liquids	water	1 000
	alcohol (ethanol)	790
	mercury	13 600

	Material	Density / kg/m^3
solids	ice	920
	wood	400–1 200
	polythene	910–970
	glass	2 500–4 200
	steel	7 500–8 100
	lead	11 340
	silver	10 500
	gold	19 300

Table 1.3 Densities of some substances. For gases, these are given at a temperature of 0 °C and a pressure of 1.0×10^5 Pa.

Values of density

Some values of density are shown in Table 1.3. Here are some points to note:

- Gases have much lower densities than solids or liquids.
- Density is the key to floating. Ice is less dense than water. This explains why icebergs float in the sea, rather than sinking to the bottom.
- Many materials have a range of densities. Some types of wood, for example, are less dense than water and will float. Others (such as mahogany) are more dense and sink. The density depends on the composition.
- Gold is denser than silver. Pure gold is a soft metal, so jewellers add silver to make it harder. The amount of silver added can be judged by measuring the density.
- It is useful to remember that the density of water is 1000 kg/m^3, 1.0 g/cm^3 or 1 kg/litre.

Calculating density

To calculate the density of a material, we need to know the mass and volume of a sample of the material.

Worked example 1

A sample of ethanol has a volume of 240 cm^3. Its mass is found to be 190.0 g. What is the density of ethanol?

Step 1: Write down what you know and what you want to know.

mass $M = 190.0$ g
volume $V = 240$ cm^3
density $D = ?$

Step 2: Write down the equation for density, substitute values and calculate D.

$$D = \frac{M}{V} = \frac{190}{240} = 0.79\ \text{g/cm}^3$$

Measuring density

The easiest way to determine the density of a substance is to find the mass and volume of a sample of the substance.

For a solid with a regular shape, find its volume by measurement (see page 4). Find its mass using a balance. Then calculate the density.

Figure 1.10 shows one way to find the density of a liquid. Place a measuring cylinder on a balance. Set the balance to zero. Now pour liquid into the cylinder. Read the volume from the scale on the cylinder. The balance shows the mass.

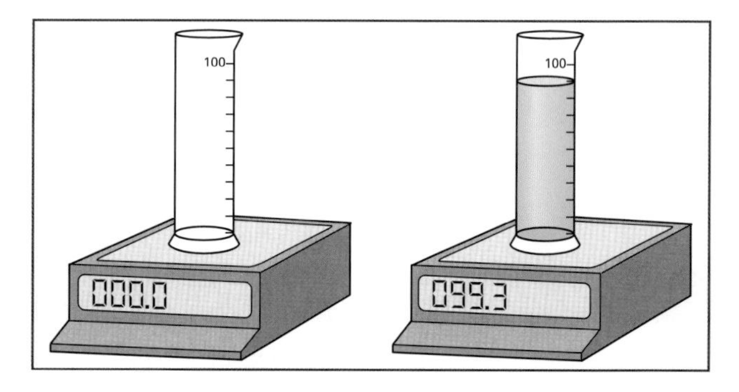

Figure 1.10 Measuring the density of a liquid.

Activity 1.2 Measuring density

Make measurements to find the densities of some blocks of different materials.

QUESTIONS

5 Calculate the density of mercury if 500 cm^3 has a mass of 6.60 kg. Give your answer in g/cm^3.

6 A steel block has mass 40 g. It is in the form of a cube. Each edge of the cube is 1.74 cm long. Calculate the density of the steel.

7 A student measures the density of a piece of steel. She uses the method of displacement to find its volume. Figure **1.11** shows her measurements. Calculate the volume of the steel and its density.

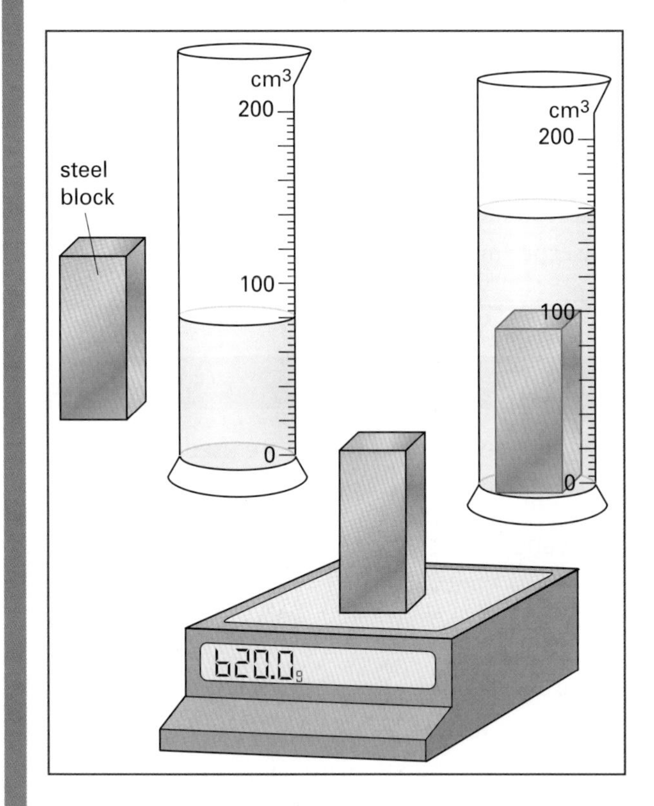

Figure 1.11 For Question 7.

1.4 Measuring time

The athletics coach in Figure **1.12** is using her stopwatch to time a sprinter. For a sprinter, a fraction of a second (perhaps just 0.01 s) can make all the difference between winning and coming second or third. It is different in a marathon, where the race lasts for more than two hours and the runners are timed to the nearest second.

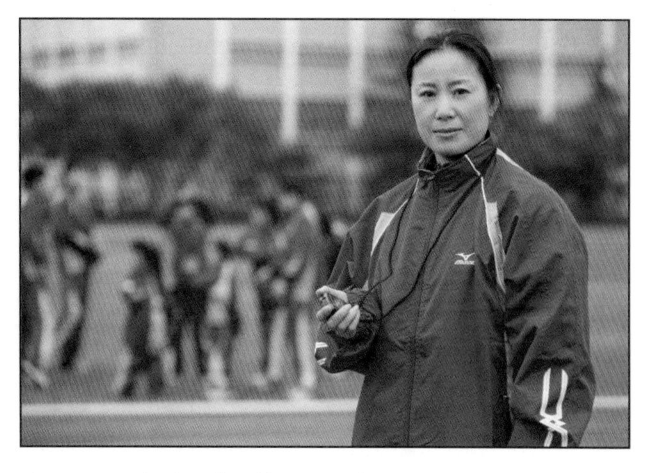

Figure 1.12 The female athletics coach uses a stopwatch to time a sprinter, who can then learn whether she has improved.

In the lab, you might need to record the temperature of a container of water every minute, or find the time for which an electric current is flowing. For measurements like these, stopclocks and stopwatches can be used.

When studying motion, you may need to measure the time taken for a rapidly moving object to move between two points. In this case, you might use a device called a light gate connected to an electronic timer. This is similar to the way in which runners are timed in major athletics events. An electronic timer starts when the marshal's gun is fired, and stops as the runner crosses the finishing line.

There is more about how to use electronic timing instruments in Chapter **2**.

Measuring short intervals of time

The time for one swing of a pendulum (from left to right and back again) is called its **period**. A single period is usually too short a time to measure accurately. However, because a pendulum swings at a steady rate, you can use a stopwatch to measure the time for a large number of swings (perhaps 20 or 50), and calculate the average time per swing. Any inaccuracy in the time at which the stopwatch is started and stopped will be much less significant if you measure the total time for a large number of swings.

E

Activity 1.3 The period of a pendulum

Figure **1.13** shows a typical lab pendulum. Devise a means of testing Galileo's idea that the period of a pendulum does not depend on the size of its swing.

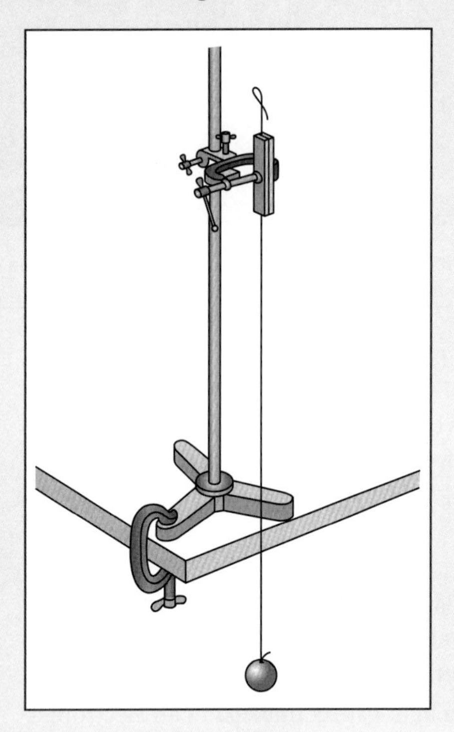

Figure 1.13 A simple pendulum.

E

QUESTIONS

8 Many television sets show 25 images, called 'frames', each second. What is the time interval between one frame and the next?

9 A pendulum is timed, first for 20 swings and then for 50 swings:

time for 20 swings = 17.4 s
time for 50 swings = 43.2 s

Calculate the average time per swing in each case. The answers are slightly different. Can you suggest any experimental reasons for this?

Summary

Rules and measuring cylinders are used to measure length and volume.

Clocks and electronic timers are used to measure intervals of time.

$$\text{Density} = \frac{\text{mass}}{\text{volume}}$$

E

Measurements of small quantities can be improved using special instruments (for example, vernier callipers and micrometer screw gauge) or by making multiple measurements.

End-of-chapter questions

1.1 An ice cube has the dimensions shown in Figure **1.14**. Its mass is 340 g. Calculate:

a its volume [3]

b its density. [3]

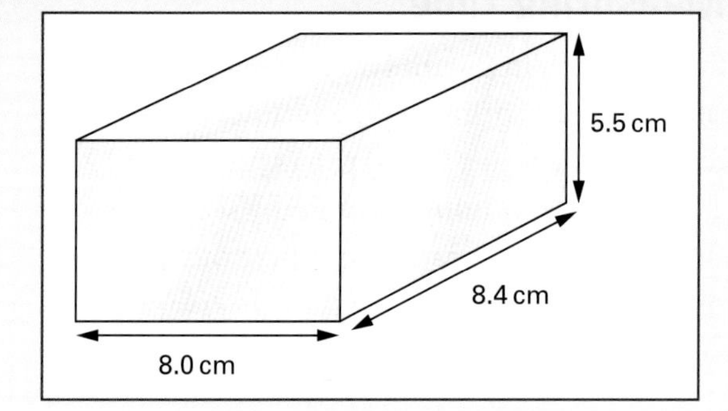

Figure 1.14 A block of ice – for Question **1.1**.

1.2 A student is collecting water as it runs into a measuring cylinder. She uses a clock to measure the time interval between measurements. Figure **1.15** shows the level of water in the cylinder at two times, together with the clock at these times. Calculate:

a the volume of water collected between these two times [2]

b the time interval. [2]

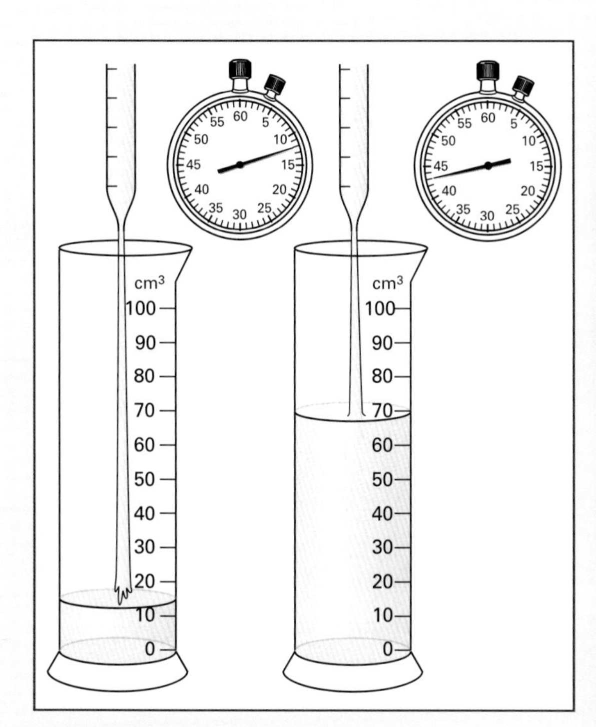

Figure 1.15 For Question **1.2**.

1.3 A student is measuring the density of a liquid. He places a measuring cylinder on a balance and records its mass. He then pours liquid into the cylinder and records the new reading on the balance. He also records the volume of the liquid.

Mass of empty cylinder = 147 g
Mass of cylinder + liquid = 203 g
Volume of liquid = 59 cm^3

Using the results shown above, calculate the density of the liquid. [5]

1.4 The inside of a sports hall measures 80 m long by 40 m wide by 15 m high. The air in it has a density of 1.3 kg/m^3 when it is cool.

a Calculate the volume of the air in the sports hall, in m^3. [3]

b Calculate the mass of the air. State the equation you are using. [3]

E

1.5 A geologist needs to measure the density of an irregularly shaped pebble.

a Describe how she can find its volume by the method of displacement. [4]

b What other measurement must she make if she is to find its density? [1]

1.6 An IGCSE student thinks it may be possible to identify different rocks (A, B and C) by measuring their densities. She uses an electronic balance to measure the mass of each sample and uses the 'displacement method' to determine the volume of each sample. Figure **1.16** shows her displacement results for sample A.

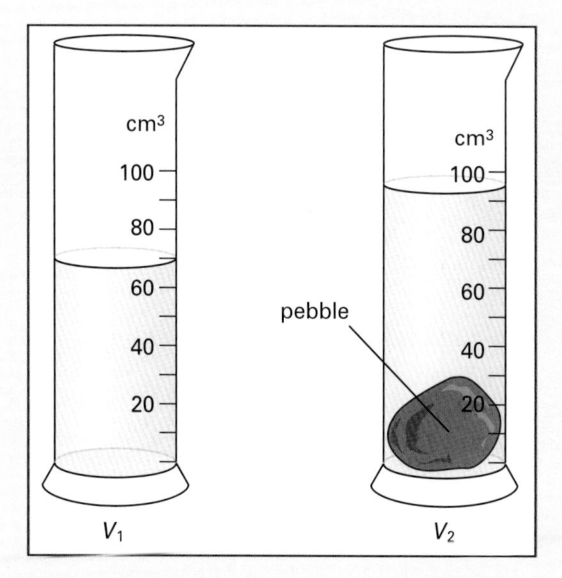

Figure 1.16 For Question **1.6**.

Sample	m / g	…… /……	…… /……	V /……	Density /……
B	144	80	44	……	……
C	166	124	71	……	……

Table 1.4 For Question **1.6**.

a State the volume shown in each measuring cylinder. [2]

b Calculate the volume V of the rock sample A. [2]

c Sample A has a mass of 102 g. Calculate its density. [3]

Table **1.4** shows the student's readings for samples B and C.

d Copy and complete the table by inserting the appropriate column headings and units, and calculating the densities. [12]

2 Describing motion

Core Interpreting distance against time and speed against time graphs
Core Calculating speed and distance
E Extension Calculating acceleration

Figure 2.1 Traffic engineers use sophisticated cameras and computers to monitor traffic. Understanding how drivers behave is important not only for safety, but also to improve the flow of traffic.

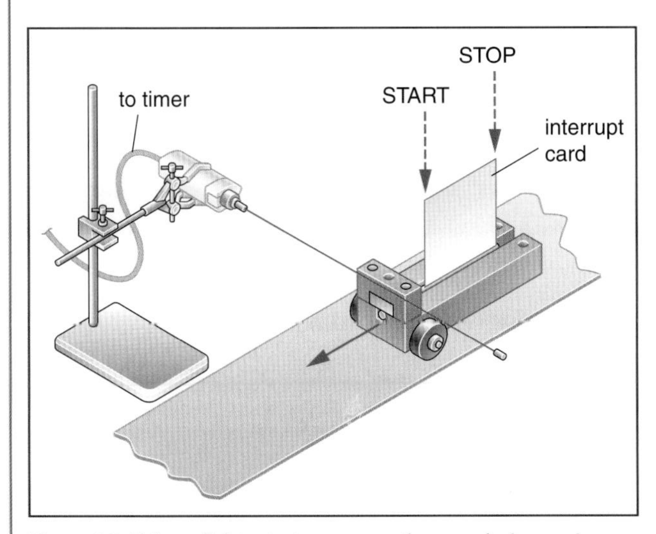

Figure 2.2 Using a light gate to measure the speed of a moving trolley in the laboratory.

Measuring speed

If you travel on a major highway or through a large city, the chances are that someone is watching you (see Figure 2.1). Cameras on the verge and on overhead gantries keep an eye on traffic as it moves along. Some cameras are there to monitor the flow, so that traffic managers can take action when blockages develop, or when accidents occur. Other cameras are equipped with sensors to spot speeding motorists, or those who break the law at traffic lights. In some busy places, traffic police may observe the roads from helicopters.

Drivers should know how fast they are moving – they have a speedometer to tell them their speed at any instant in time. Traffic police can use a radar speed 'gun' to give them an instant readout of another vehicle's speed (such 'guns' use the Doppler effect to measure a car's speed). Alternatively, they may time a car between two fixed points on the road. Knowing the distance between the two points, they can calculate the car's speed.

In the laboratory, the speed of a moving trolley can be measured using a **light gate** connected to an electronic timer (see Figure 2.2). A piece of card, called an **interrupt card**, is mounted on the trolley. The light gate has a beam of (invisible) infrared radiation. As the trolley passes through the gate, the front edge of the card breaks the beam and starts the timer. When the trailing edge passes the gate, the beam is no longer broken and the timer stops. The faster the trolley is moving, the shorter the time for which the beam is broken. Given the length of the card, the trolley's speed can be calculated.

2.1 Understanding speed

In this chapter, we will look at ideas of motion and speed. In Chapter 3, we will look at how physicists came to understand the forces involved in motion, and how to control them to make our everyday travel possible.

Distance, time and speed

As we have seen, there is more than one way to determine the speed of a moving car or aircraft. Several methods rely on making two measurements:

- the **distance travelled** between two points;
- the **time taken** to travel between these two points.

Then we can work out the average speed between the two points:

$$\textbf{average speed} = \frac{\text{distance travelled}}{\text{time taken}}$$

$$\text{or} \qquad \textbf{speed} = \frac{\text{distance}}{\text{time}}$$

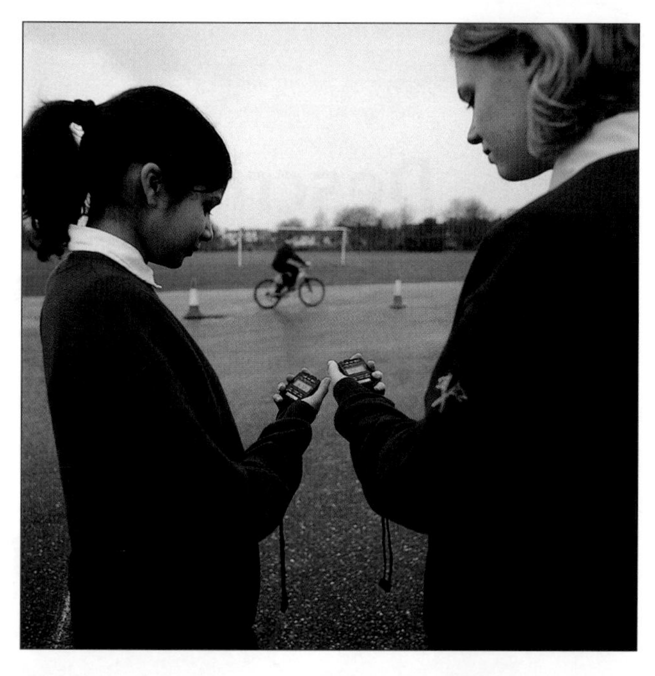

Figure 2.3 Timing a cyclist over a fixed distance. Using a stopwatch involves making judgements as to when the cyclist passes the starting and finishing lines. This can introduce an error into the measurements. An automatic timing system might be better.

Notice that this equation tells us the vehicle's average speed. We cannot say whether it was travelling at a steady speed, or if its speed was changing. For example, you could use a stopwatch to time a friend cycling over a fixed distance – say, 100 m (see Figure 2.3). Dividing distance by time would tell you their average speed, but they might have been speeding up or slowing down along the way.

Table 2.1 shows the different units that may be used in calculations of speed. SI units are the 'standard' units used in physics (SI is short for *Système International* or International System). In practice, many other units are used. In US space programmes, heights above the Earth are often given in feet, while the spacecraft's speed is given in knots (nautical miles per hour). These awkward units did not prevent them from reaching the Moon!

Quantity	SI unit	Other units	
distance	metre, m	kilometre, km	miles
time	second, s	hour, h	hour, h
speed	metres per second, m/s, $m\,s^{-1}$	kilometres per hour, km/h	miles per hour, mph

Table 2.1 Quantities, symbols and units in measurements of speed.

Worked example 1

A cyclist completed a 1500 m stage of a race in 37.5 s. What was her average speed?

Step 1: Start by writing down what you know, and what you want to know.

distance = 1500 m
time = 37.5 s
speed = ?

Step 2: Now write down the equation.

$$\text{speed} = \frac{\text{distance}}{\text{time}}$$

Step 3: Substitute the values of the quantities on the right-hand side.

$$\text{speed} = \frac{1500\,\text{m}}{37.5\,\text{s}}$$

Step 4: Calculate the answer.

speed = 40 m/s

So the cyclist's average speed was 40 m/s.

QUESTIONS

1 If you measured the distance travelled by a snail in inches and the time it took in minutes, what would be the units of its speed?

2 Which of the following could not be a unit of speed?
km/h, s/m, mph, m/s, m s.

3 Table **2.2** shows information about three cars travelling on a motorway.
- **a** Which car is moving fastest?
- **b** Which car is moving slowest?

Vehicle	Distance travelled / km	Time taken / minutes
car A	80	50
car B	72	50
car C	85	50

Table 2.2 For Question **3**.

Rearranging the equation

The equation

$$\text{speed} = \frac{\text{distance}}{\text{time}}$$

allows us to calculate speed from measurements of distance and time. We can rearrange the equation to allow us to calculate distance or time.

For example, a railway signalman might know how fast a train is moving, and need to be able to predict where it will have reached after a certain length of time:

$$\text{distance} = \text{speed} \times \text{time}$$

Similarly, the crew of an aircraft might want to know how long it will take for their aircraft to travel between two points on its flight path:

$$\text{time} = \frac{\text{distance}}{\text{speed}}$$

Worked example 2

A spacecraft is orbiting the Earth at a steady speed of 8 km/s (see Figure **2.4**). How long will it take to complete a single orbit, a distance of 40 000 km?

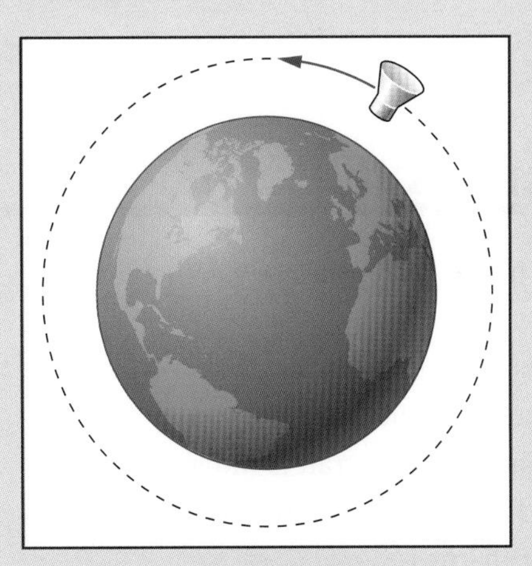

Figure 2.4 A spacecraft orbiting Earth.

Step 1: Start by writing down what you know, and what you want to know.

speed = 8 km/s
distance = 40 000 km
time = ?

Step 2: Choose the appropriate equation, with the unknown quantity 'time' as the subject (on the left-hand side).

$$\text{time} = \frac{\text{distance}}{\text{speed}}$$

Step 3: Substitute values – it can help to include units.

$$\text{time} = \frac{40\,000\,\text{km}}{8\,\text{km/s}}$$

Step 4: Perform the calculation.

$$\text{time} = 5000\,\text{s}$$

This is about 83 minutes. So the spacecraft takes 83 minutes to orbit the Earth once.

Worked example 2 illustrates the importance of keeping an eye on units. Because speed is in km/s and distance is in km, we do not need to convert to m/s and m. We would get the same answer if we did the conversion:

$$\text{time} = \frac{40\,000\,000\,\text{m}}{8000\,\text{m/s}}$$

$$= 5000\,\text{s}$$

QUESTIONS

4 An aircraft travels 1000 m in 4.0 s. What is its speed?

5 A car travels 150 km in 2 hours. What is its speed? (Show the correct units.)

6 An interplanetary spacecraft is moving at 20 000 m/s. How far will it travel in one day? (Give your answer in km.)

7 How long will it take a coach travelling at 90 km/h to travel 300 km along a highway?

2.2 Distance against time graphs

You can describe how something moves in words: 'The coach pulled away from the bus stop. It travelled at a steady speed along the main road, heading out of town. After five minutes, it reached the highway, where it was able to speed up. After ten minutes, it was forced to stop because of congestion.'

We can show the same information in the form of a distance against time graph, as shown in Figure **2.5**. This graph is in three sections, corresponding to the three sections of the coach's journey:

A The graph slopes up gently, showing that the coach was travelling at a slow speed.

B The graph becomes steeper. The distance of the coach from its starting point is increasing more rapidly. It is moving faster.

C The graph is flat (horizontal). The distance of the coach from its starting point is not changing. It is stationary.

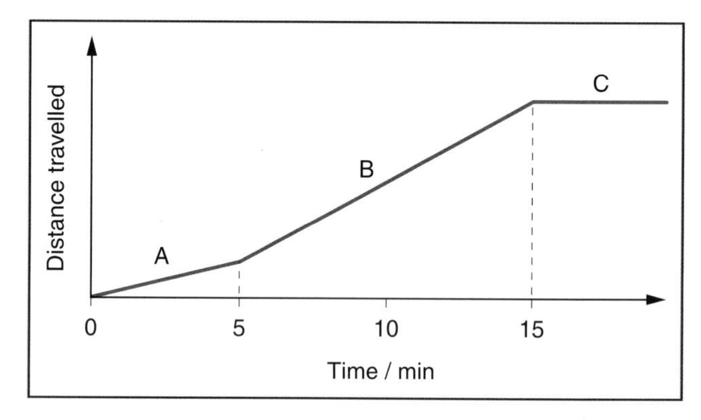

Figure 2.5 A graph to represent the motion of a coach, as described in the text. The slope of the graph tells us about the coach's speed. The steepest section (B) corresponds to the greatest speed. The horizontal section (C) shows that the coach was stationary.

The slope of the distance against time graph tells us how fast the coach was moving. The steeper the graph, the faster it was moving (the greater its speed). When the graph becomes horizontal, its slope is zero. This tells us that the coach's speed was zero. It was not moving.

QUESTION

8 Sketch a distance against time graph to show this: 'The car travelled along the road at a steady speed. It stopped suddenly for a few seconds. Then it continued its journey, at a slower speed than before.'

Activity 2.1 Story graphs

Sketch a distance against time graph. Then ask your partner to write a description of it on a separate sheet of paper.

Choose four graphs and their descriptions. Display them separately and challenge the class to match them up.

Calculating speed

Table 2.3 shows information about a car journey between two cities. The car travelled more slowly at some times than at others. It is easier to see this if we present the information as a graph (see Figure 2.6).

From the graph, you can see that the car travelled slowly at the start of its journey, and also at the end, when it was travelling through the city. The graph is steeper in the middle section, when it was travelling on the open road between the cities.

The graph of Figure 2.6 also shows how to calculate the car's speed. Here, we are looking at the straight section of the graph, where the car's speed was constant. We need to find the value of the gradient (or slope) of the graph, which will tell us the speed:

speed = gradient of distance against time graph

Distance travelled / km	Time taken / h
0	0.0
10	0.5
20	0.8
100	1.8
110	2.3

Table 2.3 Distance and time data for a car journey. This data is represented by the graph in Figure **2.6**

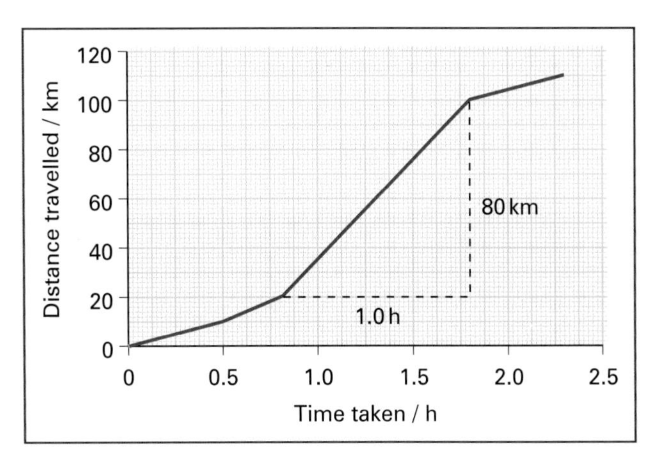

Figure 2.6 Distance against time graph for a car journey, for the data from Table **2.3**.

Worked example 3

These are the steps you take to find the gradient:

Step 1: Identify a straight section of the graph.

Step 2: Draw horizontal and vertical lines to complete a right-angled triangle.

Step 3: Calculate the lengths of the sides of the triangle.

Step 4: Divide the vertical height by the horizontal width of the triangle ('up divided by along').

Here is the calculation for the triangle shown in Figure 2.6:

$$\text{vertical height} = 80\,\text{km}$$
$$\text{horizontal width} = 1.0\,\text{h}$$
$$\text{gradient} = \frac{80\,\text{km}}{1.0\,\text{h}} = 80\,\text{km/h}$$

So the car's speed was 80 km/h for this section of its journey. It helps to include units in this calculation. Then the answer will automatically have the correct units – in this case, km/h.

QUESTION

9 Table 2.4 shows information about a train journey. Use the data in the table to plot a distance against time graph for the train. Find the train's average speed between Beeston and Deeville. Give your answer in km/h.

Station	Distance travelled / km	Time taken / minutes
Ayton	0	0
Beeston	20	30
Seatown	28	45
Deeville	36	60
Eton	44	70

Table 2.4 For Question **9**.

Activity 2.2 Measuring speed

Measure the speed of a cyclist or runner in the school grounds.

Activity 2.3 Measuring speed in the lab

Use lab equipment to measure the speed of a moving trolley or toy car.

Express trains, slow buses

An express train is capable of reaching high speeds, perhaps more than 300 km/h. However, when it sets off on its journey, it may take several minutes to reach this top speed. Then it takes a long time to slow down when it approaches its destination. The famous French TGV trains (Figure 2.7) run on lines that are reserved solely for their operation, so that their high-speed journeys are not disrupted by slower, local trains. It takes time to accelerate (speed up) and decelerate (slow down).

A bus journey is full of accelerations and decelerations (Figure 2.8). The bus accelerates away from the stop. Ideally, the driver hopes to travel at a steady speed until the next stop. A steady speed means that you can sit comfortably in your seat. Then there is a rapid deceleration as the bus slows to a halt. A lot of accelerating and decelerating means that you are likely to be thrown about as the bus changes speed. The gentle acceleration of an express train will barely disturb the drink in your cup. The bus's rapid accelerations and decelerations would make it impossible to avoid spilling the drink.

Figure 2.7 France's high-speed trains, the TGVs (*Trains à Grande Vitesse*), run on dedicated tracks. Their speed has made it possible to travel 600 km from Marseille in the south to Paris in the north, attend a meeting, and return home again within a single day.

Figure 2.8 It can be uncomfortable on a packed bus as it accelerates and decelerates along its journey.

2.3 Understanding acceleration

Some cars, particularly high-performance ones, are advertised according to how rapidly they can accelerate. An advert may claim that a car goes 'from 0 to 60 miles per hour (mph) in 6 s'. This means that, if the car accelerates at a steady rate, it reaches 10 mph after 1 s, 20 mph after 2 s, and so on. We could say that it speeds up by 10 mph every second. In other words, its acceleration is 10 mph per second.

So, we say that an object **accelerates** if its speed increases. Its **acceleration** tells us the rate at which its speed is changing – in other words, the change in speed per unit time.

If an object slows down, its speed is also changing. This is sometimes described as **decelerating**.

Speed against time graphs

Just as we can represent the motion of a moving object by a distance against time graph, we can also represent it by a speed against time graph. (It is easy to get these two types of graph mixed up. Always check out any graph by looking at the axes to see what their labels say.) A speed against time graph shows how the object's speed changes as it moves.

Figure **2.9** shows a speed against time graph for a bus as it follows its route through a busy town. The graph frequently drops to zero because the bus must keep stopping to let people on and off. Then the line slopes up, as the bus accelerates away from the stop. Towards the end of its journey, it manages to move at a steady speed (horizontal graph), as it does not have to stop. Finally, the graph slopes downwards to zero again as the bus pulls into the terminus and stops.

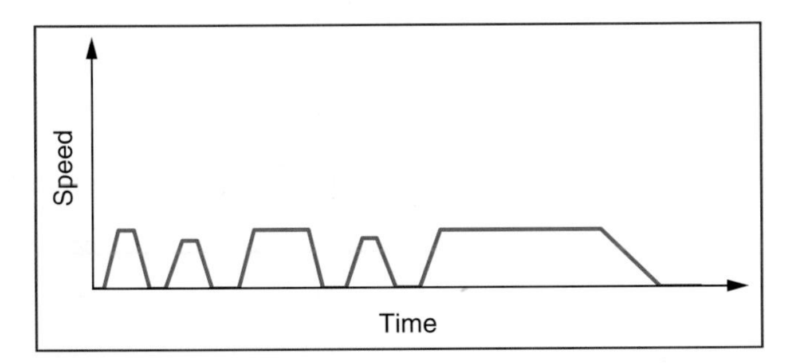

Figure 2.9 A speed against time graph for a bus on a busy route. At first, it has to halt frequently at bus stops. Towards the end of its journey, it maintains a steady speed.

The slope of the speed against time graph tells us about the bus's acceleration:

- The steeper the slope, the greater the acceleration.
- A negative slope means a deceleration (slowing down).
- A horizontal graph (slope = 0) means a constant speed.

Graphs of different shapes

Speed against time graphs can show us a lot about an object's movement. Was it moving at a steady speed, or speeding up, or slowing down? Was it moving at all? The graph shown in Figure **2.10** represents a train journey.

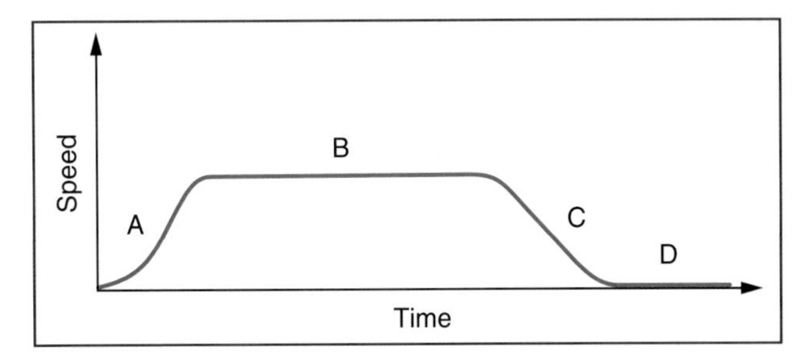

Figure 2.10 An example of a speed against time graph (for a train during part of its journey). This illustrates how such a graph can show acceleration (section A), constant speed (section B), deceleration (section C) and zero speed (section D).

If you study the graph, you will see that it is in four sections. Each section illustrates a different point.

A Sloping upwards – speed increasing – the train was accelerating.

B Horizontal – speed constant – the train was travelling at a steady speed.

C Sloping downwards – speed decreasing – the train was decelerating.

D Horizontal – speed has decreased to zero – the train was stationary.

The fact that the graph lines are curved in sections A and C tells us that the train's acceleration was changing. If its speed had changed at a steady rate, these lines would have been straight.

QUESTIONS

10 A car travels at a steady speed. When the driver sees the red traffic lights ahead, she slows down and comes to a halt. Sketch a speed against time graph for this journey.

11 Look at the speed against time graph in Figure 2.11. Name the sections that represent:
   a steady speed
   b speeding up (accelerating)
   c being stationary
   d slowing down (decelerating).

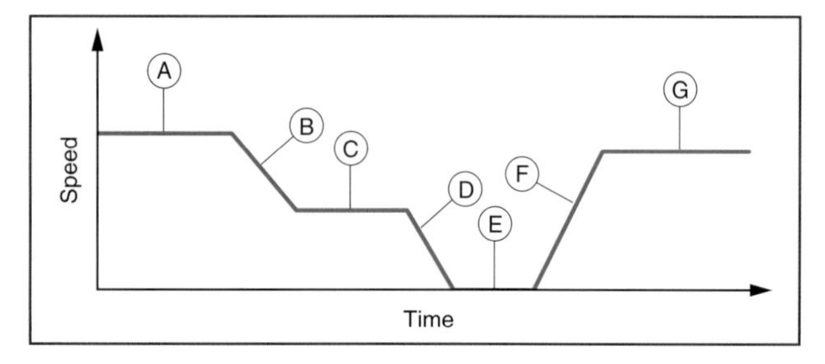

Figure 2.11 For Question **11**.

Finding distance moved

A speed against time graph represents an object's movement. It tells us about how its speed changes. We can use the graph to deduce how far the object moves. To do this, we have to make use of the equation

distance = area under speed against time graph

To understand this equation, consider these two examples.

Example 1

You cycle for 20 s at a constant speed of 10 m/s (see Figure 2.12). The distance you travel is

$$\text{distance moved} = 10\,\text{m/s} \times 20\,\text{s} = 200\,\text{m}$$

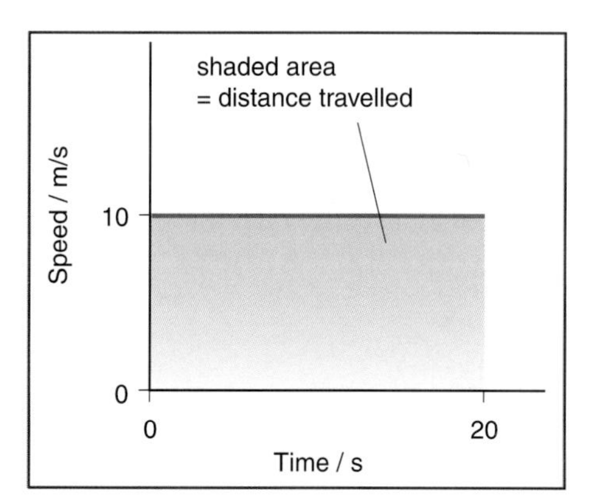

Figure 2.12 Speed against time graph for constant speed. The distance travelled is represented by the shaded area under the graph.

This is the same as the shaded area under the graph. This rectangle is 20 s long and 10 m/s high, so its area is $10\,\text{m/s} \times 20\,\text{s} = 200\,\text{m}$.

Example 2

This is a little more complicated. You set off down a steep ski slope. Your initial speed is 0 m/s. After 10 s you are travelling at 30 m/s (see Figure 2.13).

To calculate the distance moved, we can use the fact that your average speed is 15 m/s. The distance you travel is

$$\text{distance moved} = 15\,\text{m/s} \times 10\,\text{s} = 150\,\text{m}$$

Again, this is represented by the shaded area under the graph. In this case, the shape is a triangle whose height is 30 m/s and whose base is 10 s. Since area of a triangle $= \frac{1}{2} \times \text{base} \times \text{height}$, we have

$$\text{area} = \tfrac{1}{2} \times 10\,\text{s} \times 30\,\text{m/s} = 150\,\text{m}.$$

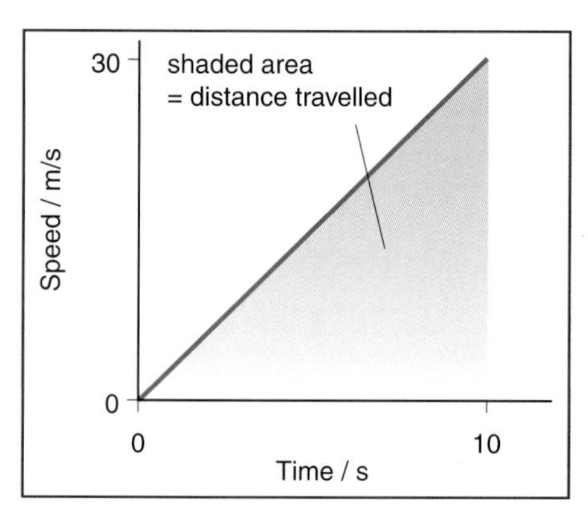

Figure 2.13 Speed against time graph for constant acceleration from rest. Again, the distance travelled is represented by the shaded area under the graph.

Worked example 4

Calculate the distance travelled in 60 s by the train whose motion is represented in Figure 2.14.

The graph in Figure 2.14 has been shaded to show the area we need to calculate to find the distance moved by the train. This area is in two parts:

- a rectangle (pink) of height 6 m/s and width 60 s

  $$\text{area} = 6\,\text{m/s} \times 60\,\text{s} = 360\,\text{m}$$

(this tells us how far the train would have travelled if it had maintained a constant speed of 6 m/s)

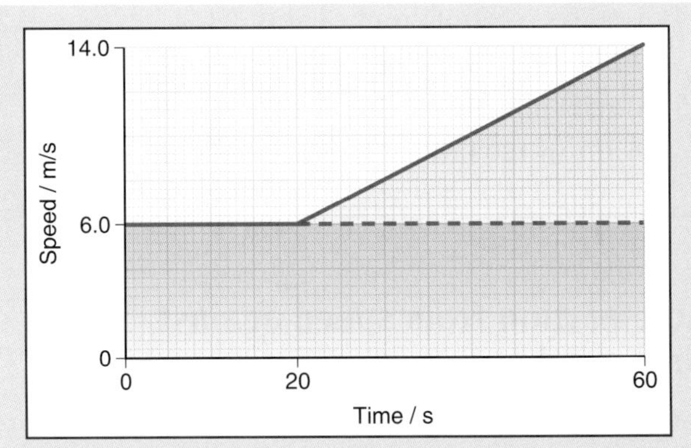

Figure 2.14 Calculating the distance travelled by a train – see Worked example **4**. Distance travelled is represented by the area under the graph. To make the calculation possible, this area is divided up into a rectangle and a triangle, as shown.

- a triangle (orange) of base 40 s and height (14 m/s – 6 m/s) = 8 m/s

$$\text{area} = \tfrac{1}{2} \times \text{base} \times \text{height}$$
$$= \tfrac{1}{2} \times 40\,\text{s} \times 8\,\text{m/s}$$
$$= 160\,\text{m}$$

(this tells us the extra distance travelled by the train because it was accelerating).

We can add these two contributions to the area to find the total distance travelled:

$$\text{total distance travelled} = 360\,\text{m} + 160\,\text{m}$$
$$= 520\,\text{m}$$

So, in 60 s, the train travelled 520 m. We can check this result using an alternative approach. The train travelled for 20 s at a steady speed of 6 m/s, and then for 40 s at an average speed of 10 m/s. So

$$\text{distance travelled} = (6\,\text{m/s} \times 20\,\text{s}) + (10\,\text{m/s} \times 40\,\text{s})$$
$$= 120\,\text{m} + 400\,\text{m}$$
$$= 520\,\text{m}$$

QUESTION

12 a Draw a speed against time graph to show the following motion. A car accelerates uniformly from rest for 5 s. Then it travels at a steady speed of 6 m/s for 5 s.

b On your graph, shade the area that shows the distance travelled by the car in 10 s.

c Calculate the distance travelled in this time.

Activity 2.4 Speed against time graphs

Solve some more problems involving speed against time graphs.

E

2.4 Calculating acceleration

An express train may take 300 s to reach a speed of 300 km/h. Its speed has increased by 1 km/h each second, and so we say that its acceleration is 1 km/h per second.

These are not very convenient units, although they may help to make it clear what is happening when we talk about acceleration. To calculate an object's acceleration, we need to know two things:

- its change in speed (how much it speeds up)
- the time taken (how long it takes to speed up).

Then the acceleration of the object is given by

$$\text{acceleration} = \frac{\text{change in speed}}{\text{time taken}}$$

Worked example 5

An aircraft accelerates from 100 m/s to 300 m/s in 100 s. What is its acceleration?

Step 1: Start by writing down what you know, and what you want to know.

$$\text{initial speed} = 100\,\text{m/s}$$
$$\text{final speed} = 300\,\text{m/s}$$
$$\text{time} = 100\,\text{s}$$
$$\text{acceleration} = ?$$

Step 2: Now calculate the change in speed.

$$\text{change in speed} = 300\,\text{m/s} - 100\,\text{m/s}$$
$$= 200\,\text{m/s}$$

Step 3: Substitute into the equation.

$$\text{acceleration} = \frac{\text{change in speed}}{\text{time taken}}$$
$$= \frac{200\,\text{m/s}}{100\,\text{s}} = 2\,\text{m/s}^2$$

Units of acceleration

In Worked example 5, the units of acceleration are given as m/s^2 (metres per second squared). These are the standard units of acceleration. The calculation shows that the aircraft's speed increased by 2 m/s every second, or 2 metres per second per second. It is simplest to write this as $2\,m/s^2$, but you may prefer to think of it as 2 m/s per second, as this emphasises the meaning of acceleration.

Other units for acceleration are possible. Earlier we saw examples of acceleration in mph per second and km/h per second, but these are unconventional. It is usually best to work in m/s^2.

QUESTIONS

13 Which of the following could **not** be a unit of acceleration?
km/s^2, mph/s, km/s, m/s^2

14 A car sets off from traffic lights. It reaches a speed of 27 m/s in 18 s. What is its acceleration?

15 A train, initially moving at 12 m/s, speeds up to 36 m/s in 120 s. What is its acceleration?

Acceleration from speed against time graphs

A speed against time graph with a steep slope shows that the speed is changing rapidly – the acceleration is greater. It follows that we can find the acceleration of an object by calculating the gradient of its speed against time graph:

acceleration = gradient of speed against time graph

If the speed against time graph is curved (rather than a straight line), the acceleration is changing.

Worked example 6

A train travels slowly as it climbs up a long hill. Then it speeds up as it travels down the other side. Table 2.5 shows how its speed changes. Draw a speed against time graph to show this data. Use the graph to calculate the train's acceleration during the second half of its journey.

Before starting to draw the graph, it is worth looking at the data in the table. The values of speed are given at equal intervals of time (every 10 s). The speed is constant at first (6.0 m/s). Then it increases in equal steps (8.0, 10.0, and so on). In fact, we can see that the speed increases by 2.0 m/s every 10 s. This is enough to tell us that the train's acceleration is $0.2\,m/s^2$. However, we will follow through the detailed calculation to illustrate how to work out acceleration from a graph.

Step 1: Figure 2.15 shows the speed against time graph drawn using the data in the table.

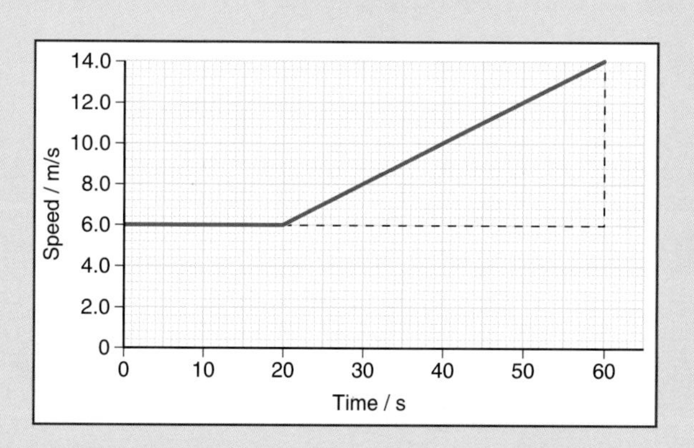

Figure 2.15 Speed against time graph for a train, based on the data in Table **2.5**. The triangle is used to calculate the slope of the second section of the graph. This tells us the train's acceleration.

Speed / m/s	6.0	6.0	6.0	8.0	10.0	12.0	14.0
Time / s	0	10	20	30	40	50	60

Table 2.5 Speed against time data for a train.

You can see that it falls into two parts.

- the initial horizontal section shows that the train's speed was constant (zero acceleration)
- the sloping section shows that the train was then accelerating.

Step 2: The triangle shows how to calculate the slope of the graph. This gives us the acceleration.

$$\text{acceleration} = \frac{14.0\,\text{m/s} - 6.0\,\text{m/s}}{60\,\text{s} - 20\,\text{s}}$$

$$= \frac{8.0\,\text{m/s}}{40\,\text{s}}$$

$$= 0.2\,\text{m/s}^2$$

So, as we expected, the train's acceleration down the hill is $0.2\,\text{m/s}^2$

QUESTION

16 A car travels for 10 s at a steady speed of 20 m/s along a straight road. The traffic lights ahead change to red, and the car slows down with a constant deceleration, so that it halts after a further 8 s.

a Draw a speed against time graph to represent the car's motion during the 18 s described.

b Use the graph to deduce the car's deceleration as it slows down.

c Use the graph to deduce how far the car travels during the 18 s described.

Activity 2.5 Acceleration problems

Solve some more problems involving acceleration.

Speed and velocity, vectors and scalars

In physics, the words **speed** and **velocity** have different meanings, although they are closely related. **Velocity** is an object's speed in a particular direction.

So, we could say that an aircraft has a speed of 200 m/s but a velocity of 200 m/s due north. We must give the direction of the velocity or the information is incomplete.

Velocity is an example of a **vector quantity**. Vectors have both magnitude (size) and direction. Another example of a vector is weight – your weight is a force that acts downwards, towards the centre of the Earth.

Speed is an example of a **scalar quantity**. Scalars only have magnitude. Temperature is an example of another scalar quantity.

There is more about vectors and scalars in Chapter **3**.

Summary

We can represent an object's motion using distance against time and speed against time graphs.

$$\text{Speed} = \frac{\text{distance}}{\text{time}}$$

Speed = gradient of distance against time graph

Distance travelled = area under speed against time graph

$$\text{Acceleration} = \frac{\text{change in speed}}{\text{time taken}}$$

Acceleration = gradient of speed against time graph

End-of-chapter questions

2.1 A runner travels 400 m in 50 s. What is her average speed? [3]

2.2 Figure **2.16** represents the motion of a bus. It is in two sections, A and B. What can you say about the motion of the bus during these two sections? [2]

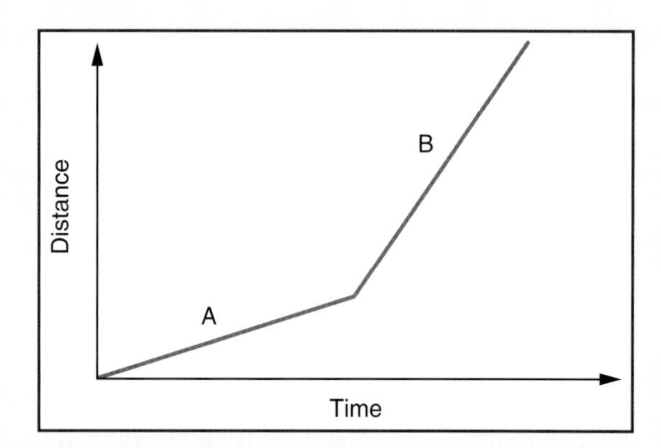

Figure 2.16 For Question **2.2**.

2.3 How far will a bus travel in 30 s at a speed of 15 m/s? [3]

2.4 Table **2.6** shows the distance travelled by a car at intervals during a short journey.

a Draw a distance against time graph to represent this data. [4]

b What does the shape of the graph tell you about the car's speed? [2]

Distance / m	0	200	400	600	800
Time / s	0	10	20	30	40

Table 2.6 For Question **2.4**.

2.5 Figure **2.17** shows the distance travelled by a car on a rollercoaster ride, at different times along its trip. It travels along the track, and then returns to its starting position. Study the graph and decide which point best fits the following descriptions. In each case, give a reason to explain why you have chosen that point.

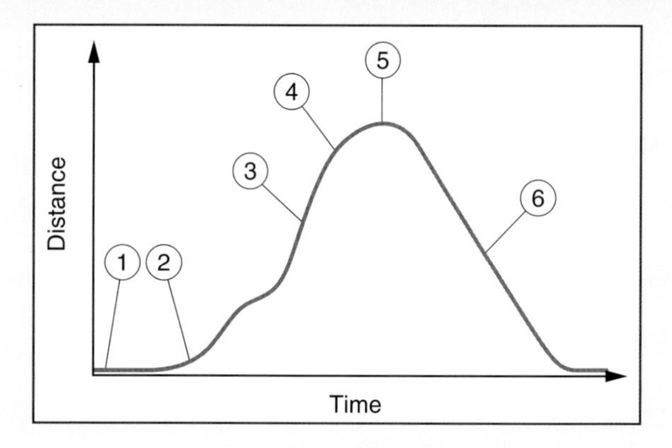

Figure 2.17 Distance against time graph for a rollercoaster car – for Question **2.5**.

a The car is stationary. [2]

b The car is travelling its fastest. [2]

c The car is speeding up. [2]

d The car is slowing down. [2]

e The car starts on its return journey. [2]

2.6 Scientists have measured the distance between the Earth and the Moon by reflecting a beam of laser light off the Moon. They measure the time taken for light to travel to the Moon and back.

a What other piece of information is needed to calculate the Earth–Moon distance? [1]

b How would the distance be calculated? [1]

2.7 Table **2.7** shows information about the motion of a number of objects. Copy and complete the table. [4]

Object	Distance travelled	Time taken	Speed
bus	20 km	0.8 h	……
taxi	6 km	……	30 m/s
aircraft	……	5.5 h	900 km/h
snail	3 mm	10 s	……

Table 2.7 For Question **2.7**.

2.8 The speed against time graph for part of a train journey is a horizontal straight line. What does this tell you about the train's speed, and about its acceleration? [2]

2.9 Sketch speed against time graphs to represent the following two situations.

a An object starts from rest and moves with constant acceleration. [3]

b An object moves at a steady speed. Then it slows down and stops. [3]

E

2.10 A runner accelerates from rest to 8 m/s in 2 s. What is his acceleration? [3]

2.11 A runner accelerates from rest with an acceleration of 4 m/s^2 for 2.3 s. What will her speed be at the end of this time? [4]

2.12 A car can accelerate at 5.6 m/s^2. Starting from rest, how long will it take to reach a speed of 24 m/s? [3]

2.13 Table **2.8** shows how the speed of a car changed during a section of a journey.

a Draw a speed against time graph to represent this data. [4]

Use your graph to calculate:

b the car's acceleration during the first 30 s of the journey [3]

c the distance travelled by the car during the journey. [5]

E

Speed / m/s	0	9	18	27	27	27
Time / s	0	10	20	30	40	50

Table 2.8 For Question **2.13**.

2.14 Figure **2.18** shows how a car's speed changed as it travelled along.

a In which section(s) was its acceleration zero? [2]

b In which section(s) was its acceleration constant? [2]

c What can you say about its acceleration in the other section(s)? [2]

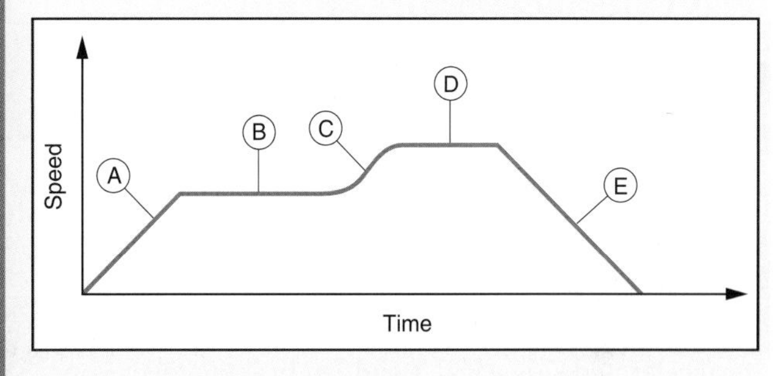

Figure 2.18 A speed against time graph for a car – for Question **2.14**.

2.15 A bus travels 1425 m in 75 s.

a What is its speed? [3]

b What other piece of information do we need in order to state its velocity? [1]

3 Forces and motion

Core Identifying the forces acting on an object
Core Describing how a resultant force changes the motion of an object
E Extension Describing how a resultant force can give rise to motion in a circle
Extension Using the relationship between force, mass and acceleration
Core Explaining the difference between mass and weight
E Extension Describing the effect of drag on a moving object
Extension Calculating the resultant of two or more vectors

Roller-coaster forces

Some people get a lot of pleasure out of sudden acceleration and deceleration. Many fairground rides involve sudden changes in speed. On a roller-coaster (Figure 3.1), you may speed up as the car runs downhill. Then, suddenly, you veer off to the left – you are accelerated sideways. A sudden braking gives you a large, negative acceleration (a deceleration). You will probably have to be fastened in to your seat to avoid being thrown out of the car by these sudden changes in speed.

Figure 3.1 A roller-coaster ride involves many rapid changes in speed. These accelerations and decelerations give the ride its thrill. The ride's designers have calculated the accelerations carefully to ensure that the car will not come off its track, and the riders will stay in the car.

What are the forces at work in a roller-coaster? If you are falling downwards, it is gravity that affects you. This gives you an acceleration of about $10\,m/s^2$. We say that the G-force acting on you is 1 (that is, one unit of gravity). When the brakes slam on, the G-force may be greater, perhaps as high as 4. The brakes make use of the force of **friction**.

Changing direction also requires a force. So when you loop the loop or veer to the side, there must be a force acting. This is simply the force of the track, whose curved shape pushes you round. Again, the G-force may reach as high as 4.

Roller-coaster designers have learned how to surprise you with sudden twists and turns. You can be scared or exhilarated. However you feel, you can release the tension by screaming.

3.1 We have lift-off

It takes an enormous force to lift the giant space shuttle off its launch pad, and to propel it into space (Figure 3.2). The booster rockets that supply the initial thrust provide a force of several million newtons. As the spacecraft accelerates upwards, the crew experience the sensation of being pressed firmly back into their seats. That is how they know that their craft is accelerating.

Figure 3.2 The space shuttle accelerating away from its launch pad. The force needed is provided by several rockets. Once each rocket has used all its fuel, it will be jettisoned, to reduce the mass that is being carried up into space.

Forces change motion. One moment, the shuttle is sitting on the ground, stationary. The next moment, it is accelerating upwards, pushed by the force provided by the rockets.

In this chapter, we will look at how forces – pushes and pulls – affect objects as they move. You will be familiar with the idea that forces are measured in **newtons (N)**. To give an idea of the sizes of various forces, here are some examples:

- You lift an apple. The force needed to lift an apple is roughly 1 newton (1 N).
- You jump up in the air. Your leg muscles provide the force needed to do this, about 1000 N.
- You reach the motorway in your high-performance car, and 'put your foot down'. The car accelerates forwards. The engine provides a force of about 5000 N.
- You are crossing the Atlantic in a Boeing 777 jumbo jet. The four engines together provide a thrust of about 500 000 N. In total, that is about half the thrust provided by each of the space shuttle's booster rockets.

Some important forces

Forces appear when two objects interact with each other. Figure **3.3** shows some important forces. Each force is represented by an arrow to show its direction.

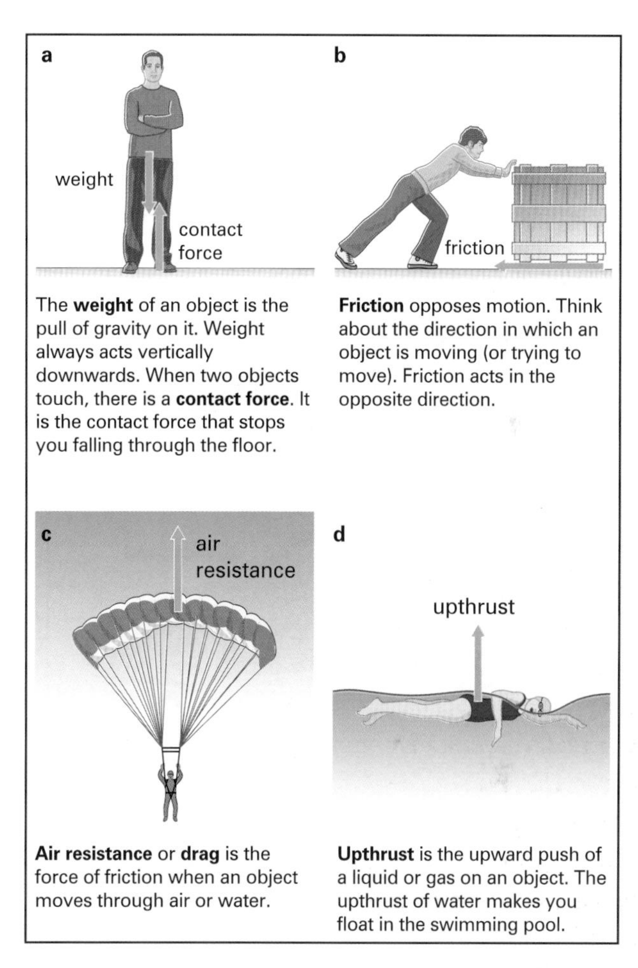

Figure 3.3 Some common forces.

Forces produce acceleration

The car driver in Figure **3.4a** is waiting for the traffic lights to change. When they go green, he moves forwards. The force provided by the engine causes the car to accelerate. In a few seconds, the car is moving quickly along the road. The arrow in the diagram shows the force pushing the car forwards. If the driver wants to get away from the lights more quickly, he can press harder on the accelerator. The forward force is then bigger, and the car's acceleration will be greater.

The driver reaches another junction, where he must stop. He applies the brakes. This provides another force to slow down the car (see Figure **3.4b**). The car is moving forwards, but the force needed to make it decelerate is directed backwards. If the driver wants to stop in a hurry, a bigger force is needed. He must press hard on the brake pedal, and the car's deceleration will be greater.

Finally, the driver wants to turn a corner. He turns the steering wheel. This produces a sideways force on the car (Figure **3.4c**), so that the car changes direction.

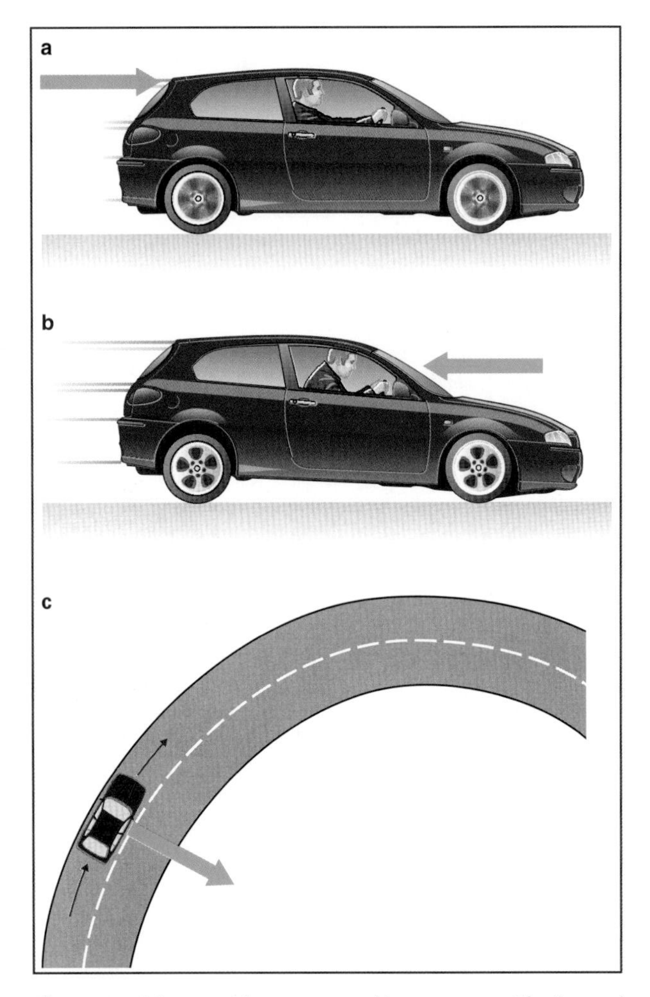

Figure 3.4 A force can be represented by an arrow. **a** The forward force provided by the engine causes the car to accelerate forwards. **b** The backward force provided by the brakes causes the car to decelerate. **c** A sideways force causes the car to change direction.

To summarise, we have seen several things about forces:

- They can be represented by arrows. A force has a direction, shown by the direction of the arrow.
- A force can make an object change speed (accelerate). A forward force makes it speed up, while a backward force makes it slow down.
- A force can change the direction in which an object is moving.

QUESTION

1 Figure **3.5** shows three objects that are moving. A force acts on each object. For each, say how its movement will change.

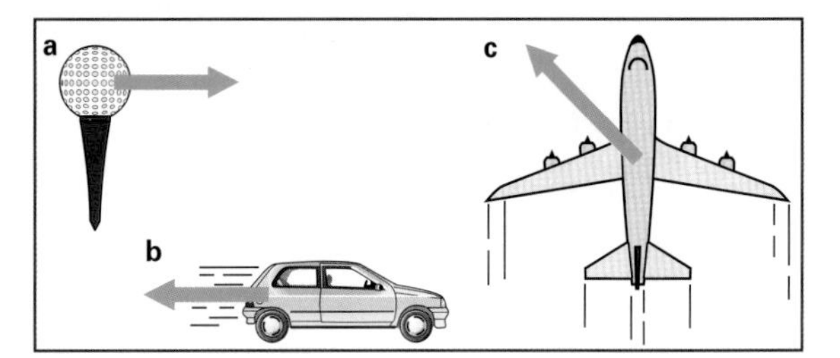

Figure 3.5 Three moving objects – for Question **1**.

Two or more forces

The car shown in Figure **3.6a** is moving rapidly. The engine is providing a force to accelerate it forwards, but there is another force acting, which tends to slow down the car. This is **air resistance**, a form of friction caused when an object moves through the air. (This frictional force is also called **drag**, especially for motion through fluids other than the air.) The air drags on the object, producing a force that acts in the opposite direction to the object's motion. In Figure **3.6a**, these two forces are:

- push of engine = 600 N to the right
- drag of air resistance = 400 N to the left.

We can work out the combined effect of these two forces by subtracting one from the other to give the **resultant force** acting on the car.

> The resultant force is the single force that has the same effect as two or more forces.

So in Figure **3.6a**:

$$\text{resultant force} = 600\,\text{N} - 400\,\text{N}$$
$$= 200\,\text{N to the right}$$

This resultant force will make the car accelerate to the right, but not as much as if there was no air resistance.

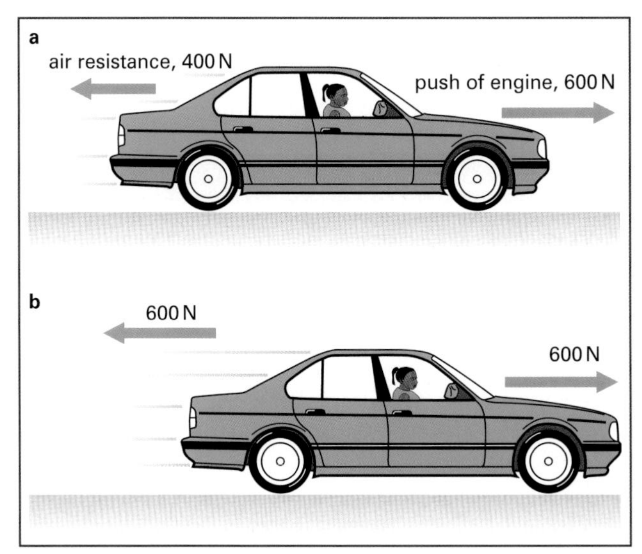

Figure 3.6 A car moves through the air. Air resistance acts in the opposite direction to its motion.

In Figure **3.6b**, the car is moving even faster, and air resistance is greater. Now the two forces cancel each other out. So in Figure **3.6b**:

resultant force = 600 N − 600 N = 0 N

We say that the forces on the car are **balanced**, and it no longer accelerates.

QUESTION

2 Figure **3.7** shows the forces acting on three objects. For each, say whether the forces are balanced or unbalanced. If the forces are unbalanced, calculate the resultant force and give its direction. Say how the object's motion will change.

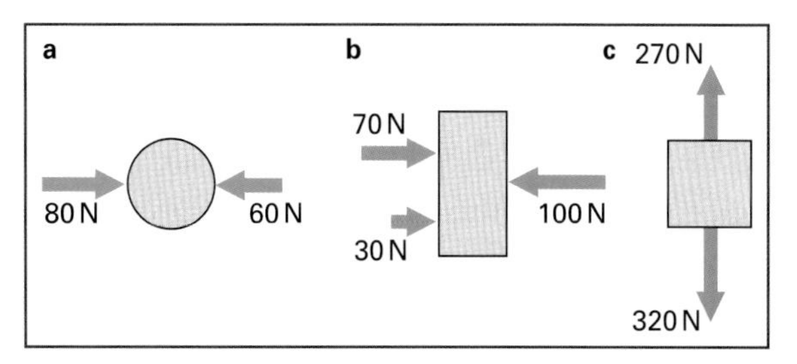

Figure 3.7 For Question **2**.

Activity 3.1 Balanced forces

Solve some problems involving two or more forces acting on an object.

Going round in circles

When a car turns a corner, it changes direction. Any object moving along a circular path is changing direction as it goes. A force is needed to do this. Figure **3.8** shows three objects following curved paths, together with the forces that act to keep them on track.

a The boy is whirling an apple around on the end of a piece of string. The tension in the string pulls on the apple, keeping it moving in a circle.

b An aircraft 'banks' (tilts) to change direction. The lift force on its wings provides the necessary force.

c The Moon is held in its orbit around the Earth by the pull of the Earth's gravity.

For an object following a circular path, the object is acted on by a force at right angles to its velocity.

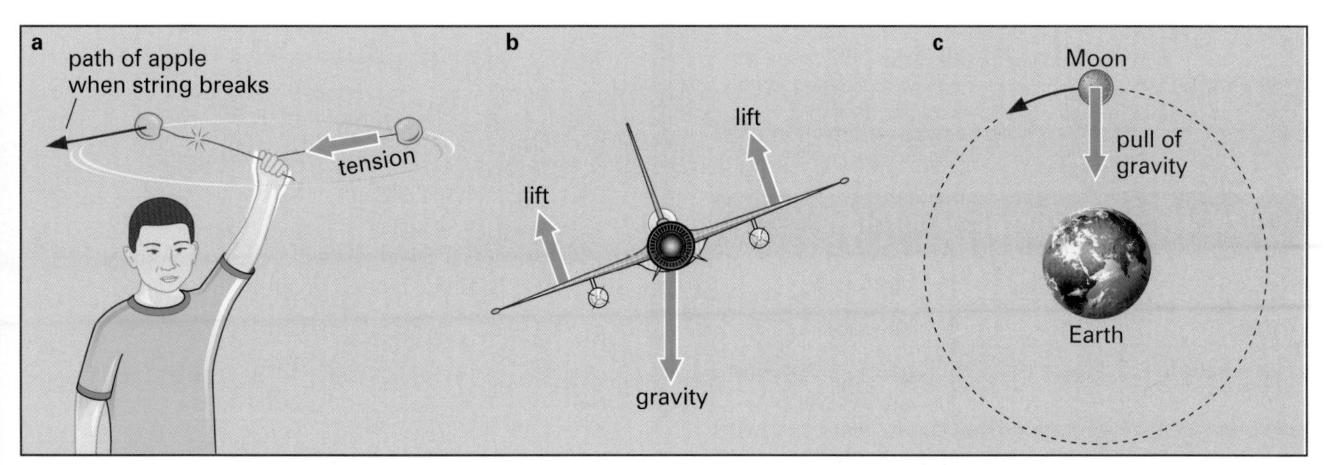

Figure 3.8 Examples of motion along a curved path. In each case, there is a sideways force holding the object in its circular path.

3.2 Force, mass and acceleration

A car driver uses the accelerator pedal to control the car's acceleration. This alters the force provided by the engine. The bigger the force acting on the car, the bigger the acceleration it gives to the car. Doubling the force produces twice the acceleration, three times the force produces three times the acceleration, and so on.

There is another factor that affects the car's acceleration. Suppose the driver fills the boot with a lot of heavy boxes and then collects several children from college. He will notice the difference when he moves away from the traffic lights. The car will not accelerate so readily, because its mass has been increased. Similarly, when he applies the brakes, it will not decelerate as readily as before. The mass of the car affects how easily it can be accelerated or decelerated. Drivers learn to take account of this.

The greater the mass of an object, the smaller the acceleration it is given by a particular force.

So big (more massive) objects are harder to accelerate than small (less massive) ones. If we double the mass of the object, its acceleration for a given force will be halved. We need double the force to give it the same acceleration.

This tells us what we mean by **mass**. It is the property of an object that resists changes in its motion.

Force calculations

These relationships between force, mass and acceleration can be combined into a single, very useful, equation:

$$\text{force} = \text{mass} \times \text{acceleration}$$
$$F = ma$$

Quantity	Symbol	SI unit
force	F	newton, N
mass	m	kilogram, kg
acceleration	a	metres per second squared, m/s^2

Table 3.1 The three quantities related by the equation force = mass × acceleration.

The quantities involved in this equation, and their units, are summarised in Table **3.1**. Worked examples **1** and **2** show how to use the equation.

Worked example 1

When you strike a tennis ball that another player has hit towards you, you provide a large force to reverse its direction of travel and send it back towards your opponent. You give the ball a large acceleration. What force is needed to give a ball of mass 0.1 kg an acceleration of 500 m/s^2?

Step 1: We have

$$\text{mass} = 0.1\,\text{kg}$$
$$\text{acceleration} = 500\,\text{m/s}^2$$
$$\text{force} = ?$$

Step 2: Substituting in the equation to find the force gives

$$\begin{aligned}\text{force} &= \text{mass} \times \text{acceleration}\\ &= 0.1\,\text{kg} \times 500\,\text{m/s}^2\\ &= 50\,\text{N}\end{aligned}$$

Worked example 2

A Boeing 777 jumbo jet (Figure **3.9**) has four engines, each capable of providing 250 000 N of thrust. The mass of the aircraft is 250 000 kg. What is the greatest acceleration that the aircraft can achieve?

Step 1: The greatest force provided by all four engines working together is 4 × 250 000 N = 1 000 000 N.

Step 2: Now we have

$$\text{force} = 1\,000\,000\,\text{N}$$
$$\text{mass} = 250\,000\,\text{kg}$$
$$\text{acceleration} = ?$$

Step 3: The greatest acceleration the engines can produce is then given by

$$\begin{aligned}\text{acceleration} &= \frac{\text{force}}{\text{mass}}\\ &= \frac{1\,000\,000\,\text{N}}{250\,000\,\text{kg}}\\ &= 4.0\,\text{m/s}^2\end{aligned}$$

Figure 3.9 A jumbo jet has four engines, each capable of providing a quarter of a million newtons of thrust. When the aircraft lands, the engines are put into 'reverse thrust' mode, so that they provide a decelerating force to bring it to a halt.

QUESTIONS

3 What force is needed to give a car of mass 600 kg an acceleration of 2.5 m/s^2?

4 A stone of mass 0.2 kg falls with an acceleration of 10.0 m/s^2. How big is the force that causes this acceleration?

5 What acceleration is produced by a force of 2000 N acting on a person of mass 80 kg?

6 One way to find the mass of an object is to measure its acceleration when a force acts on it. If a force of 80 N causes a box to accelerate at 0.1 m/s^2, what is the mass of the box?

Activity 3.2 *F*, *m* and *a*

Change the force acting on an object, and see how its acceleration changes.

Change the mass of an object, and see how its acceleration changes.

3.3 Mass, weight and gravity

If you drop an object, it falls to the ground. It is difficult to see how a falling object moves. However, a multi-flash photograph can show the pattern of movement when an object falls.

Figure 3.10 shows a ball falling. There are seven images of the ball, taken at equal intervals of time. The ball falls further in each successive time interval. This shows that its speed is increasing – it is accelerating.

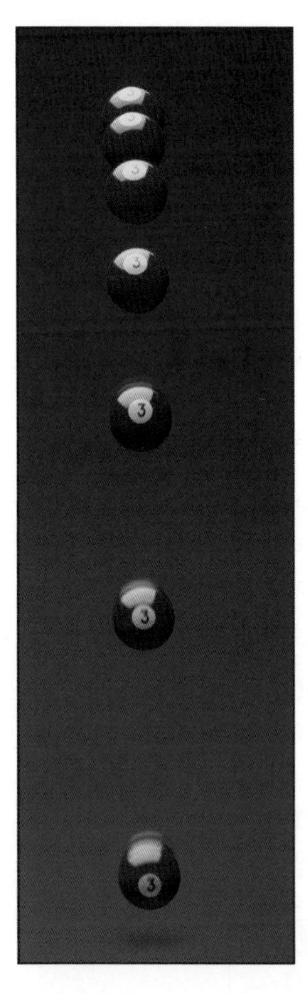

Figure 3.10 The increasing speed of a falling ball is captured in this multi-flash image.

If an object accelerates, there must be a force that is causing it to do so. In this case, the force of **gravity** is pulling the ball downwards. The name given to the force of gravity acting on an object is its **weight**. Because weight is a force, it is measured in newtons (N).

Every object on or near the Earth's surface has weight. This is caused by the attraction of the Earth's gravity. The Earth pulls with a force of 10 N (approximately) on each kilogram of matter, so an object of mass 1 kg has a weight of 10 N:

weight of 1 kg mass = 10 N

Because the Earth pulls with the same force on every kilogram of matter, every object falls with the same acceleration close to the Earth's surface. If you drop a 5 kg ball and a 1 kg ball at the same time, they will reach the ground at the same time.

The acceleration caused by the pull of the Earth's gravity is called the **acceleration of free fall** or the **acceleration due to gravity**, and its value is $10\,m/s^2$ close to the surface of the Earth:

acceleration of free fall = $10\,m/s^2$

Distinguishing mass and weight

It is important to understand the difference between the two quantities, mass and weight.

- The **mass** of an object, measured in kilograms, tells you how much matter it is composed of.
- The **weight** of an object, measured in newtons, is the force of gravity that acts on it.

If you take an object to the Moon, it will weigh less, because the Moon's gravity is weaker than the Earth's. However, its mass will be **unchanged** because it is made of just as much matter as when it was on Earth.

When we weigh an object using a balance, we are comparing its weight with that of standard weights on the other side of the balance (Figure 3.11). We are making use of the fact that, if two objects weigh the same, their masses will be the same.

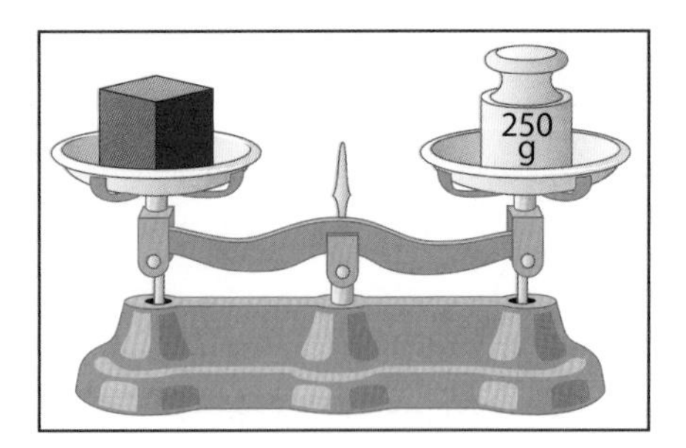

Figure 3.11 When the balance is balanced, we know that the weights on opposite sides are equal, and so the masses must also be equal.

Activity 3.3 Comparing masses

You can compare the masses of two objects by holding them. How good are you at judging mass?

QUESTION

7 A book is weighed on Earth. It is found to have a mass of 1 kg. So its weight on the Earth is 10 N. What can you say about its mass and its weight if you take it:

a to the Moon, where gravity is weaker than on Earth?

b to Jupiter, where gravity is stronger?

3.4 Falling through the air

The Earth's gravity is equally strong at all points close to the Earth's surface. If you climb to the top of a tall building, your weight will stay the same. We say that there is a **uniform gravitational field** close to the Earth's surface. This means that all objects fall with the same acceleration as the ball shown above in Figure 3.10, provided there is no other force acting to reduce their acceleration. For many objects, the force of air resistance can affect their acceleration.

Parachutists make use of air resistance. A free-fall parachutist (Figure 3.12) jumps out of an aircraft and accelerates downwards. Figure 3.13 shows the forces on a parachutist at different points in his fall. At first, air resistance has little effect. However, air resistance increases as he falls, and eventually this force balances his weight. Then the parachutist stops accelerating – he falls at a steady rate known as the **terminal velocity**.

Figure 3.12 Free-fall parachutists, before they open their parachutes. They can reach a terminal velocity of more than 50 m/s.

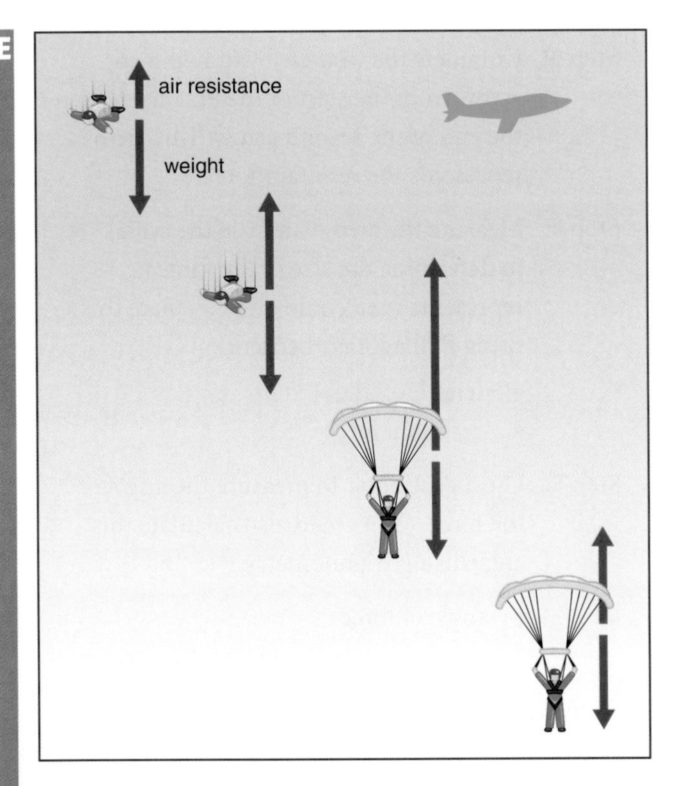

Figure 3.13 The forces on a falling parachutist. Notice that his weight is constant. When air resistance equals weight, the forces are balanced and the parachutist reaches a steady speed. The parachutist is always falling (velocity downwards), although his acceleration is upwards when he opens his parachute.

Opening the parachute greatly increases the area and hence the air resistance. Now there is a much bigger force upwards. The forces on the parachutist are again unbalanced, and he slows down. The idea is to reach a new, slower, terminal velocity of about 10 m/s, at which speed he can safely land. At this point, weight = drag, and so the forces on the parachutist are balanced.

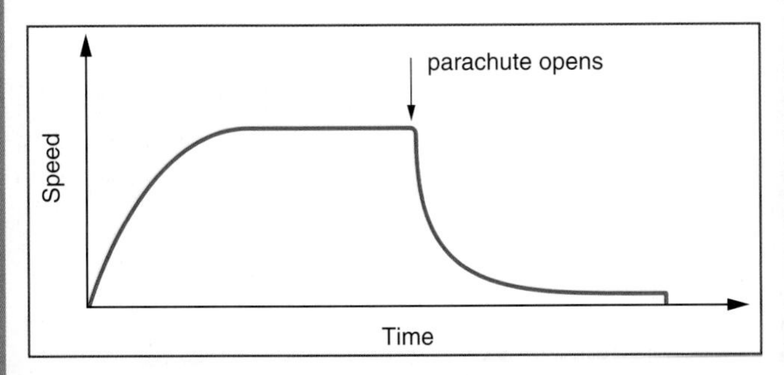

Figure 3.14 A speed against time graph for a falling parachutist.

The graph in Figure **3.14** shows how the parachutist's speed changes during a fall.

- When the graph is horizontal, speed is constant and forces are balanced.
- When the graph is sloping, speed is changing. The parachutist is accelerating or decelerating, and forces are unbalanced.

QUESTION

8 Look at the graph of Figure **3.14**. Find a point where the graph is sloping upwards.

a Is the parachutist accelerating or decelerating?

b Which of the two forces acting on the parachutist is greater?

c Explain the shape of the graph after the parachute has opened.

3.5 More about scalars and vectors

We can represent forces using arrows because a force has a **direction** as well as a **magnitude**. This means that force is a **vector quantity** (see Chapter 2). Table **3.2** lists some scalar and vector quantities.

Scalar quantities	Vector quantities
speed	velocity
mass	force
energy	weight
density	acceleration
temperature	

Table 3.2 Some scalar and vector quantities.

Adding forces

What happens if an object is acted on by two or more forces? Figure **3.15a** shows someone pushing a car. Friction opposes their pushing force. Because the forces are acting in a straight line, it is simple to calculate the resultant force, provided we take into account the directions of the forces:

resultant force = 500 N − 350 N
= 150 N to the right

Note that we must give the direction of the resultant force, as well as its magnitude. The car will accelerate towards the right.

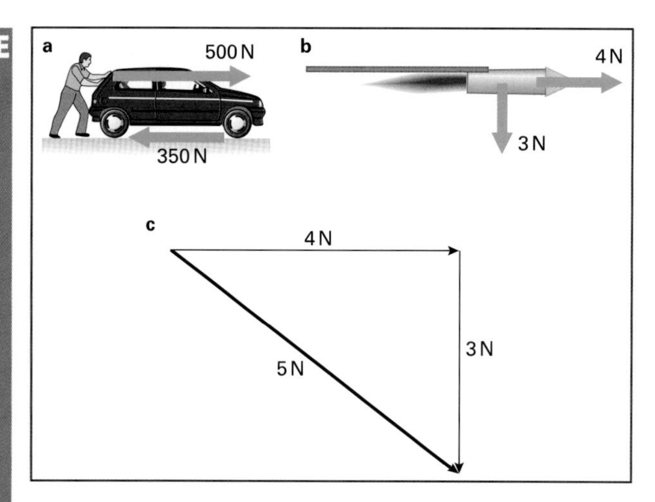

Figure 3.15 Adding forces: **a** two forces in a straight line; **b** two forces in different directions; **c** a vector triangle shows how to add the forces in **b**.

Figure **3.15b** shows a more difficult situation. A firework rocket is acted on by two forces.

- The thrust of its burning fuel pushes it towards the right.
- Its weight acts vertically downwards.

Worked example **3** shows how to find the resultant force by the method of drawing a **vector triangle** (graphical representation of vectors).

Worked example 3

Find the resultant force acting on the rocket shown in Figure **3.15b**. What effect will the resultant force have on the rocket?

Step 1: Look at the diagram. The two forces are 4 N horizontally and 3 N vertically.

Step 2: Draw a scale diagram to represent these forces, as follows (see Figure **3.15c**). In Figure **3.15c** we are using a scale of 1 cm to represent 1 N.

- Draw a horizontal arrow, 4 cm long, to represent the 4 N force. Mark it with an arrow to show its direction.
- Using the end of this arrow as the start of the next arrow, draw a vertical arrow, 3 cm long, to represent the 3 N force.

Step 3: Complete the triangle by drawing an arrow from the start of the first arrow to the end of the second arrow. This arrow represents the resultant force.

Step 4: Measure the arrow, and use the scale to determine the size of the force it represents (you could also calculate this using Pythagoras' theorem).

- length of line = 5 cm
- resultant force = 5 N

Step 5: Use a protractor to measure the angle of the force. (You could also calculate this angle using trigonometry.)

angle of force
= 37° below the horizontal

So the resultant force acting on the rocket is 5 N acting at 37° below the horizontal. The rocket will be given an acceleration in this direction.

Rules for vector addition

You can add two or more forces by the following method – simply keep adding arrows end-to-end:

- Draw arrows end-to-end, so that the end of one is the start of the next.
- Choose a scale that gives a large triangle.
- Join the start of the first arrow to the end of the last arrow to find the resultant.

Other vector quantities (for example, two velocities) can be added in this way. Imagine that you set out to swim across a fast-flowing river. You swim towards the opposite bank, but the river's velocity carries you downstream. Your resultant velocity will be at an angle to the bank.

Airline pilots must understand vector addition. Aircraft fly at high speed, but the air they are moving through is also moving fast. If they are to fly in a straight line towards their destination, the pilot must take account of the wind speed.

QUESTION

9 An aircraft can fly at a top speed of 600 km/h.

a What will its speed be if it flies into a head-wind of 100 km/h? (A head-wind blows in the opposite direction to the aircraft.)

b The pilot directs the aircraft to fly due north at 600 km/h. A side-wind blows at 100 km/h towards the east. What will be the aircraft's resultant velocity? (Give both its speed and direction.)

Summary

A force can cause a body to accelerate, decelerate, or change direction.

The resultant force on a body is the single force that has the same effect as all of the forces acting on it.

When combining forces, we must take account of their directions.

The force of gravity on an object is its weight.

Force, mass and acceleration are related by

force = mass × acceleration

$F = ma$

An object following a circular path is acted on by a force at right angles to its velocity.

End-of-chapter questions

3.1 What are the units of **a** mass, **b** force and **c** acceleration? [3]

3.2 a Why is it sensible on diagrams to represent a force by an arrow? [1]

b Why should mass not be represented by an arrow? [1]

3.3 Which will produce a bigger acceleration: a force of 10 N acting on a mass of 5 kg, or a force of 5 N acting on a mass of 10 kg? [2]

3.4 An astronaut is weighed before he sets off to the Moon. He has a mass of 80 kg.

a What will his weight be on Earth? [3]

b When he arrives on the Moon, will his mass be more, less, or the same? [1]

c Will his weight be more, less, or the same? [1]

3.5 Figure **3.16** shows the forces acting on a lorry as it travels along a flat road.

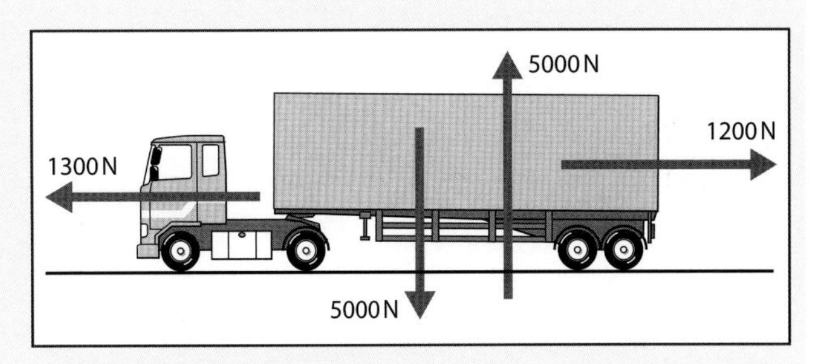

Figure 3.16 For Question **3.5**.

a Two of the forces have effects that cancel each other out. Which two? Explain your answer. [2]

b What is the resultant force acting on the lorry? Give its magnitude and direction. [3]

c What effect will this resultant force have on the speed at which the lorry is travelling? [1]

E

3.6 What force is needed to give a mass of 20 kg an acceleration of 5 m/s^2? [3]

3.7 A train of mass 800 000 kg is slowing down. What acceleration is produced if the braking force is 1 400 000 N? [3]

3.8 A car speeds up from 12 m/s to 20 m/s in 6.4 s. If its mass is 1200 kg, what force must its engine provide? [6]

3.9 The gravitational field of the Moon is weaker than that of the Earth. It pulls on each kilogram of mass with a force of 1.6 N. What will be the weight of a 50 kg mass on the Moon? [3]

E

3.10 Figure **3.17** shows a diver underwater.

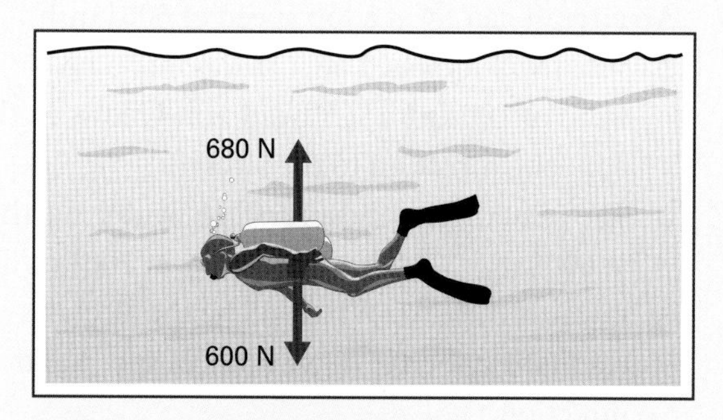

Figure 3.17 For Question **3.10**.

a Calculate the resultant force on the diver. [3]

b Explain how his motion will change. [1]

Turning effects of forces

Core Describing the turning effect of a force
Core Stating the conditions for equilibrium of an object
E **Extension** Calculating moments, forces and distances
Core Understanding centre of mass and stability

Keeping upright

Human beings are inherently unstable. We are tall and thin and walk upright. Our feet are not rooted into the ground. So you might expect us to keep toppling over. Human children learn to stand and walk at the age of about 12 months. It takes a lot of practice to get it right. We have to learn to coordinate our muscles so that our legs, body and arms move correctly. There is a special organ in each of our ears (the semicircular canals) that keeps us aware of whether we are vertical or tilting. Months of practice and many falls are needed to develop the skill of walking.

Figure 4.1 This cyclist must balance with great care because the load he is carrying on his head makes him even more unstable.

We have the same experience later in life if we learn to ride a bicycle (Figure 4.1). A bicycle is even more unstable than a person. If you ride a bicycle, you are constantly adjusting your position to maintain your stability and to remain upright. If the bicycle tilts slightly to the left, you automatically lean slightly to the right to provide a force that tips it back again. You make these adjustments unconsciously. You know intuitively that if you let the bicycle tilt too far, you will not be able to recover the situation, and you will end up sprawling on the ground.

4.1 The moment of a force

Figure 4.2 shows a boy who is trying to open a heavy door by pushing on it. He must make the **turning effect** of his force as big as possible. How should he push?

First of all, look for the **pivot** – the fixed point about which the door will turn. This is the hinge of the door. To open the door, push with as big a force as possible, and as far as possible from the pivot – at the other edge of the door. (That's why the door handle is fitted there.)

Figure 4.2 Opening a door – how can the boy have a big turning effect?

To have a big turning effect, the person must push hard at **right angles** to the door. Pushing at a different angle gives a smaller turning effect.

The quantity that tells us the turning effect of a force about a pivot is its **moment**.

- The moment of a force is bigger if the force is bigger.
- The moment of a force is bigger if it acts further from the pivot.
- The moment of a force is greatest if it acts at 90° to the object it acts on.

Making use of turning effects

Figure **4.3** shows how understanding moments can be useful.

- Using a crowbar to lift a heavy paving slab – pull near the end of the bar, and at 90°, to have the biggest possible turning effect.
- Lifting a load in a wheelbarrow – the long handles help to increase the moment of the lifting force.

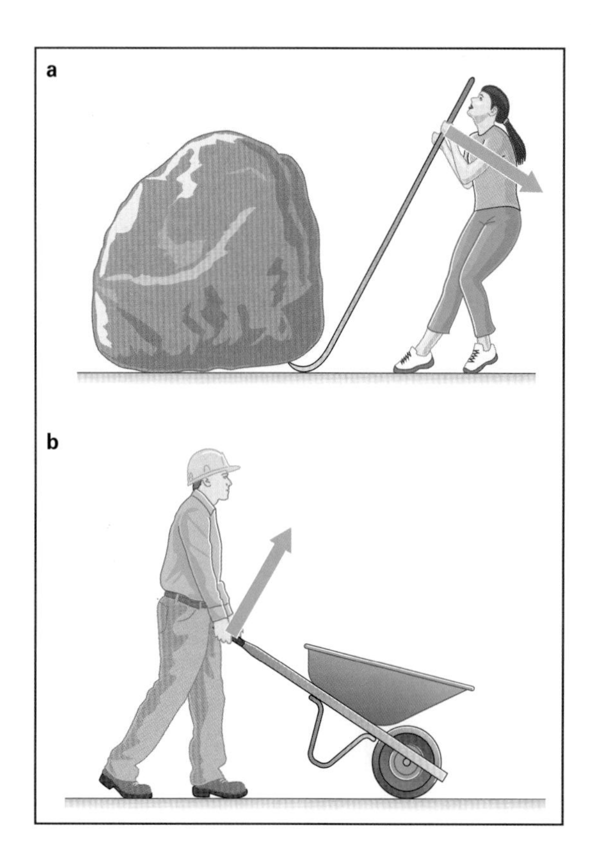

Figure 4.3 Understanding moments can help in some difficult tasks.

Balancing a beam

Figure **4.4** shows a small child sitting on the left-hand end of a see-saw. Her weight causes the see-saw to tip down on the left. Her father presses down on the other end. If he can press with a force greater than her weight, the see-saw will tip to the right and she will come up in the air.

Now, suppose the father presses down closer to the pivot. He will have to press with a greater force if the turning effect of his force is to overcome the turning effect of his daughter's weight. If he presses at half the distance from the pivot, he will need to press with twice the force to balance her weight.

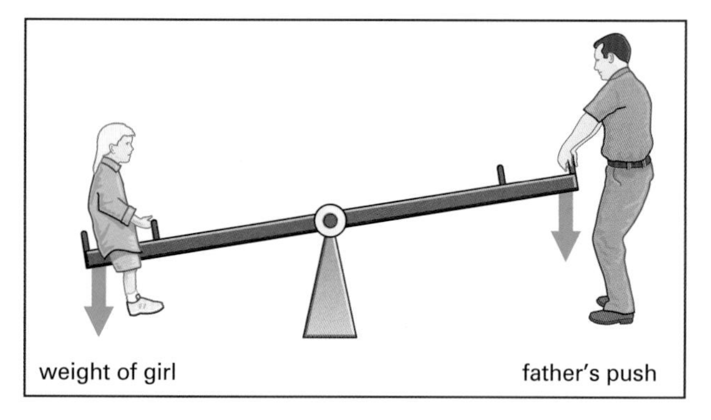

Figure 4.4 Two forces are causing this see-saw to tip. The girl's weight causes it to tip to the left, while her father provides a force to tip it to the right. He can increase the turning effect of his force by increasing the force, or by pushing down at a greater distance from the pivot.

A see-saw is an example of a **beam**, a long, rigid object that is pivoted at a point. The girl's weight is making the beam tip one way. The father's push is making it tip the other way. If the beam is to be balanced, the moments of the two forces must cancel each other out.

Equilibrium

When a beam is balanced, we say that it is in **equilibrium**. If an object is in equilibrium:

- the forces on it must be balanced (no resultant force)
- the turning effects of the forces on it must also be balanced (no resultant turning effect).

If a resultant force acts on an object, it will start to move off in the direction of the resultant force. If there is a resultant turning effect, it will start to rotate.

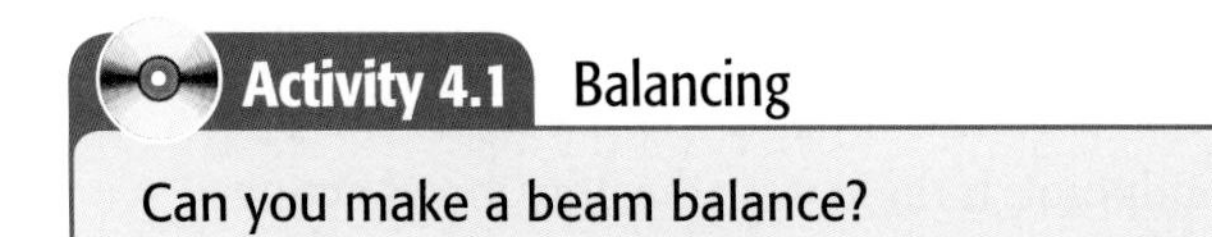

QUESTIONS

1 Figure **4.5** shows a heavy trapdoor. Three different forces are shown pulling on the trapdoor. Which force will have the biggest turning effect? Explain your answer.

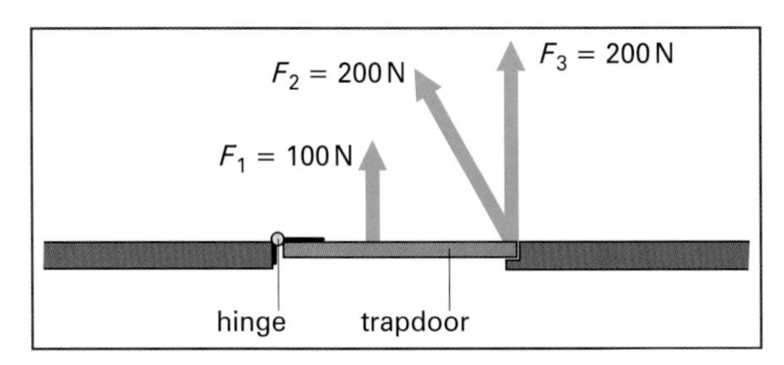

Figure 4.5 For Question **1**.

2 A tall tree can survive a gentle breeze but it may be blown over by a high wind. Explain why a tall tree is more likely to blow over than a short tree.

4.2 Calculating moments

We have seen that, the greater a force and the further it acts from the pivot, the greater is its moment. We can write an equation for calculating the moment of a force:

> moment of a force = force × perpendicular distance from pivot to force

Units: since moment is a force (in N) multiplied by a distance (in m), its unit is simply the newton metre (N m). There is no special name for this unit in the SI system.

Figure **4.6** shows an example. The 40 N force is 2.0 m from the pivot, so:

$$\text{moment of force} = 40\,\text{N} \times 2.0\,\text{m} = 80\,\text{N m}$$

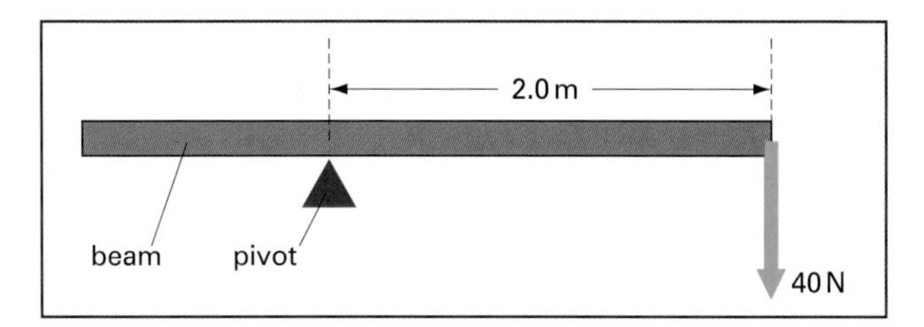

Figure 4.6 Calculating the moment of a force.

Balancing moments

The three children in Figure **4.7** have balanced their see-saw – it is in equilibrium. The weight of the child on the left is tending to turn the see-saw anticlockwise. So the weight of the child on the left has an anticlockwise moment. The weights of the two children on the right have clockwise moments.

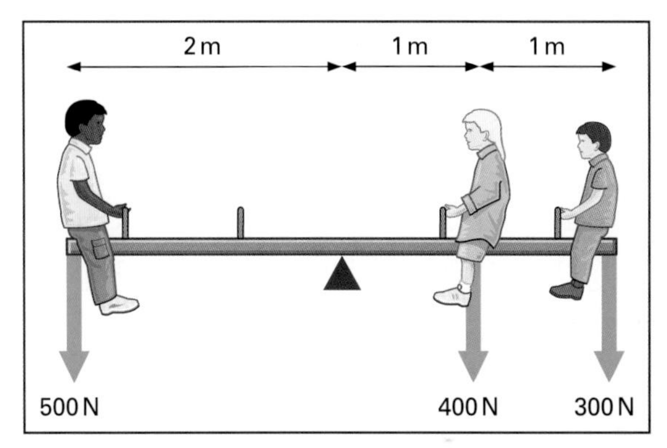

Figure 4.7 A balanced see-saw. On her own, the child on the left would make the see-saw turn anticlockwise; her weight has an anticlockwise moment. The weight of each child on the right has a clockwise moment. Since the see-saw is balanced, the sum of the clockwise moments must equal the anticlockwise moment.

From the data in Figure **4.7**, we can calculate these moments:

$$\begin{aligned}\text{anticlockwise moment} &= 500 \times 2.0 = 1\,000\,\text{N m}\\ \text{clockwise moments} &= (300 \times 2.0) + (400 \times 1.0)\\ &= 600\,\text{N m} + 400\,\text{N m}\\ &= 1000\,\text{N m}\end{aligned}$$

(The brackets are included as a reminder to perform the multiplications before the addition.) We can see that in this situation:

> total clockwise moment = total anticlockwise moment

So the see-saw in Figure **4.7** is balanced.

We can use this idea to find the value of an unknown force or distance, as shown in Worked example **1**.

E

Worked example 1

The beam shown in Figure **4.8** is 2.0 m long and has a weight of 20 N. It is pivoted as shown. A force of 10 N acts downwards at one end. What force F must be applied downwards at the other end to balance the beam?

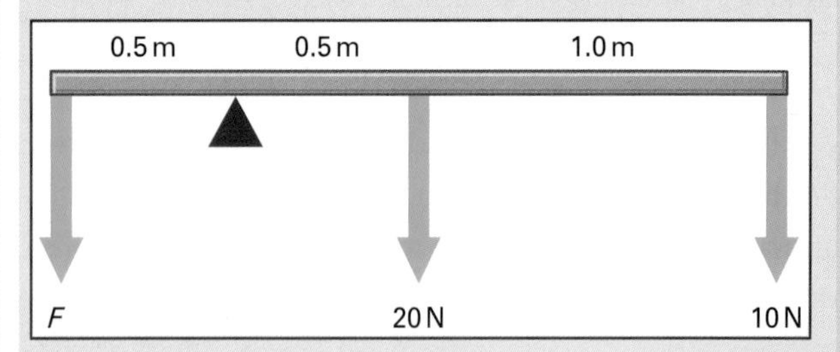

Figure 4.8 A balanced beam. Note that the weight of the beam (20 N) is represented by a downward arrow at its midpoint.

Step 1: Identify the clockwise and anticlockwise forces. Two forces act clockwise: 20 N at a distance of 0.5 m, and 10 N at 1.5 m. One force acts anticlockwise: the force F at 0.5 m.

Step 2: Since the beam is in equilibrium, we can write

total clockwise moment
= total anticlockwise moment

Step 3: Substitute in the values from Step 1, and solve.

$$(20\,\text{N} \times 0.5\,\text{m}) + (10\,\text{N} \times 1.5\,\text{m}) = F \times 0.5\,\text{m}$$

$$10\,\text{N m} + 15\,\text{N m} = F \times 0.5\,\text{m}$$

$$25\,\text{N m} = F \times 0.5\,\text{m}$$

$$F = \frac{25\,\text{N m}}{0.5\,\text{m}} = 50\,\text{N}$$

So a force of 50 N is needed.

(You might have been able to work this out in your head, by looking at the diagram. The 20 N weight requires 20 N to balance it, and the 10 N at 1.5 m needs 30 N at 0.5 m to balance it. So the total force needed is 50 N.)

In equilibrium

In the drawing of the three children on the see-saw (Figure **4.7**), three forces are shown acting downwards.

E

There is also the weight of the see-saw itself, 200 N, to consider, which also acts downwards, through its midpoint. If these were the **only** forces acting, they would make the see-saw accelerate downwards. Another force acts to prevent this from happening. There is an upward **contact force** where the see-saw sits on the pivot. Figure **4.9** shows all five forces.

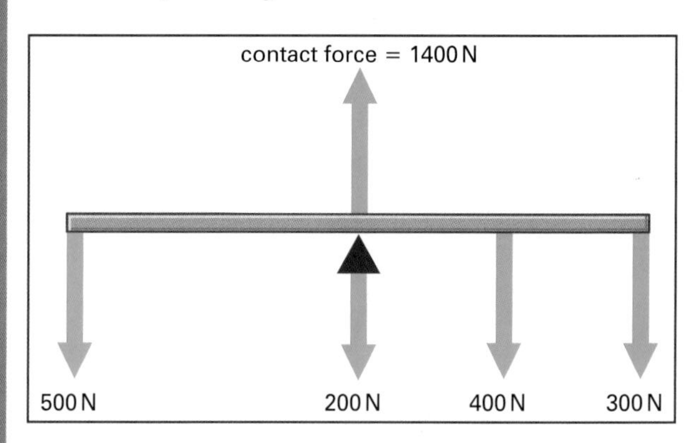

Figure 4.9 A force diagram for the see-saw shown in Figure **4.7**. The upward contact force of the pivot on the see-saw balances the downward forces of the children's weights and the weight of the see-saw itself. The contact force has no moment about the pivot because it acts through the pivot. The weight of the see-saw is another force that acts through the pivot, so it also has no moment about the pivot.

Because the see-saw is in equilibrium, we can calculate this contact force. It must balance the four downwards forces, so its value is $(500 + 200 + 400 + 300)\,\text{N} = 1400\,\text{N}$, upwards. This force has no turning effect because it acts through the pivot. Its distance from the pivot is zero, so its moment is zero.

Now we have satisfied the two conditions that must be met if an object is to be in equilibrium:

- there must be no resultant force acting on it
- total clockwise moment = total anticlockwise moment.

You can use these two rules to solve problems concerning the forces acting on objects in equilibrium.

Activity 4.2 A question of balance

Predict the forces on a balanced beam.

Activity 4.3 Balancing problems

Solve some more problems for systems in equilibrium.

QUESTIONS

3 Figure **4.10** shows a balanced beam. Calculate the unknown forces X and Y.

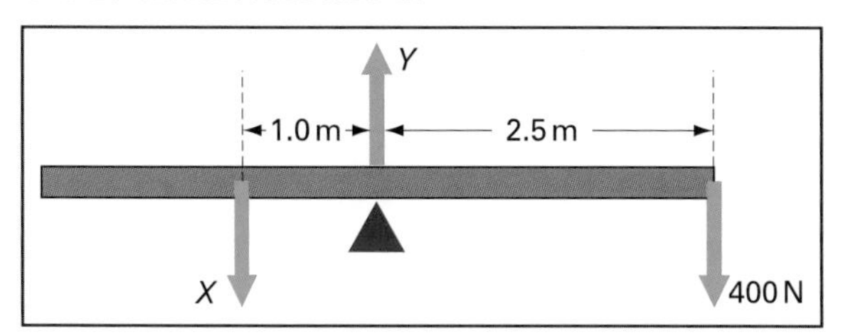

Figure 4.10 For Question **3**.

4 Figure **4.11** shows a beam, balanced at its midpoint. The weight of the beam is 40 N. Calculate the unknown force Z, and the length of the beam.

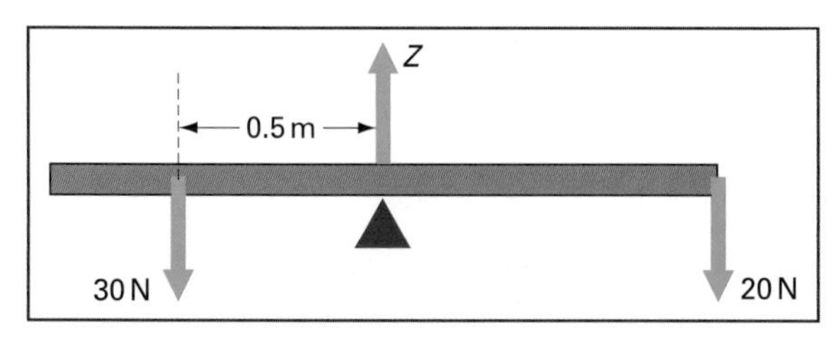

Figure 4.11 For Question **4**.

4.3 Stability and centre of mass

People are tall and thin, like a pencil standing on end. Unlike a pencil, we do not topple over when touched by the slightest push. We are able to remain upright, and to walk, because we make continual adjustments to the positions of our limbs and body. We need considerable brain power to control our muscles for this. The advantage is that, with our eyes about a metre higher than if we were on all-fours, we can see much more of the world.

Circus artistes such as tightrope walkers (Figure **4.12**) have developed the skill of remaining upright to a high degree. They use items such as poles or parasols to help them maintain their balance. The idea of moments can help us to understand why some objects are stable while others are more likely to topple over.

A tall glass is easily knocked over – it is unstable. It could be described as top-heavy, because most of its mass is concentrated high up, above its stem. Figure **4.13** shows what happens if the glass is tilted.

Figure 4.12 This high-wire artiste is using a long pole to maintain her stability on the wire. If she senses that her weight is slightly too far to the left, she can redress the balance by moving the pole to the right. Frequent, small adjustments allow her to walk smoothly along the wire.

a When the glass is upright, its weight acts downwards and the contact force of the table acts upwards. The two forces are in line, and the glass is in equilibrium.

b If the glass is tilted slightly to the right, the forces are no longer in line. There is a pivot at the point where the base of the glass is in contact with the table. The line of the glass's weight is to the left of this pivot, so it has an anticlockwise moment, which tends to tip the glass back to its upright position.

c Now the glass is tipped further. Its weight acts to the right of the pivot, and has a clockwise moment, which makes the glass tip right over.

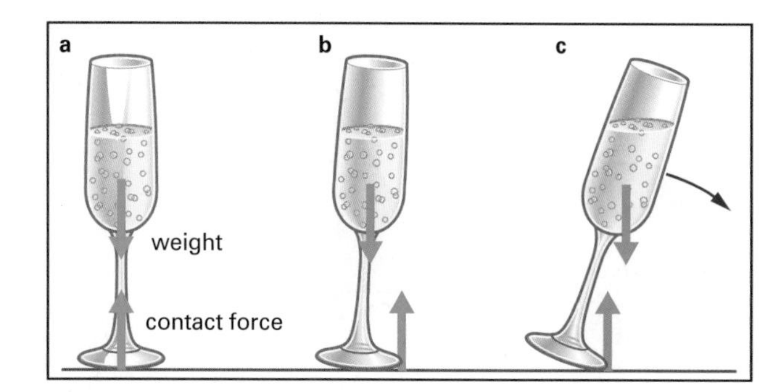

Figure 4.13 A tall glass is easily toppled. Once the line of action of its weight is beyond the edge of the base, as in **c**, the glass tips right over.

Centre of mass

In Figure **4.13**, the weight of the glass is represented by an arrow starting at a point inside the liquid in the bowl of the glass. Why is this? The reason is that the glass behaves as if all of its mass were concentrated at this point, known as the **centre of mass**. The glass is top-heavy because its centre of mass is high up. The force of gravity acts on the mass of the glass – each bit of the glass is pulled by the Earth's gravity. However, rather than drawing lots of weight arrows, one for each bit of the glass, it is simpler to draw a single arrow acting through the centre of mass. (Because we can think of the weight of the glass acting at this point, it is sometimes known as the centre of gravity.)

Figure **4.14** shows the position of the centre of mass for several objects. A person is fairly symmetrical, so their centre of mass must lie somewhere on the axis of symmetry. (This is because half of their mass is on one side of the axis, and half on the other.) The centre of mass is in the middle of the body, roughly level with the navel. A ball is much more symmetrical, and its centre of mass is at its centre.

For an object to be stable, it should have a low centre of mass and a wide base. The pyramid in Figure **4.14** is an example of this. (The Egyptian pyramids are among the Wonders of the World. It has been suggested that, if they had been built the other way up, they would have been even greater wonders!) The tightrope walker shown in Figure **4.12** has to adjust her position so that her centre of mass remains above her 'base' – the point where her feet make contact with the rope.

Finding the centre of mass

Balancing is the clue to finding an object's centre of mass. A metre rule balances at its midpoint, so that is where its centre of mass must lie.

The procedure for finding the centre of mass of a more irregularly shaped object is shown in Figure **4.15**. In this case, the object is a piece of card, described as a plane **lamina**. The card is suspended from a pin. If it is free to move, it hangs with its centre of mass below the point of suspension. (This is because its weight pulls it round until the weight and the contact force at the pin are lined up. Then there is no moment about the pin.) A plumb-line is used to mark a vertical line below the pin. The centre of mass must lie on this line.

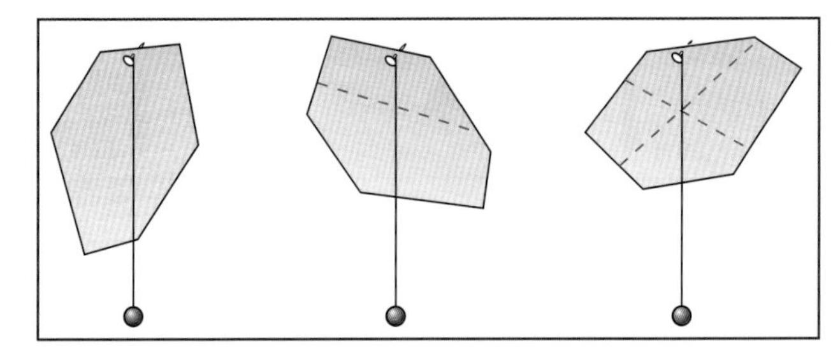

Figure 4.15 Finding the centre of mass of an irregularly shaped piece of card. The card hangs freely from the pin. The centre of mass must lie on the line indicated by the plumb-line hanging from the pin. Three lines are enough to find the centre of mass.

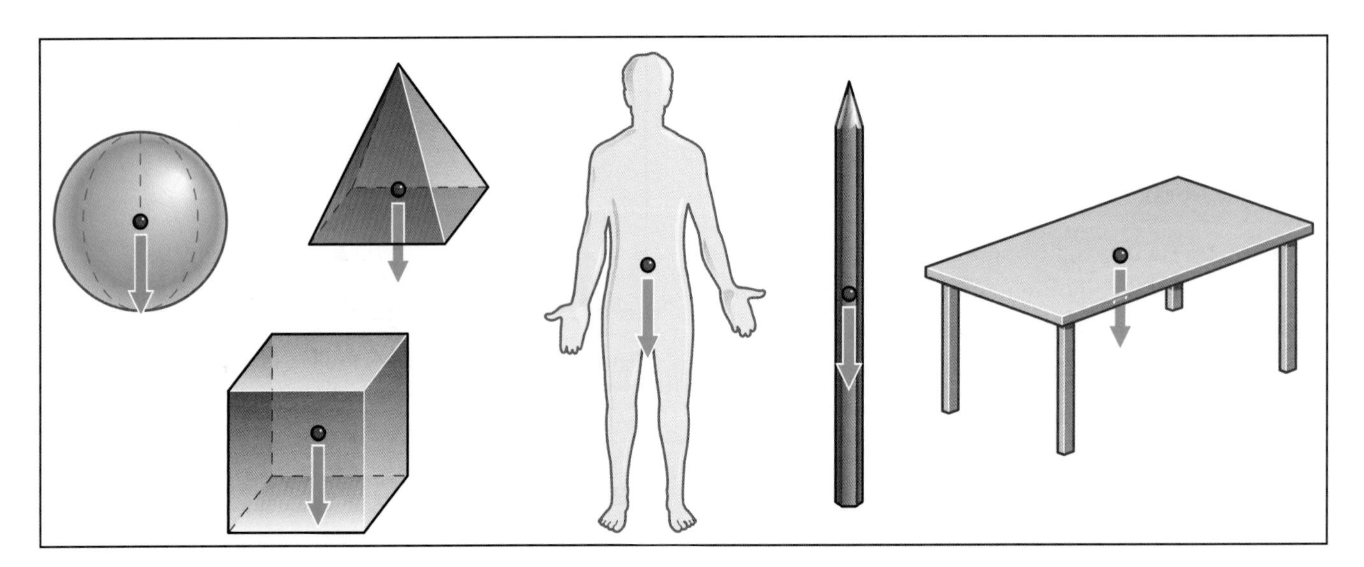

Figure 4.14 The weight of an object acts through its centre of mass. Symmetry can help to judge where the centre of mass lies. An object's weight can be considered to act through this point. Note that, for the table, its centre of mass is in the air below the tabletop.

The process is repeated for two more pinholes. Now there are three lines on the card, and the centre of mass must lie on all of them, that is, at the point where they intersect. (Two lines might have been enough, but it is advisable to use at least three points to show up any inaccuracies.)

Activity 4.4 Centre of mass of a plane lamina

Use the method described in the text to find the centre of mass of a sheet of card.

QUESTIONS

5 Use the ideas of **stability** and **centre of mass** to explain the following.

  a Double-decker buses have heavy weights attached to their undersides.

  b The crane shown in Figure **4.16** has a heavy concrete block attached to one end of its arm, and others placed around its base.

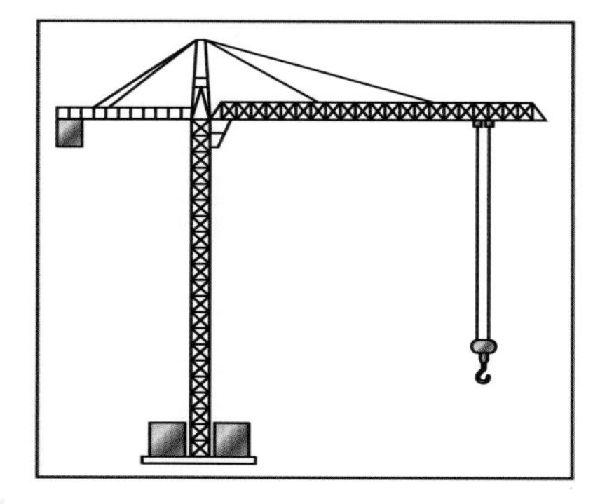

Figure 4.16 For Question **5**.

6 Figure **4.17** shows the forces acting on a cyclist. Look at part **a** of the diagram.

  a Explain how you can tell that the cyclist shown in part **a** is in equilibrium.

  Now look at part **b** of the diagram.

  b Are the forces on the cyclist balanced now? How can you tell?

  c Would you describe the cyclist as **stable** or **unstable**? Explain your answer.

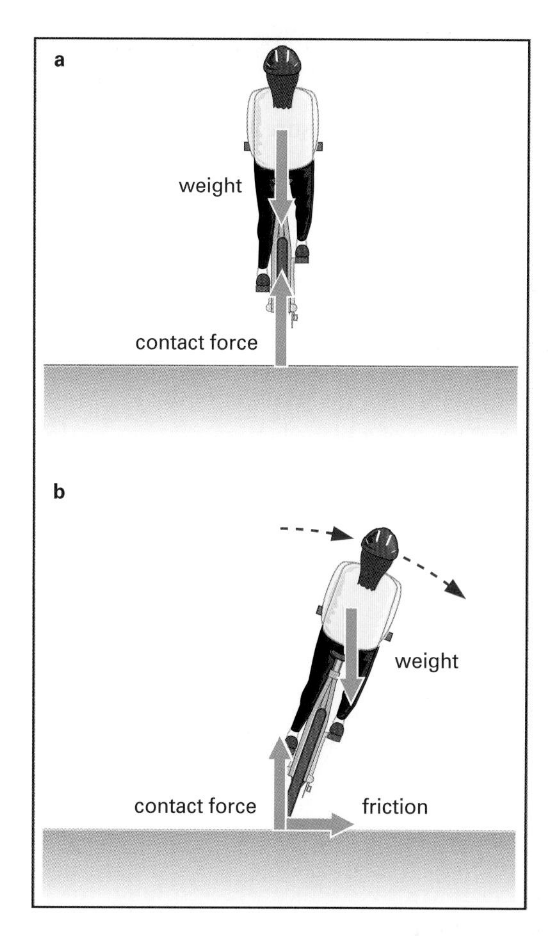

Figure 4.17 Forces on a cyclist – for Question **6**.

Summary

The moment of a force is a measure of its turning effect.

When a system is in equilibrium, the resultant force is zero and the resultant turning effect is zero.

For an object to be stable, its centre of mass must be low down and it must have a large base.

E

Moment of a force = force × perpendicular distance from pivot to force

When an object is in equilibrium:
total clockwise moment = total anticlockwise moment

End-of-chapter questions

4.1 What quantity is a measure of the turning effect of a force? [1]

4.2 What **two** conditions must be met if an object is to be in equilibrium? [2]

4.3 Write out step-by-step instructions for an experiment to find the position of the centre of mass of a plane lamina. [5]

E

4.4 The diagram (Figure **4.18**) shows a 3 m uniform beam AB, pivoted 1.0 m from the end A. The weight of the beam is 200 N.

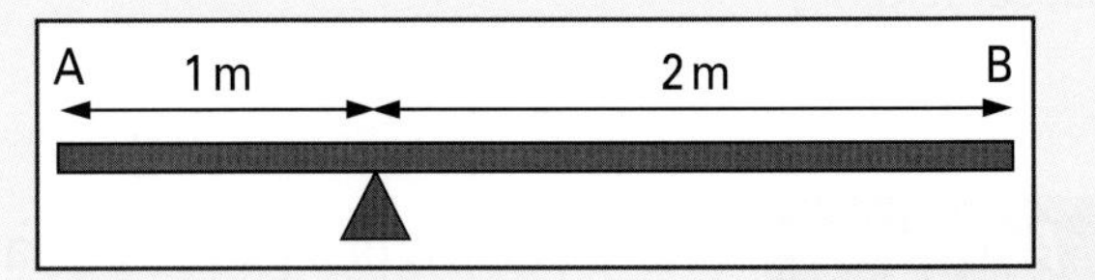

Figure 4.18 For Question **4.4**.

a Copy the diagram and mark the beam's centre of mass. [1]

b Add arrows to show the following forces: the weight of the beam; the contact force on the beam at the pivot. [2]

c A third force *F* presses down on the beam (at end point A). What value of *F* is needed to balance the beam? [5]

d When this force is applied, what is the value of the contact force that the pivot exerts on the beam? [3]

5 Forces and matter

Core	Using forces to change the shape and size of a body
Core	Carrying out experiments to produce extension against load graphs
E Extension	Interpreting extension against load graphs
Extension	Using Hooke's law
Core	Understanding the factors that affect pressure
E Extension	Calculating pressure

5.1 Forces acting on solids

Forces can change the size and shape of an object. They can stretch, squash, bend or twist it. Figure 5.1 shows the forces needed for these different ways of deforming an object. You could imagine holding a cylinder of foam rubber, which is easy to deform, and changing its shape in each of these ways.

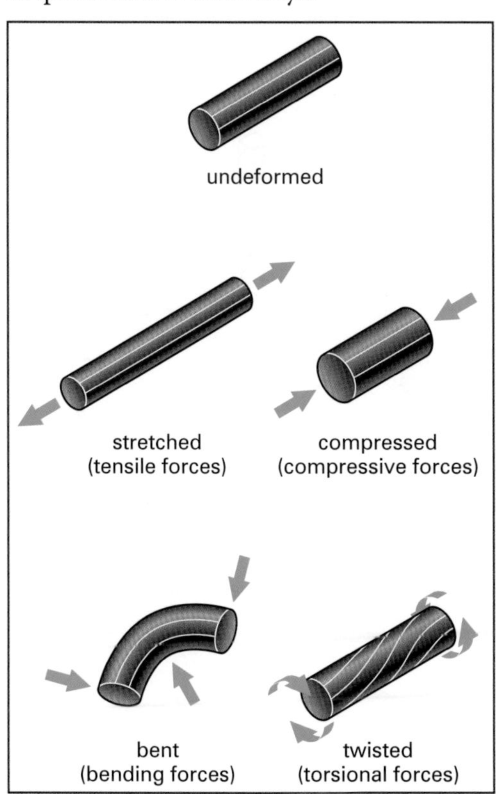

Figure 5.1 Forces can change the size and shape of a solid object. These diagrams show four different ways of deforming a solid object.

Foam rubber is good for investigating how things deform because, when the forces are removed, it springs back to its original shape. Here are two more examples of materials that deform in this way:

- When a football is kicked, it is compressed for a short while (see Figure 5.2). Then it springs back to its original shape as it pushes itself off the foot of the player who has kicked it. The same is true for a tennis ball when struck by a racket.
- Bungee jumpers rely on the springiness of the rubber rope, which breaks their fall when they jump from a height. If the rope became permanently stretched, they would stop suddenly at the bottom of their fall, rather than bouncing up and down and gradually coming to a halt.

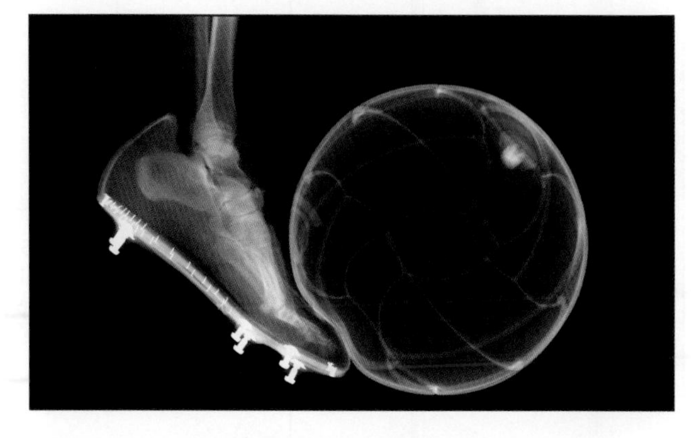

Figure 5.2 This remarkable X-ray image shows how a football is compressed when it is kicked. It returns to its original shape as it leaves the player's boot. (This is an example of an elastic deformation.) The boot is also compressed slightly but, because it is stiffer than the ball, the effect is less noticeable.

Some materials are less springy. They become permanently deformed when forces act on them.

- When two cars collide, the metal panels of their bodywork are bent. In a serious crash, the solid metal sections of the car's chassis are also bent.
- Gold and silver are metals that can be deformed by hammering them (see Figure 5.3). People have known for thousands of years how to shape rings and other ornaments from these precious metals.

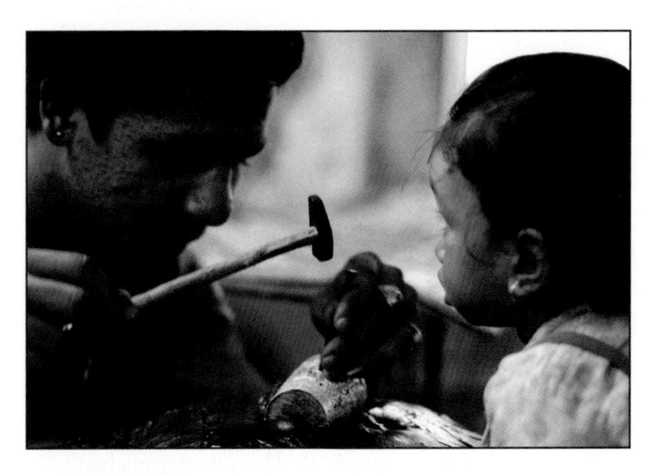

Figure 5.3 A Tibetan silversmith making a wrist band. Silver is a relatively soft metal at room temperature, so it can be hammered into shape without the need for heating.

5.2 Stretching springs

To investigate how objects deform, it is simplest to start with a spring. Springs are designed to stretch a long way when a small force is applied, so it is easy to measure how their length changes.

Figure 5.4 shows how to carry out an investigation on stretching a spring. The spring is hung from a rigid clamp, so that its top end is fixed. Weights are hung on the end of the spring – these are referred to as the **load**. As the load is increased, the spring stretches and its length increases.

Figure 5.5 shows the pattern observed as the load is increased in regular steps. The length of the spring increases (also in regular steps). At this stage the spring will return to its original length if the load is removed. However, if the load is increased too far, the spring becomes permanently stretched and will not return to its original length. It has been **inelastically deformed**.

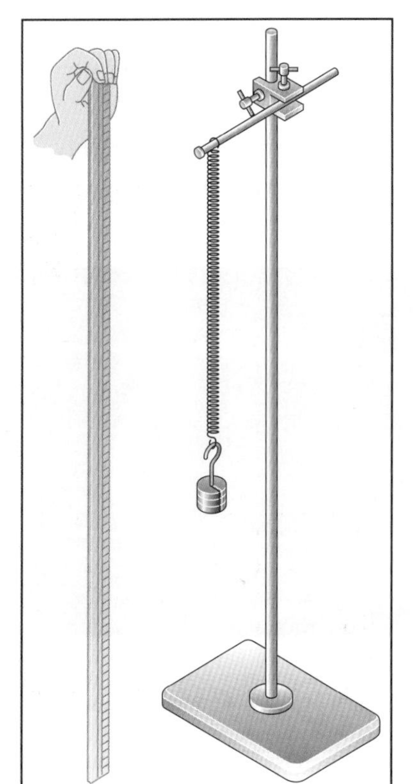

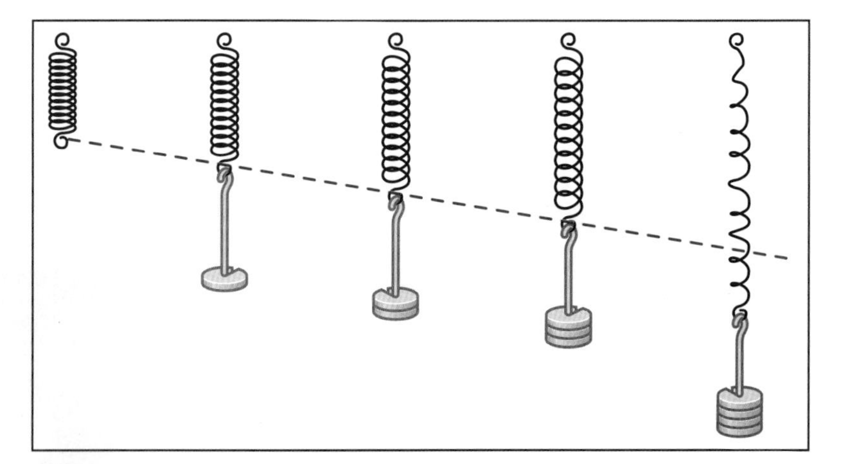

Figure 5.5 Stretching a spring. At first, the spring deforms elastically. It will return to its original length when the load is removed. Eventually, however, the load is so great that the spring is damaged.

Figure 5.4 Investigating the stretching of a spring.

Extension of a spring

As the force stretching the spring increases, it gets longer. It is important to consider the increase in length of the spring. This quantity is known as the **extension**.

length of stretched spring
= original length + extension

Table **5.1** shows how to use a table with three columns to record the results of an experiment to stretch a spring. The third column is used to record the value of the extension, calculated by subtracting the original length from the value in the second column.

Load / N	Length / cm	Extension / cm
0.0	24.0	0.0
1.0	24.6	0.6
2.0	25.2	1.2
3.0	25.8	1.8
4.0	26.4	2.4
5.0	27.0	3.0
6.0	27.6	3.6
7.0	28.6	4.6
8.0	29.5	5.6

Table 5.1 Results from an experiment to find out how a spring stretches as the load on it is increased.

To see how the extension depends on the load, we draw an extension against load graph (Figure **5.6**). You can see that the graph is in two parts.

- At first, the graph slopes up steadily. This shows that the extension increases in equal steps as the load increases.
- Then the graph bends. This happens when the load is so great that the spring has become permanently damaged. It will not return to its original length.

(You can see the same features in Table **5.1**. Look at the third column. At first, the numbers go up in equal steps. The last two steps are bigger.)

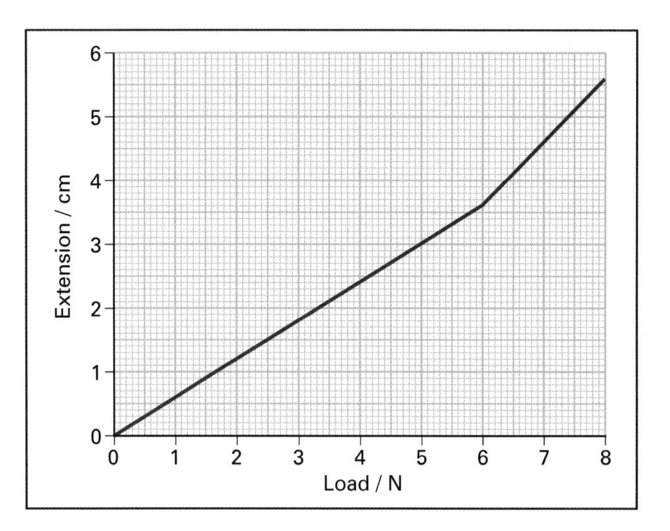

Figure 5.6 An extension against load graph for a spring, based on the data in Table **5.1**.

Activity 5.1 Investigating springs

Use weights to stretch a spring, and then plot a graph to show the pattern of your results.

QUESTIONS

1 A piece of elastic cord is 80 cm long. When it is stretched, its length increases to 102 cm. What is its extension?

2 Table **5.2** shows the results of an experiment to stretch an elastic cord. Copy and complete the table, and draw a graph to represent this data.

Load / N	Length / mm	Extension / mm
0.0	50	0
1.0	54	
2.0	58	
3.0	62	
4.0	66	
5.0	70	
6.0	73	
7.0	75	
8.0	76	

Table 5.2 For Question **2**.

5.3 Hooke's law

The mathematical pattern of the stretching spring was first described by the English scientist Robert Hooke. He realised that, when the load on the spring was doubled, the extension also doubled. Three times the load gave three times the extension, and so on. This shows up in the graph in Figure 5.7. The graph shows how the extension depends on the load. At first, the graph is a straight line, leading up from the origin. This shows that the extension is proportional to the load.

At a certain point, the graph bends and the line slopes up more steeply. This point is called the **limit of proportionality**. If the spring is stretched beyond this point, it will be permanently damaged. If the load is removed, the spring will not return all the way to its original, undeformed length. (This point is also known as the **elastic limit**.)

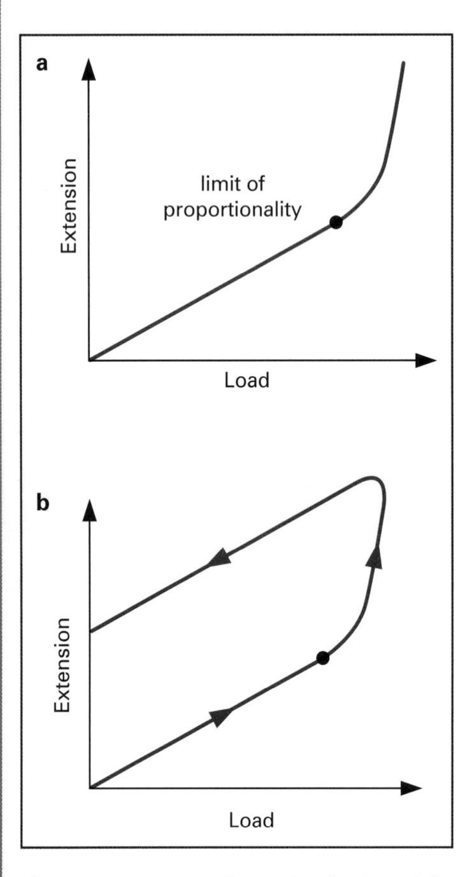

Figure 5.7 **a** An extension against load graph for a spring. Beyond the limit of proportionality, the graph is no longer a straight line, and the spring is permanently deformed. **b** This graph shows what happens when the load is removed. The extension does not return to zero, showing that the spring is now longer than at the start of the experiment.

The behaviour of the spring is represented by the graph of Figure 5.7**a** and is summed up by **Hooke's law**:

> The extension of a spring is proportional to the load applied to it, provided the limit of proportionality is not exceeded.

We can also write Hooke's law as an equation:

$$F = kx$$

In this equation, F is the load (force) stretching the spring, k is the stiffness of the spring, and x is the extension of the spring.

Worked example 1

A spring has a stiffness $k = 20\,\text{N/cm}$. What load is needed to produce an extension of 2.5 cm?

Step 1: Write down what you know and what you want to find out.

load $F = ?$
stiffness $k = 20\,\text{N/cm}$
extension $= 2.5\,\text{cm}$

Step 2: Write down the equation linking these quantities, substitute values and calculate the result.

$F = kx$
$F = 20 \times 2.5 = 50\,\text{N}$

So a load of 50 N will stretch the spring by 2.5 cm.

How rubber behaves

A rubber band can be stretched in a similar way to a spring. As with a spring, the bigger the load, the bigger the extension. However, if the weights are added with great care, and then removed one by one without releasing the tension in the rubber, the following can be observed:

- The graph obtained is not a straight line. Rather, it has a slightly S-shaped curve. This shows that the extension is not exactly proportional to the load. Rubber does not obey Hooke's law.

- Eventually, increasing the load no longer produces any extension. The rubber feels very stiff. When the load is removed, the graph does not come back exactly to zero.

Activity 5.2 Stretching rubber

Carry out an investigation into the stretching of a rubber band. This is a good test of your experimental skills. You will need to work carefully if you are to see the effects described above.

Hooke and springs

Why was Robert Hooke so interested in springs? Hooke was a scientist, but he was also a great inventor. He was interested in springs for two reasons:

- Springs are useful in making weighing machines, and Hooke wanted to make a weighing machine that was both very sensitive (to weigh very light objects) and very accurate (to measure very precise quantities).
- He also realised that a spiral spring could be used to control a clock or even a wristwatch.

Figure 5.8 shows a set of diagrams drawn by Hooke, including a long spring and a spiral spring, complete with pans for carrying weights. You can also see some of his graphs.

For scientists, it is important to publish results so that other scientists can make use of them. Hooke was very secretive about some of his findings, because he did not want other people to use them in their own inventions. For this reason, he published some of his findings in code. For example, instead of writing his law of springs as given above, he wrote this: "ceiiinosssttuv". Later, when he felt that it was safe to publish his ideas, he revealed that this was an anagram of a sentence in Latin. Decoded, it said:

Ut tensio, sic vis.

In English, this is:

As the extension increases, so does the force.

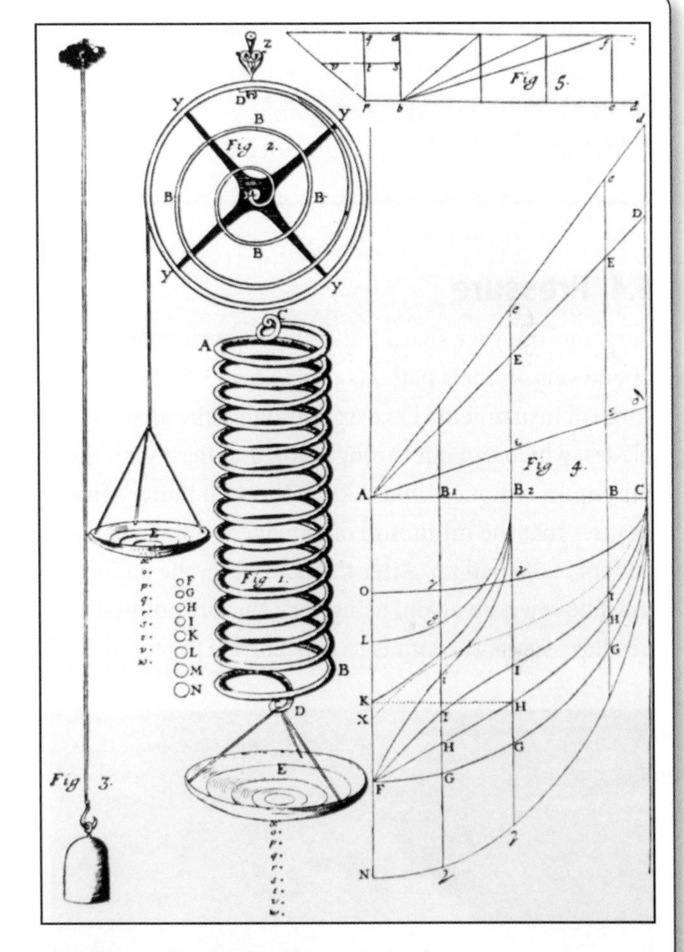

Figure 5.8 Robert Hooke's diagrams of springs.

In other words, the extension is proportional to the force producing it. You can see Hooke's straight-line graph in Figure 5.8.

QUESTIONS

3 A spring requires a load of 2.5 N to increase its length by 4 cm. The spring obeys Hooke's law. What load will give it an extension of 12 cm?

4 A spring has an unstretched length of 12.0 cm. Its stiffness k is 8 N/cm. What load is needed to stretch the spring to a length of 15.0 cm?

5 Table 5.3 shows the results of an experiment to stretch a spring. Use the results to plot an extension against load graph. On your graph, mark the limit of proportionality and state the value of the load at this point.

Load / N	Length / m
0.0	0.800
2.0	0.815
4.0	0.830
6.0	0.845
8.0	0.860
10.0	0.880
12.0	0.905

Table 5.3 For Question **5**.

5.4 Pressure

Surgeons use very sharp knives when performing operations on their patients (see Figure 5.9). Today's surgical instruments have exceedingly thin, sharp blades, which can cut through skin and flesh with the minimum of force – 'like a knife through butter'. This ensures that the minimum of damage is done when a patient is opened up. After the operation, the patient must be sewn up again, using very sharply pointed needles. Again, the aim is to concentrate the force pushing the needle onto a very small area. Then a small force is needed to pass through the skin.

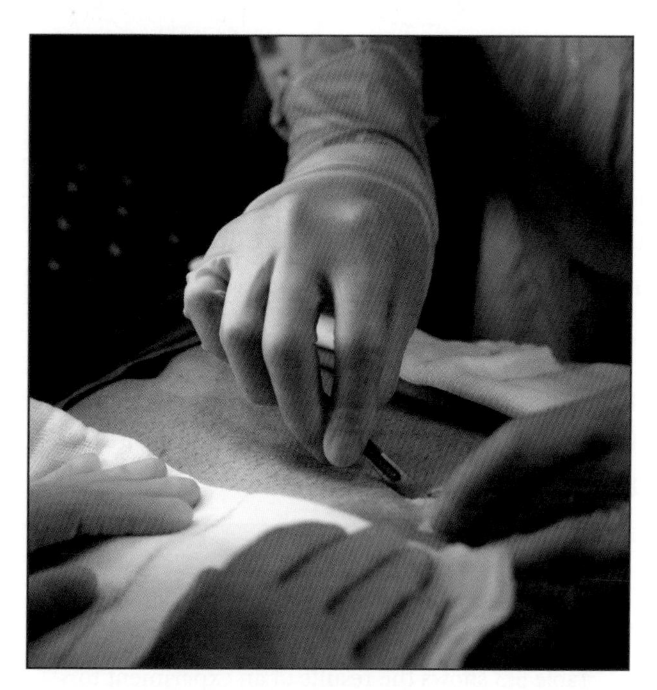

Figure 5.9 Surgeons aim to minimise the damage to a patient during an operation. Surgical instruments are made from special steels, which can be sharpened to give very fine points and edges. This means that less force is needed to cut through a patient's skin, or to cut into an organ.

Knives and needles have sharp blades and tips in order to concentrate the force that is being applied onto a small area. This means that the **pressure** that is being applied is as large as possible. Pressure tells us how concentrated a force is. If a force is spread out over a large area, its pressure is low. If a force is concentrated on a small area, its pressure is high.

High pressure, low pressure

Knives have sharp blades to give a high pressure. Ice skates also have 'blades', the part that is in contact with the ice. This is narrow, so that the skater's weight is concentrated on a small area. The effect of this high pressure is to melt the ice just below the blade. This gives a thin film of water, which provides lubrication for the skate as it skims over the ice. As the skate moves on, the water re-freezes. On very cold days, the pressure may not be enough to melt the ice, and skating is impossible.

If you are unlucky enough to find yourself standing on thin ice, it is advisable to lie down on it to spread out your weight. This reduces the pressure, and the ice is less likely to break. The same principle is used elsewhere. 'Crawling boards' are often used to climb over a glass roof. These spread out the weight of anyone needing to climb on the roof. Some vehicles are fitted with very wide tyres (Figure 5.10) so that they exert less pressure on soft ground.

Figure 5.10 This truck has wide tyres to spread its weight as it travels over the sand dunes. Camels have big, flat feet for the same reason, to reduce the pressure on the sand so that they are less likely to sink in and become bogged down.

QUESTION

6 Use the idea of pressure to explain the following.
  a Sharks and crocodiles have sharp teeth.
  b Camels have wide, flat feet.
  c If you walk on a wooden floor wearing stilettos (shoes with very narrow heels), you may damage the floor.

E

Calculating pressure

A large force pressing on a small area gives a high pressure. We can think of **pressure** as the force per unit area acting on a surface:

$$\text{pressure} = \frac{\text{force}}{\text{area}} \qquad p = \frac{F}{A}$$

Units: if force F is measured in newtons (N) and area A in square metres (m^2), pressure p is in newtons per square metre (N/m^2). In the SI system of units, this is given the name **pascal (Pa)**.

Worked example 2

Shoes with stiletto heels go in and out of fashion. ('Stiletto' is an Italian word meaning a small and murderous dagger.) Such heels can damage floors, and dance halls often have notices requiring them to be removed. Calculate the pressure exerted by a woman dancer weighing 600 N standing on a single heel of area 1 cm^2. If the surface of the dancefloor is broken by pressures over five million pascals (5 MPa), will it be damaged?

Step 1: To calculate the pressure, we need to know the force, and the area on which the force acts, in m^2.

force $F = 600\,\text{N}$

area $A = 1\,\text{cm}^2 = 0.0001\,\text{m}^2 = 10^{-4}\,\text{m}^2$

Step 2: Now we can calculate the pressure p.

$$p = \frac{F}{A} = \frac{600\,\text{N}}{0.0001\,\text{m}^2}$$

$$= 6\,000\,000\,\text{Pa} = 6\,\text{MPa}$$

E

The pressure is thus 6×10^6 Pa, or 6 MPa. This is more than the minimum pressure needed to break the surface of the floor, so it will be damaged.

QUESTIONS

7 Write down an equation that defines pressure.

8 What are the SI units of pressure?

9 Which exerts a greater pressure, a force of 100 N acting on 1 cm^2, or the same force acting on 2 cm^2?

10 What pressure is exerted by a force of 40 000 N acting on 2 m^2?

11 A swimming pool has a level, horizontal, bottom of area 10.0 m by 4.0 m. If the pressure of the water on the bottom is 15 000 Pa, what total force does the water exert on the bottom of the pool?

Pressure in fluids

In a fluid such as water or air, pressure does not simply act downwards – it acts equally in all directions. This is because the molecules of the fluid move around in all directions, causing pressure on every surface they collide with.

If you dive into a swimming pool, you will experience the pressure of the water on you. It provides the upthrust on you, which pushes you back to the surface. The deeper you go, the greater the pressure acting on you. Deep-sea divers have to take account of this. They wear protective suits, which will stop them being crushed by the pressure. Submarines and marine exploring vehicles (Figure **5.11**) must be designed to withstand very great pressures. They have curved surfaces, which are less likely to buckle under pressure, and they are made of thick metal.

This pressure comes about because any object under water is being pressed down on by the weight of water above it. The deeper you go, the greater the amount of water pressing down on you (see Figure **5.12a**). In a similar way, the atmosphere exerts pressure on us, although we are not normally conscious of this. The Earth's gravity pulls it downwards, so that

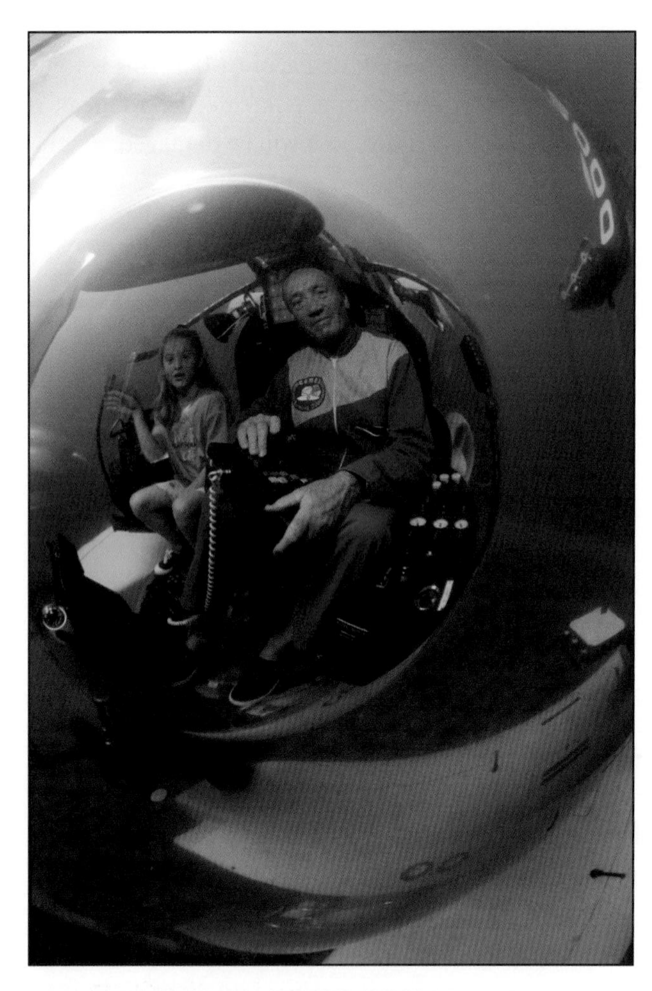
Figure 5.11 This underwater exploring vehicle is used to carry tourists to depths of 600 m, where the pressure is 60 times that at the surface. The design makes use of the fact that spherical and cylindrical surfaces stand up well to pressure. The viewing window is made of acrylic plastic and is 9.5 cm thick.

the atmosphere presses downwards on our heads. Mountaineers climbing to the top of Mount Everest rise through two-thirds of the atmosphere, so the pressure is only about one-third of the pressure down at sea-level. There is much less air above them, pressing down.

The pressure caused by water is much greater than that caused by air because water is much denser than air. Figure **5.12b** shows how a dam is designed to withstand the pressure of the water behind it. Because the pressure is greatest at the greatest depth, the dam must be made thickest at its base.

Pressure measurements

A **manometer** is a simple instrument for showing the difference in pressure between two gases or liquids.

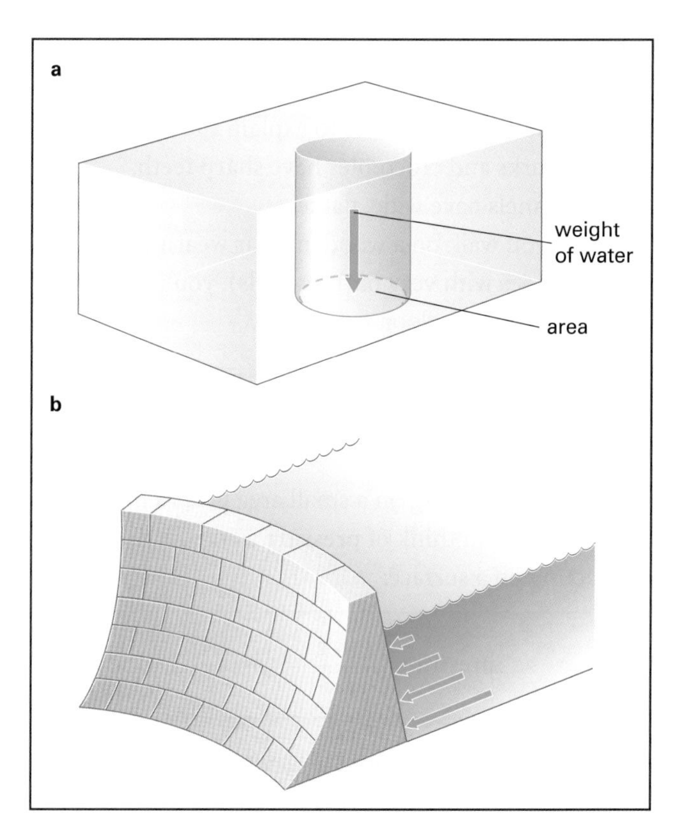

Figure 5.12 **a** Pressure is caused by the weight of water (or other fluid) above an object, pressing down on it. **b** This dam is thickest near its base, because that is where the pressure is greatest.

Figure **5.13** shows how a manometer is used to measure the pressure of the laboratory gas supply. This pressure must be higher than atmospheric pressure, or gas would not flow out of the pipe.

- A manometer is a U-shaped tube, holding a small amount of liquid.
- When both ends are open, the levels of the liquid in the two sides are the same.
- If one side is connected to the gas supply, the gas pushes down on the liquid and forces it round the bend. The levels are now unequal, showing that there is a difference in pressure.

A **barometer** can be used to measure atmospheric pressure. One simple type, the mercury barometer, is shown in Figure **5.14**. It consists of a long glass tube, at least 80 cm in length. The tube is filled with mercury and then carefully inverted into a trough containing mercury. This must be done carefully, so that no air enters the tube.

Once the tube is safely inverted, the level of mercury in the tube drops. The length l of the mercury column,

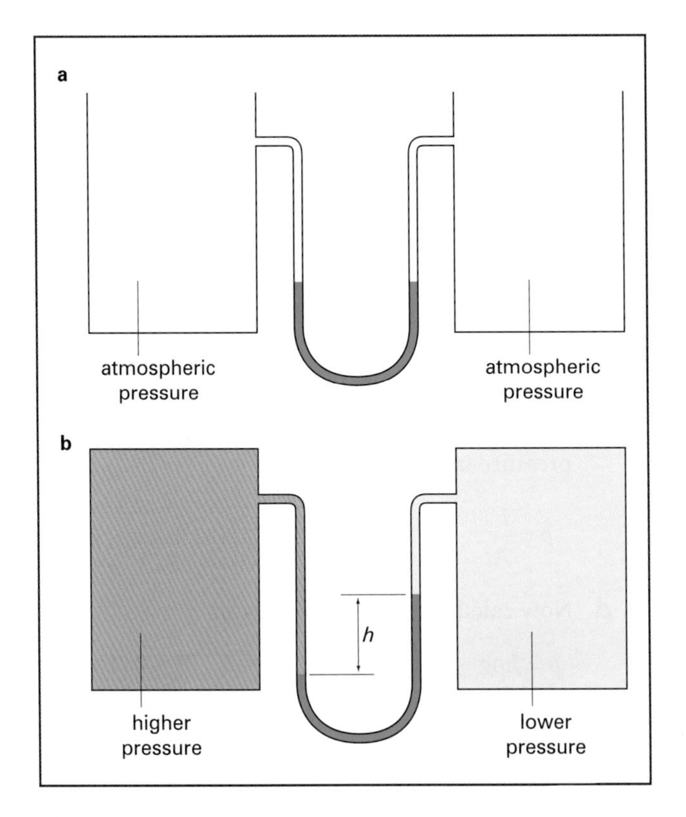

Figure 5.13 Using a manometer to measure the pressure difference between two gases. **a** With atmospheric pressure on both sides of the U-tube, the liquid is at the same level in both sides. **b** With higher pressure on one side, the liquid is pushed round. The greater the pressure difference, the greater is the difference in levels, *h*.

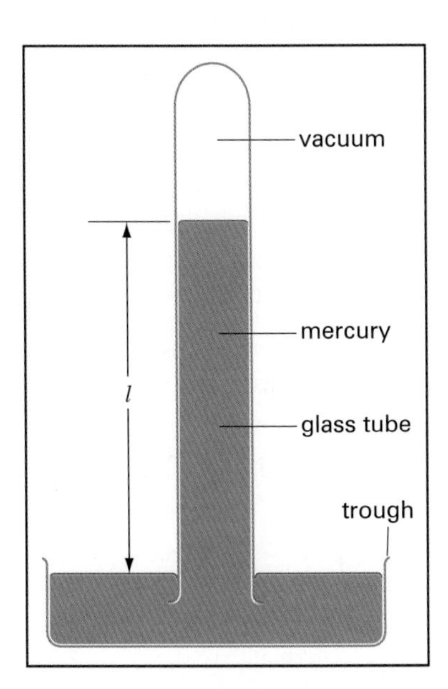

Figure 5.14 A mercury barometer is used to measure atmospheric pressure.

measured from the surface of the mercury in the trough, is about 76 cm. The space above the mercury column is a vacuum (with a small amount of mercury vapour).

The column length *l* depends on the atmospheric pressure. On a day when the atmospheric pressure is high, the air presses more strongly on the mercury in the trough, so that it rises further in the tube. If the pressure falls, the force on the mercury decreases, and the level in the tube decreases.

Mercury is used in barometers like this because it has a high density (more than 13 times the density of water). A barometer made using water would require a much taller tube, over 10 m in height!

Activity 5.3 Pressure experiments

Try out some simple experiments to explore the idea of pressure.

QUESTIONS

12 Name an instrument used to measure:

a atmospheric pressure

b differences in pressure.

13 Figure 5.15 shows two tanks, A and B. Each tank contains gas and is fitted with a manometer to show how the pressure compares with atmospheric pressure outside the tank.

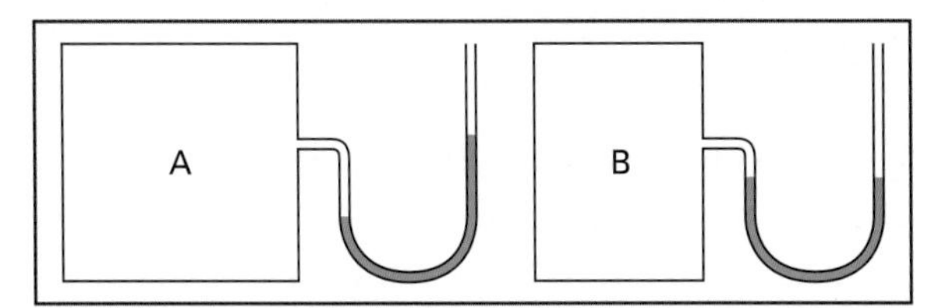

Figure 5.15 For Question **13**.

a In which tank is the gas pressure greater than atmospheric pressure? Explain how you can tell.

b What can you say about the pressure of the gas in the other tank?

E Pressure, depth and density

We have seen that the deeper one dives into water, the greater the pressure. Pressure *p* is proportional to depth *h* (we use the letter *h*, for height). Twice the depth means twice the pressure. Pressure also depends on the density ρ of the material (here ρ is the Greek letter rho). If you dive into mercury, which is more than ten times as dense as water, the pressure will be more than ten times as great.

We can write an equation for the pressure at a depth h in a fluid of density ρ:

pressure = depth × density × acceleration due to gravity

$p = h\rho g$

Worked example 3

Calculate the pressure on the bottom of a swimming pool 2.5 m deep. How does the pressure compare with atmospheric pressure, 10^5 Pa? (Density of water = 1000 kg/m^3.)

Step 1: Write down what you know, and what you want to know.

$h = 2.5\,\text{m}$
$\rho = 1000\,\text{kg/m}^3$
$g = 10\,\text{m/s}^2$
$p = ?$

Step 2: Write down the equation for pressure, substitute values and calculate the answer.

$$p = h\rho g = 2.5\,\text{m} \times 1000\,\text{kg/m}^3 \times 10\,\text{m/s}^2 = 2.5 \times 10^4\,\text{Pa}$$

This is one-quarter of atmospheric pressure. We live at the bottom of the atmosphere. There is about 10 km of air above us, pressing downwards on us – that is the origin of atmospheric pressure.

QUESTIONS

14 A water tank holds water to a depth of 80 cm. What is the pressure on the bottom of the tank? (Density of water = 1000 kg/m^3.)

15 Figure **5.16** shows a tank that is filled with oil. The density of the oil is 920 kg/m^3.

a Calculate the volume of the tank from the dimensions shown in the diagram.

b Calculate the weight of the oil in the tank.

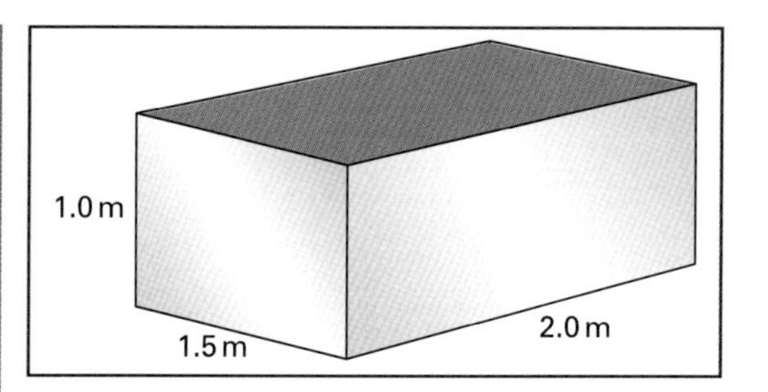

Figure 5.16 For Question **15**.

c The pressure on the bottom of the tank is caused by the weight of the oil. Calculate the pressure using

$$p = \frac{F}{A}$$

d Now calculate the pressure using

$$p = h\rho g$$

Do you find the same answer?

Summary

Forces can change the size and shape of a body.

An extension against load graph shows how a body stretches when a load is applied to it.

Hooke's law: The extension of a spring is proportional to the load applied to it, provided the limit of proportionality is not exceeded.

Hooke's law: $F = kx$.

An extension against load graph is a straight line up to the limit of proportionality.

Pressure is greater when a large force acts on a small area.

$$\text{Pressure} = \frac{\text{force}}{\text{area}} \qquad p = \frac{F}{A}$$

The pressure in a fluid is greater at greater depths, and when the fluid has a greater density.

Pressure = depth × density × acceleration due to gravity

$p = h\rho g$

End-of-chapter questions

5.1 When a spring is stretched, its length increases from 58 cm to 66 cm. Calculate its extension. [3]

5.2 A student has a short spring. He is required to investigate how the length of the spring changes as the load stretching it increases. Describe the experimental procedure he should follow, stating the equipment he should use and the measurements he should make. [6]

5.3 Table **5.4** shows the results of an experiment in which a long piece of plastic foam was stretched by hanging weights from one end.

Load / N	Length / cm	Extension / cm
0.0	83.0	0.0
5.0	87.0	……
10.0	91.0	……
15.0	95.0	……
20.0	99.0	……

Table 5.4 For Question **5.3**.

a Copy the table and complete the third column to show the value of the extension produced by each load. [4]

b Use your completed table to plot an extension against load graph. [3]

5.4 a Draw a labelled diagram to show a simple mercury barometer. [3]

b Describe how such a barometer shows changes in atmospheric pressure. [1]

5.5 Your friend has fallen through the thin ice on a frozen pond. You come to the rescue by laying a ladder across the ice and crawling along the ladder to reach your friend. Use the idea of **pressure** to explain why it is safer to use the ladder than to walk on the ice. [3]

E

5.6 An unstretched spring is 12 cm long. A load of 5 N stretches it to 15 cm. How long will it be under a load of 15 N? (Assume that the spring obeys Hooke's law.) [3]

5.7 A group of students carried out an experiment in which they stretched a length of wire by hanging weights on the end. For each value of the load, they measured the length of the wire. Table **5.5** shows their results.

a Copy the table and add a row showing the extension for each load. [4]

b Use the data in your table to draw an extension against load graph for the wire. [4]

c From your graph, determine the extension produced by a load of 25 N. [2]

d Determine the value of the load at the limit of proportionality. [2]

5.8 The pressure of the atmosphere is 100 000 Pa.

a Calculate the force with which the atmosphere presses on the outside of a large window 2.0 m high and 1.25 m wide. [3]

b Explain why this force does not break the window. [1]

5.9 On a particular day, the height of the mercury column in a simple barometer is 760 mm. Calculate the atmospheric pressure on this day. (Density of mercury = 13 600 kg/m^3, g = 10 m/s^2.) [3]

E

Load / N	0	10	20	30	40	50	60	70
Length / m	3.200	3.207	3.215	3.222	3.230	3.242	3.255	3.270

Table 5.5 For Question **5.7**.

6 Energy transformations and energy transfers

Core Identifying forms of energy
Core Describing energy conversions
Core Applying the principle of conservation of energy
Core Explaining energy efficiency
E **Extension** Calculating percentage efficiency
Extension Calculating kinetic and potential energy

Energy for life

Crocodiles (Figure **6.1**) are efficient creatures. Their jaws snap down on their prey, and there is no escape. You might imagine that a crocodile has a big appetite, but that is not so. A crocodile needs very little food. It can exist on just one-quarter of its own body weight each year. For a human being, this is equivalent to surviving on fish and chips once a week!

There are several reasons for this. It does not take much energy to lie in wait in a water-hole. The water supports your weight, and you do not have to move around a lot. Also, crocodiles (like all reptiles) are cold-blooded, so that their body temperature is close to that of their surroundings. On a cold day, they are sluggish and much more approachable. On hot days, their system is more active, and they are much more agile and dangerous. Finally, their bodies make good use of the food they consume. Unlike humans, they do not have much of a brain (which uses a lot of a human's energy supply). Instead, their energy is stored efficiently and only released when it is time to grab a snack.

In this chapter, we will look at how energy is used in various forms, and how we can use energy efficiently to avoid wasting it.

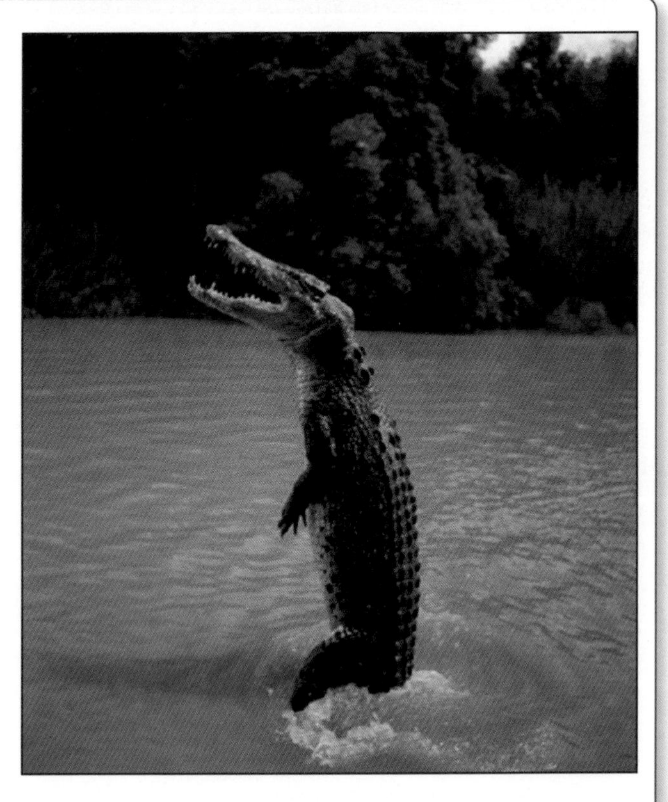

Figure 6.1 Crocodiles are cold-blooded creatures, so it is relatively safe to approach them on a cold day. On a hot day, they are much more active. Crocodiles are not big eaters, but they make very efficient use of the energy supplied by their food.

6.1 Forms of energy

Energy, and energy changes, are involved in all sorts of activities. We will look at two examples and see how we can describe them in terms of energy. We need to have the idea of forms of **energy**.

Example 1: running

At the start of a race, you are stationary, waiting for the starter's pistol. Energy is stored in your toned-up muscles, ready to be released. As you set off, the energy from your muscles gets you moving. If you are running a marathon, you will need to make use of the energy in the longer-term stores in the fatty tissues of your body.

The energy changes involved are shown in Figure 6.2. Your muscles store **chemical energy**. The energy is stored by chemicals in your muscles, ready to be released at a moment's notice. Your muscles start you moving, and you then have **kinetic energy**. Running makes you hot. This tells us that some of the energy released in your muscles is wasted as **thermal (heat) energy**, rather than becoming useful kinetic energy. Fitness training helps people to reduce this waste.

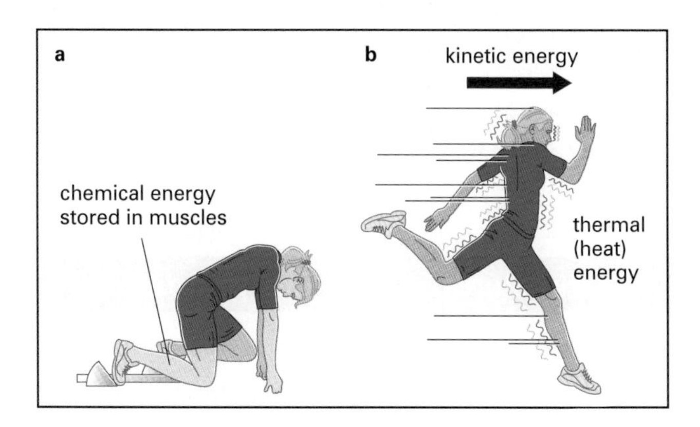

Figure 6.2 a At the start of a race, the runner's muscles are stores of chemical energy. **b** As the runner starts to move, chemical energy is transformed to kinetic energy and thermal (heat) energy.

Example 2: switching on a light

It is evening, and the daylight is fading. You switch on the light. Your electricity meter starts to turn a little faster, recording the fact that you are drawing more energy from the distant power station.

The energy changes involved are shown in Figure 6.3. Electricity is useful because it brings energy, available at the flick of a switch. We can think of the energy

Figure 6.3 Switching on the light requires a supply of electrical energy. In the light bulb, electrical energy is transformed to light energy and thermal (heat) energy.

it brings as **electrical energy**. In the light bulb, this energy is transformed into **light energy**. Every light bulb also produces **thermal (heat) energy**.

Naming forms of energy

The examples above highlight some of the various forms of energy. We now take a brief look at further examples of all of these forms.

A moving object has **kinetic energy (k.e.)**. The faster an object moves, the greater its k.e. We know this because we need to transfer energy to an object to get it moving, and we need to transfer more energy to get it moving faster. Also, if you stand in the path of a moving object so that it runs into you, it will move more slowly. It has transferred some of its energy to you.

If you lift an object upwards, you give it **gravitational potential energy (g.p.e.)**. The higher an object is above the ground, the greater its g.p.e. If you let the object fall, you can get the energy back again. This is exploited in many situations. The water stored behind a hydro-electric dam has g.p.e. As the water falls, it can be used to drive a turbine to generate electricity. A grandfather clock has weights that must be pulled upwards once a week. Then, as they gradually fall, they drive the pendulum to operate the clock's mechanism.

Fuels such as coal or petrol are stores of **chemical energy**. We know that a fuel is a store of energy because, when the fuel burns, the stored energy is released, usually as heat and light. There are many other stores of chemical energy (see Figure 6.4). As we saw above,

Figure 6.4 Some stores of chemical energy – petrol, batteries and bread. Our bodies have long-term stores of energy in the form of fatty tissues.

energy is stored by chemicals in our bodies. Batteries are also stores of energy. When a battery is part of a complete circuit, the chemicals start to react with one another and an electric current flows. The current carries energy to the other components in the circuit.

An electric current is a good way of transferring energy from one place to another. It carries **electrical energy**. When the current flows through a component such as a heater, it gives up some of its energy.

A close relation of chemical energy is **nuclear energy**. Uranium is an example of a nuclear fuel, which is a store of nuclear energy. All radioactive materials are also stores of nuclear energy. In these substances, the energy is stored in the nucleus of the atoms – the tiny positively charged core of the atom. A nuclear power station is designed to release the nuclear energy stored in uranium.

If you stretch a rubber band, it becomes a store of **strain energy**. The band can give its energy to a paper pellet and send it flying across the room. Strain energy is the energy stored by an object that has been stretched or squashed in an elastic way (so that it will spring back to its original dimensions when the stretching or squashing forces are removed). The metal springs of a car are constantly storing and releasing elastic energy as the car travels along, so that the occupants have a smoother ride. A wind-up clock stores energy in a spring, which is the energy source needed to keep its mechanism operating.

If you heat an object so that it gets hotter, you are giving energy to its atoms. The energy stored in a hot object is called **internal energy**. We can picture the atoms of a hot object jiggling rapidly about – they have a lot of energy. This picture is developed further in Chapter **9**.

If you get close to a hot object, you may feel **thermal (heat) energy** coming from it. This is energy travelling from a hotter object to a colder one. The different ways in which this can happen are described in Chapter **11**.

Very hot objects glow brightly. They are giving out **light energy**. Light radiates outwards all around the hot object.

Another way in which energy can be transferred to an object's surroundings is as **sound energy**. An electric current brings electrical energy to a loudspeaker – sound energy and some thermal energy are produced (see Figure **6.5**).

Energy stores, energy transfers

Energy can be stored in an object, or it can be transferred from one object to another. Table **6.1** lists the forms of energy described above under two headings, **energy stores** and **energy transfers**. An energy transfer is 'energy on the move', from one place to another.

Energy stores	Energy transfers
kinetic energy	electrical energy
gravitational potential energy	thermal (heat) energy
chemical energy	light energy
nuclear energy	sound energy
strain energy	
internal energy	

Table 6.1 Different forms of energy can be classified as stores or transfers.

Energy can be transferred from one object to another, or from place to place. (Remember that a 'ferry' transfers people from place to place.) Here are four different ways in which energy can be transferred:

Figure 6.5 At a major rock concert, giant loudspeakers pour out sound energy to the audience. Extra generators may have to be brought on to the site to act as a source of electrical energy to power the speaker systems. Much of the energy supplied is wasted as heat energy, because only a fraction of the electrical energy is transformed into sound energy.

- **By a force.** If you lift something, you give it gravitational potential energy – you provide the force that lifts it. Alternatively, you can provide the force needed to start something moving – you give it kinetic energy. When energy is transferred from one object to another by means of a force, we say that the force is **doing work**. This is discussed in detail in Chapter **8**.
- **By heating.** We have already seen how thermal (heat) energy spreads out from hot objects. No matter how good the insulation, energy is transferred from a hot object to its cooler surroundings. This is discussed in detail in Chapter **11**.
- **By radiation.** Light reaches us from the Sun. That is how energy is transferred from the Sun to the Earth. Some of the energy is also transferred as infrared and ultraviolet radiation. These are all forms of **electromagnetic radiation** (see Chapter **15**).
- **By electricity.** An electric current is a convenient way of transferring energy from place to place. The electricity may be generated in a power station many kilometres away from where the energy is required. Alternatively, a torch battery provides the energy needed to light a bulb. Electricity transfers the energy from the battery to the bulb. This is covered in Chapter **19**.

QUESTIONS

1. What name is given to the energy of a moving object?
2. The Sun is a very hot object. Name **two** forms of energy that arrive at the Earth from the Sun.
3. What form of energy is stored by a stretched spring?
4. What do the letters g.p.e. stand for? How can an object be given g.p.e.?
5. Name a device that transforms electrical energy to sound energy. (It may also produce thermal (heat) energy.)
6. Name **three** forms of energy that are given out by a television set.
7. Look at the list of energy stores, shown in Table **6.1**. For each, give an example of an object or material that stores energy in this form.

6.2 Energy conversions

When energy changes from one form to another, we say that it has been **converted** or **transformed**. We have already mentioned several examples of **energy conversions**. Now we will look at a few more and think a little about the forms of energy that are involved.

The rocket in Figure **6.6a** is lifting off from the ground as it carries a new spacecraft up into space. Its energy comes from its store of fuel and oxygen. It carries tanks of liquid hydrogen. These are its store of chemical energy. When fuel burns, its store of energy is released.

The rocket is accelerating, so we can say that its kinetic energy is increasing. It is also rising upwards, so its gravitational potential energy is increasing. In Figure **6.6a**, you can see light coming from the burning fuel. You can also imagine that large amounts of thermal (heat) energy and sound energy are produced. This energy conversion is shown in Figure **6.6b**. We can also represent the conversion as an equation:

$$\text{chemical energy} \rightarrow \text{k.e.} + \text{g.p.e.} + \text{thermal energy} + \text{light energy} + \text{sound energy}$$

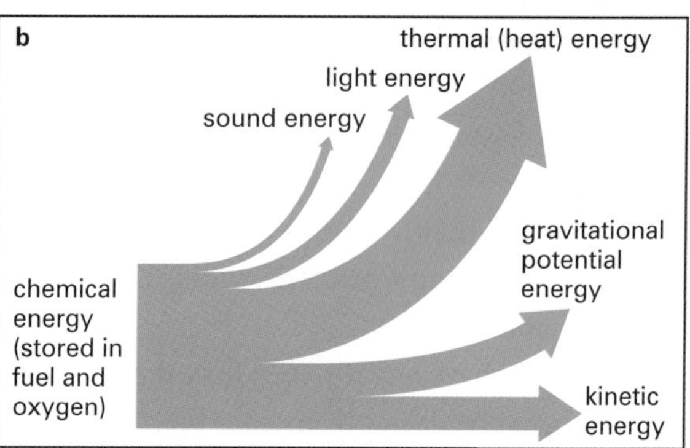

Figure 6.6 **a** This giant rocket uses rocket motors to lift it up into space. Each rocket motor burns about one tonne of fuel and oxygen every minute to provide the energy needed to move the rocket upwards. **b** This diagram represents the energy transformations going on as the rocket accelerates upwards. Chemical energy in the fuel and oxygen is transformed into five other forms of energy.

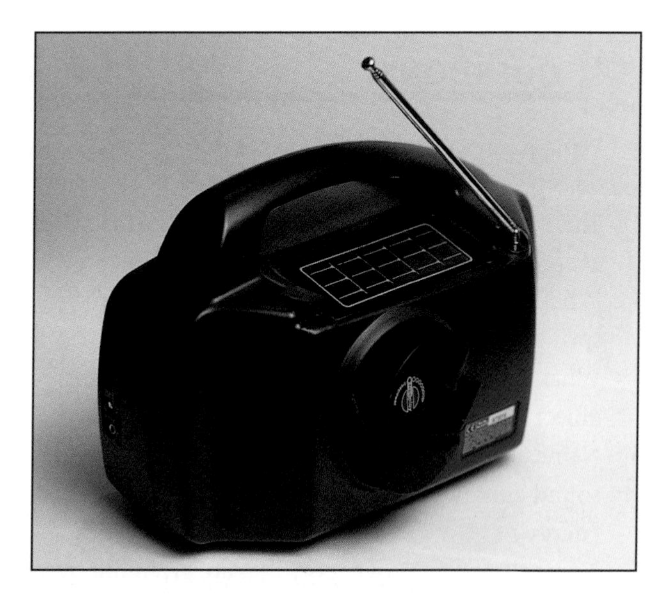

Figure 6.7 This clockwork radio is designed for use by people who do not have a ready supply of batteries or mains electricity. It operates from two alternative energy sources: a wound-up clockwork spring, and a solar cell. Since many users live in sunny parts of the world, a solar cell is a useful feature.

The clockwork radio (Figure **6.7**) is a famous invention. Where people do not have ready access to batteries or mains electricity, it allows them to listen to radio broadcasts with minimal running costs. The model shown in the photograph has an additional feature: a solar cell acts as an alternative energy source.

The wound-up spring of the clockwork mechanism is a store of elastic (strain) energy. The radio requires electrical energy to function. As the spring unwinds, it turns a generator. The elastic energy of the unwinding spring first becomes kinetic energy of the turning mechanism, and then electrical energy carried by the current to the radio. Finally, the energy is converted to sound energy. Along the way, energy is wasted as thermal (heat) energy. This is because no generator can convert all of the kinetic energy it is supplied with into electrical energy – some becomes thermal energy. Similarly, heat is produced by the electronic circuits of the radio, and by its loudspeaker. We can represent these conversions by an equation with several steps:

elastic energy → k.e.
→ electrical energy + thermal energy
→ sound energy + thermal energy

The solar cell converts light energy directly into electrical energy. Again, some energy is wasted as heat. The whole conversion then becomes:

light energy → electrical energy + thermal energy
→ sound energy + thermal energy

QUESTION

8 What energy conversions are going on in the following? In each case, write an equation to represent the conversion.

a Coal is burned to heat a room and to provide a supply of hot water.

b A student uses an electric lamp while she is doing her homework.

c A hairdryer is connected to the mains electricity supply. It blows hot air at the user's wet hair. It whirrs as it does so.

Activity 6.1 Energy conversions

Examine some devices that convert energy from one form to another. Can you decide what is going on?

6.3 Conservation of energy

When energy is transformed from one form to another, it is often the case that some of the energy ends up in a form that we do not want. The energy transformations in a light bulb were represented earlier in Figure **6.3**. The bulb produces light energy, which we do want, but also thermal (heat) energy, which we do not want. The rocket motor (see Figure **6.6**) transforms chemical energy into two forms that we do want (k.e. and g.p.e.) and three that we do not want (heat, light and sound).

Figure **6.8** shows an energy diagram for a car, driving along a flat road. Its source of energy is the petrol it burns, and the numbers show that the fuel supplies 80 kJ (kilojoules) every second. Some thermal energy escapes from the hot engine and in the exhaust gases. Some energy is wasted as heat produced by friction within the workings of the car. The rest is used in overcoming air resistance, another form of friction, so that the air is warmer after the car has passed through it.

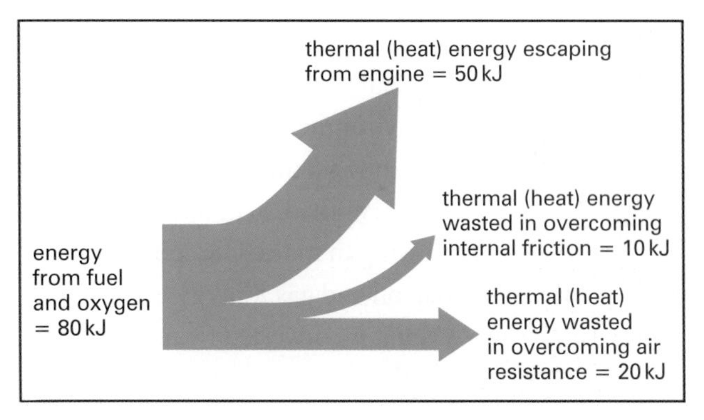

Figure 6.8 An energy diagram for a car, showing the energy converted by the car each second.

All of the energy supplied by the car's fuel ends up as thermal energy. If you add up the different amounts of thermal energy, you will see that they come to 80 kJ. This is an example of a very important idea, the **principle of conservation of energy**:

> In any energy conversion, the total amount of energy before and after the conversion is constant.

This tells us something very important about energy: it cannot be created or destroyed. The total amount of energy is constant. If we measure or calculate the amount of energy before a conversion, and again afterwards, we will always get the same result. If we find any difference, we must look for places where energy may be entering or escaping unnoticed.

Keeping an eye on the amounts of energy is rather like a form of book-keeping or accounting. Energy is like money: the amounts entering a system must equal the amounts leaving it, or stored within it.

QUESTION

9 A light bulb is supplied with 100 J of energy each second.

a How many joules of energy leave the bulb each second in the form of heat and light?

b If 10 J of energy leave the lamp each second in the form of light, how many joules leave each second in the form of heat?

Energy efficiency

Energy is expensive, and we do not want to waste it. Using more energy than necessary increases the damage we do to the environment, so it is important to avoid waste. Figure **6.9** shows a diagram that represents energy flows in the whole of the UK in one typical year. Most of the energy flowing in to the UK comes from fuels, particularly coal, oil and gas. Energy is wasted in two general ways: when it is converted (transformed) into electricity, and when it is used (for example, in light bulbs).

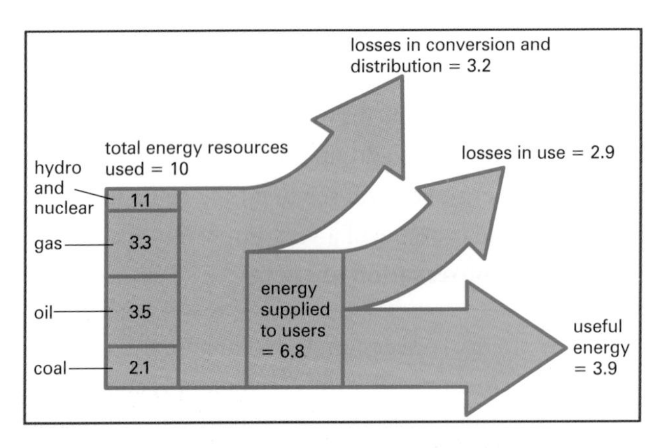

Figure 6.9 A diagram showing energy flows in the UK in a typical year, 2000. (All numbers are $\times 10^{18}$ J.) A large proportion of the energy supplied by fuels is wasted in conversion processes and in its final use. Some of this waste is inevitable, but better insulation and more efficient machines could reduce the waste and environmental damage, and save money.

Most wasted energy ends up as thermal (heat) energy. There are two main reasons for this:

- When fuels are burned (perhaps to generate electricity, or to drive a car), heat is produced as an intermediate step. Hot things readily lose energy to their surroundings, even if they are well insulated. Also, engines and boilers have to lose heat as part of the way they operate: power stations produce warm cooling water; and cars produce hot exhaust gases.
- Friction is very often a problem when things are moving. Lubrication can help to reduce friction, and a streamlined design can reduce air resistance. But it is impossible to eliminate friction entirely from machines with moving parts. Friction generates heat.

Another common form of wasted energy is sound. Noisy machinery, loud car engines and so on are all wasting energy. However, loud noises do not contain very much energy, so there is little to be gained (in terms of energy) by reducing noise. Waste energy in the form of heat and sound is sometimes referred to as low-grade energy.

Making better use of energy

It is important to make good use of the energy resources available to us. This is because energy is expensive, supplies are often limited, and our use of energy can damage the environment. So we must use resources efficiently. Here is what we mean by **efficiency**:

> The efficiency of an energy conversion is the fraction of the energy that ends up in the desired form.

Figure **6.10** shows one way to make more efficient use of electricity. It shows two types of light bulb and the energy they use each second. One is a filament lamp, and the other is an energy-efficient lamp. We use light bulbs to provide us with light. The diagrams show that each of the two bulbs produces the same amount of light energy. However, the energy-efficient lamp requires a much smaller input of electrical energy because it wastes much less energy as heat.

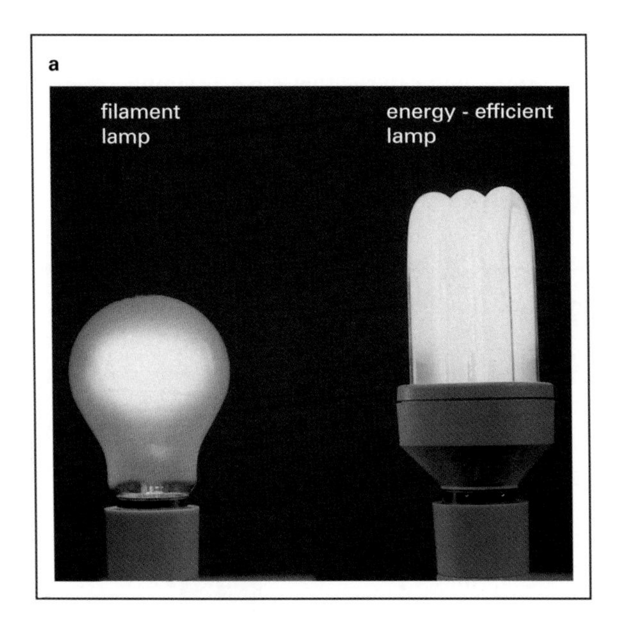

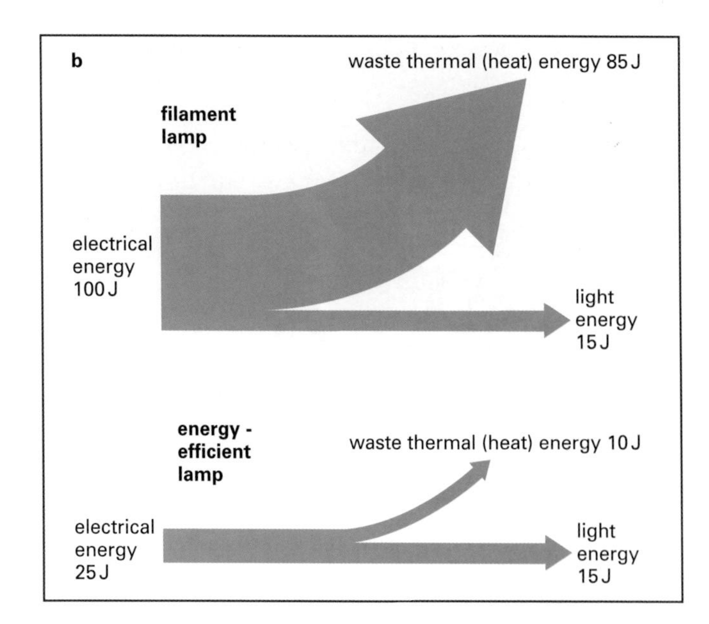

Figure 6.10 **a** Each of these two light bulbs provides the same amount of light. **b** The energy-efficient lamp (on the right) wastes much less energy as heat.

Device	Typical efficiency
electric heater	100%
large electric motor	90%
washing machine motor	70%
gas-fired power station	50%
diesel engine	40%
car petrol engine	30%
steam locomotive	10%

Table 6.2 Energy efficiencies. Most devices are less than 100% efficient because they produce waste heat. An electric heater is 100% efficient because all of the electrical energy supplied is transformed to heat. There is no problem about waste here!

Table **6.2** shows the typical efficiencies for some important devices. You can see that even the most modern gas-fired power station is only 50% efficient. Half of the energy it is supplied with is wasted.

QUESTIONS

10 a What is the most common form of waste energy?

b Name another form in which energy is often wasted.

11 Why is it important not to waste energy? Give **three** reasons.

E

Calculating efficiency

You can see from Table **6.2** that efficiency is often given as a percentage. We can calculate the percentage efficiency of an energy change as follows:

$$\text{efficiency} = \frac{\text{useful energy output}}{\text{energy input}} \times 100\%$$

When the filament lamp shown in Figure **6.10** is supplied with 100 J of electrical energy, it produces 15 J of useful light energy. Its efficiency is thus

$$\text{efficiency of filament lamp} = \frac{15}{100} \times 100\% = 15\%$$

QUESTIONS

12 Calculate the efficiency of the energy-efficient lamp shown in Figure **6.10**.

13 A coal-fired power station produces 100 MJ of electrical energy when it is supplied with 400 MJ of energy from its fuel. Calculate its efficiency.

14 A lamp is 10% efficient. How much electrical energy must be supplied to the lamp each second if it produces 20 J of light energy per second?

Figure 6.11 Astronauts on the Moon. The Moon's gravity is one-sixth that of the Earth. Experiments on the Moon have shown that a golf ball can be hit much farther than on Earth. This is because it travels a much greater distance horizontally before gravity pulls it back to the ground.

6.4 Energy calculations

Energy is not simply an idea, it is also a quantity that we can calculate.

Gravitational potential energy

Mountaineering on the Moon should be easy (see Figure **6.11**). The Moon's gravity is much weaker than the Earth's, because the Moon's mass is only one-eightieth of the Earth's. This means that the weight of an astronaut on the Moon is a fraction of his or her weight on the Earth. In principle, it is possible to jump six times as high on the Moon. Unfortunately, because an astronaut has to carry an oxygen supply and wear a cumbersome suit, this is not possible.

Earlier, we saw that an object's gravitational potential energy (g.p.e.) depends on its height above the ground. The higher it is, the greater its g.p.e. If you lift an object upwards, you provide the force needed to increase its g.p.e. The heavier the object, the greater the force needed to lift it, and hence the greater its g.p.e.

This suggests that an object's **gravitational potential energy** depends on two factors:

- the object's weight mg – the greater its weight, the greater its g.p.e.
- the object's height h above ground level – the greater its height, the greater its g.p.e.

This is illustrated in Figure **6.12**. From the numbers in the diagram, you can see that g.p.e. is simply calculated by multiplying weight by height. (Here, we are assuming that an object's g.p.e. is zero when it is at ground level.) We can write this as an equation:

gravitational potential energy = weight × height

$$\text{g.p.e.} = mg \times h$$

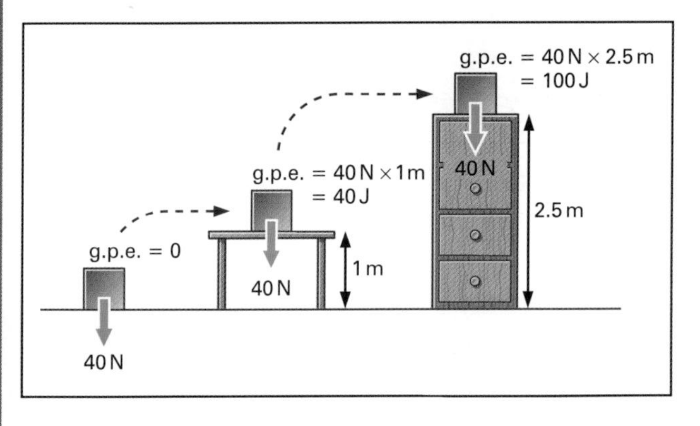

Figure 6.12 The gravitational potential energy of an object increases as it is lifted higher. The greater its weight, the greater its g.p.e.

Worked example 1

An athlete of mass 50 kg runs up a hill. The foot of the hill is 400 m above sea-level. The summit is 1200 m above sea-level. By how much does the athlete's g.p.e. increase? (Acceleration due to gravity $g = 10\,\text{m/s}^2$.)

Step 1: Assume that g.p.e. is zero at the foot of the hill. Calculate the increase in height.

$$h = 1200\,\text{m} - 400\,\text{m} = 800\,\text{m}$$

Step 2: Write down the equation for g.p.e., substitute values and solve.

$$\begin{aligned}\text{g.p.e.} &= \text{weight} \times \text{height}\\ &= mg \times h\\ &= 50\,\text{kg} \times 10\,\text{m/s}^2 \times 800\,\text{m}\\ &= 400\,000\,\text{J}\\ &= 400\,\text{kJ}\end{aligned}$$

So the athlete's g.p.e. increases by 400 kJ.

A note on height

We have to be careful when measuring or calculating the change in an object's height.

First, we have to consider the **vertical** height through which it moves. A train may travel 1 km up a long and gentle slope, but its vertical height may only increase by 10 m. A satellite may travel around the Earth in a circular orbit. It stays at a constant distance from the centre of the Earth, and so its height does not change. Its g.p.e. is constant.

Second, it is the change in height of the object's centre of gravity that we must consider. This is illustrated by the high-jumper shown in Figure **6.13**. As he jumps, he must try to increase his g.p.e. enough to get over the bar. In fact, by curving his body, he passes over the bar but his centre of gravity may pass under it.

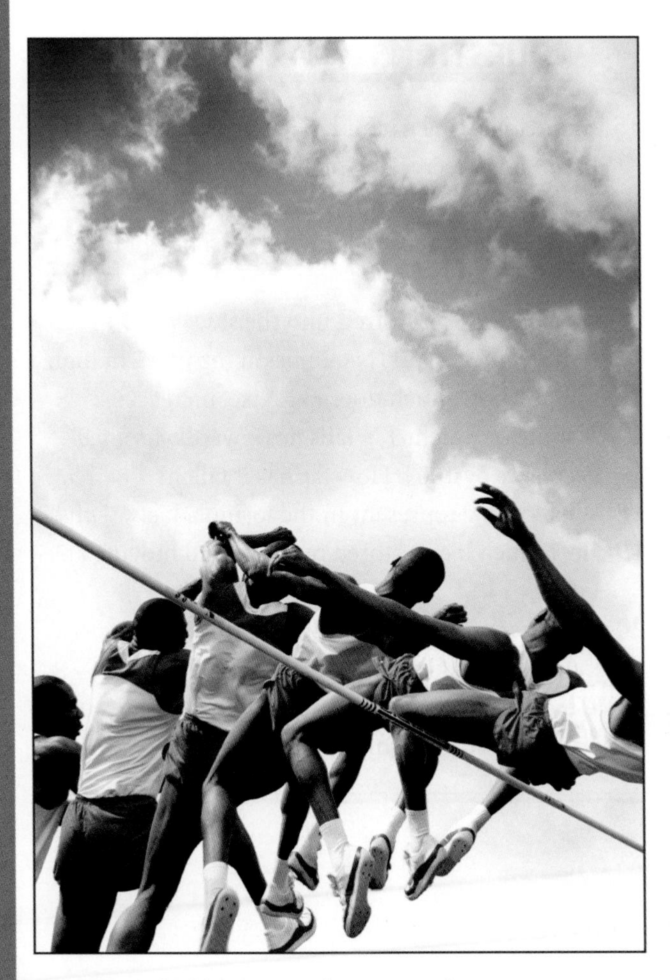

Figure 6.13 This high-jumper adopts a curved posture to get over the bar. He cannot increase his g.p.e. enough to get his whole body above the level of the bar. His centre of gravity may even pass under the bar, so that at no time is his body entirely above the bar.

Kinetic energy

It takes energy to make things move. You transfer energy to a ball when you throw it or hit it. A car uses energy from its fuel to get it moving. Elastic energy stored in a stretched piece of rubber is needed to fire a pellet from a catapult. So a moving object is a store of energy. This energy is known as kinetic energy (k.e.).

We often make use of an object's kinetic energy. To do this, we must slow it down. For example, moving air turns a wind turbine. This slows down the air, reducing its k.e. The energy extracted can be used to turn a generator to produce electricity.

This suggests that the **kinetic energy** of an object depends on two factors:

- the object's mass m – the greater the mass, the greater its k.e.
- the object's speed v – the greater the speed, the greater its k.e.

These are combined in a formula for k.e.:

$$\text{kinetic energy} = \frac{1}{2} \times \text{mass} \times \text{speed}^2$$
$$\text{k.e.} = \frac{1}{2} mv^2$$

Worked example **2** shows how to use the formula to calculate the k.e. of a moving object. Note also that kinetic energy (like all forms of energy) is a scalar quantity, despite the fact that it involves v. It is best to think of v here as **speed** rather than velocity.

Worked example 2

A van of mass 2000 kg is travelling at 10 m/s. Calculate its kinetic energy. If its speed increases to 20 m/s, by how much does its kinetic energy increase?

Step 1: Calculate the van's k.e. at 10 m/s.

$$\begin{aligned}\text{k.e.} &= \frac{1}{2} mv^2 \\ &= \frac{1}{2} \times 2000\,\text{kg} \times (10\,\text{m/s})^2 \\ &= 100\,000\,\text{J} \\ &= 100\,\text{kJ}\end{aligned}$$

E

Step 2: Calculate the van's k.e. at 20 m/s.

$$\begin{aligned}\text{k.e.} &= \tfrac{1}{2}mv^2\\ &= \tfrac{1}{2} \times 2000\,\text{kg} \times (20\,\text{m/s})^2\\ &= 400\,000\,\text{J}\\ &= 400\,\text{kJ}\end{aligned}$$

Step 3: Calculate the change in the van's k.e.

$$\begin{aligned}\text{change in k.e.} &= 400\,\text{kJ} - 100\,\text{kJ}\\ &= 300\,\text{kJ}\end{aligned}$$

So the van's k.e. increases by 300 kJ when it speeds up from 10 m/s to 20 m/s.

Comments on Worked example 2

It is worth looking at Worked example **2** in detail, since it illustrates several important points.

When calculating k.e. using $\frac{1}{2}mv^2$, take care! Only the speed is squared. Using a calculator, start by squaring the speed. Then multiply by the mass, and finally divide by 2.

When the van's speed doubles from 10 m/s to 20 m/s, its k.e. increases from 100 kJ to 400 kJ. In other words, when its speed increases by a factor of 2, its k.e. increases by a factor of 4. This is because k.e. depends on speed squared. If the speed trebled (increased by a factor of 3), the k.e. would increase by a factor of 9 (see Figure **6.14**).

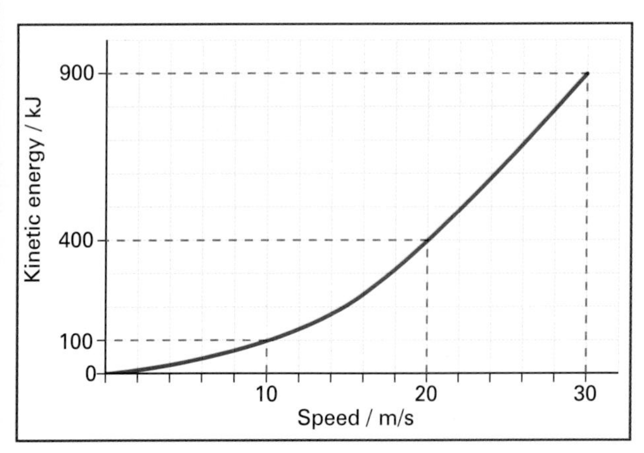

Figure 6.14 The faster the van travels, the greater its kinetic energy – see Worked example **2**. Double the speed means four times the kinetic energy, because k.e. depends on speed². The graph shows that k.e. increases more and more rapidly as the van's speed increases.

E

When the van starts moving from rest and speeds up to 10 m/s, its k.e. increases from 0 to 100 kJ. When its speed increases by the same amount again, from 10 m/s to 20 m/s, its k.e. increases by 300 kJ, three times as much. It takes a lot more energy to increase your speed when you are already moving quickly. That is why a car's fuel consumption starts to increase rapidly when the driver tries to accelerate in the fast lane of a motorway.

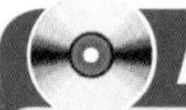

Activity 6.2 Running downhill

When a toy car runs downhill, g.p.e. changes to k.e. Can you test this idea?

QUESTIONS

15 In the following examples, is the object's g.p.e. increasing, decreasing or remaining constant?

- **a** An apple falls from a tree.
- **b** An aircraft flies horizontally at a height of 9000 m.
- **c** A sky-rocket is fired into the sky.

16 A girl of weight 500 N climbs on top of a 2 m high wall. By how much does her g.p.e. increase?

17 A stone of weight 1 N falls downwards. Its g.p.e. decreases by 100 J. How far has it fallen?

18 What does v represent in the formula k.e. $= \frac{1}{2}mv^2$?

19 How much k.e. is stored by a 1 kg ball moving at 1 m/s?

20 A runner of mass 80 kg is moving at 8 m/s. Calculate her kinetic energy.

21 Which has more k.e., a 2 g bee flying at 1 m/s, or a 1 g wasp flying at 2 m/s?

Summary

Energy can be converted from one form to another.

In any energy conversion, the total amount of energy before the conversion is equal to the total amount after the conversion. This is the principle of conservation of energy.

Energy can be transferred from place to place, or from one object to another, by a variety of means.

In energy conversions, some energy often appears in forms that are not wanted, particularly as waste heat.

Energy efficiency indicates the fraction of the input energy that ends up in a useful form.

E

Gravitational potential energy g.p.e. = $mg \times h$.

Kinetic energy k.e. = $\frac{1}{2}mv^2$.

End-of-chapter questions

6.1 What name is given to:

a the energy of a moving object? [1]

b the energy stored in a fuel? [1]

c the energy stored in a hot object? [1]

6.2 What are the energy conversions in the following? Write an equation for each.

a A glow-worm is an insect that glows in the dark. Chemicals in its body react together to produce light and heat. [2]

b An electric motor is used to start a computer's disk drive spinning round. [2]

c A wind turbine spins and generates electricity. [2]

d Friction in a car's brakes slows it down. [2]

6.3 A light bulb is supplied with 100 J of electrical energy each second. It produces 7 J of light energy and 93 J of thermal (heat) energy. Explain how this shows that energy is conserved. [3]

6.4 The girl on the skate ramp (Figure **6.15**) roller-skates down one side of the slope and up the opposite side. She cannot quite reach the top of the slope, level with her starting position.

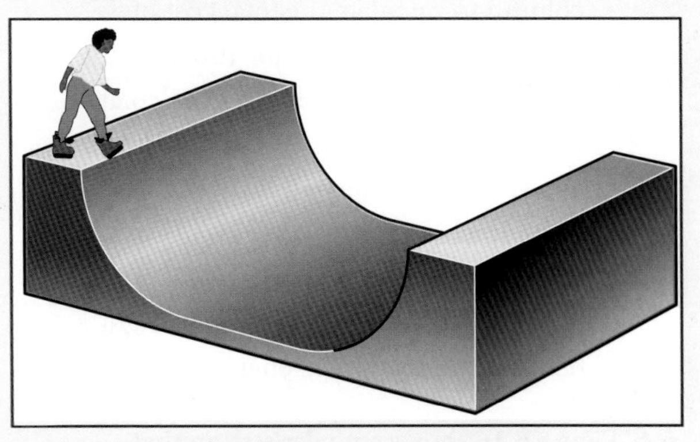

Figure 6.15 For Question **6.4**.

a What energy conversion is taking place as the girl moves downwards? [2]

b What energy conversion is taking place as the girl moves back upwards? [2]

c Explain why the girl cannot reach the top of the slope. [2]

d Suggest how the girl could reach the top of the slope. [2]

E

6.5 Low-energy light bulbs are designed to save energy, but do they also save money? An individual low-energy bulb is more expensive than the filament bulb it replaces. However, it lasts for much longer, typically 10 000 hours. Table **6.3** shows typical costs in pence (p).

	Low-energy bulb	Filament bulbs
cost of one bulb	400 p	50 p
number of bulbs required for 10 000 hours	1	10
cost of electricity for 1 hour	0.2 p	1.0 p
total cost of electricity for 10 000 hours	……	……
total cost of bulbs and electricity	……	……

Table 6.3 For Question **6.5**.

a Copy the table and complete the second column to calculate the total cost of using a low-energy bulb for 10 000 hours. [2]

b Complete the third column to calculate the cost of using filament bulbs instead of a single low-energy bulb. [2]

c How much money is saved by using a low-energy bulb? [2]

d Suggest **two** reasons that people might have for not using low-energy bulbs. [2]

6.6 Figure **6.16** shows a power station that burns rubbish to generate electricity. It also supplies hot water to nearby offices and shops.

a What **two** useful energy forms are produced? [2]

b What waste energy is produced? [1]

c Is this an efficient use of energy? Explain your answer using information from the diagram. [2]

E

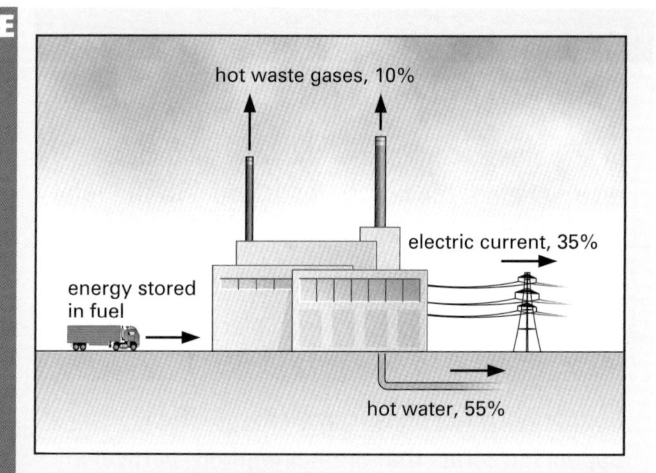

Figure 6.16 For Question **6.6**.

6.7 Figure **6.17** shows an idea for a perpetual motion machine. The car runs on electricity. As it moves along, the air moving past the car turns the generator on the roof. This generates the electricity needed to power the car.

Figure 6.17 For Question **6.7**.

a Explain the energy transformations that are going on here. [2]

b Explain why this idea will not work in practice. [2]

6.8 An astronaut on the Moon has a mass (including his spacesuit and equipment) of 180 kg. The acceleration due to gravity on the Moon's surface is 1.6 m/s^2.

a Calculate the astronaut's weight on the Moon. [3]

The astronaut climbs 100 m to the top of a crater.

b By how much does his gravitational potential energy (g.p.e.) change? [3]

c Does his g.p.e. increase or decrease? [1]

7 Energy resources

Core Identifying and describing energy resources

E Extension Understanding the Sun's energy source

7.1 The energy we use

Here on Earth, we rely on the Sun for most of the energy we use. The Sun is a fairly average star, 150 million kilometres away. The heat and light we receive from it have taken about eight minutes to travel through empty space to get here. Plants absorb this energy in the process of photosynthesis, and animals are kept warm by it.

The Earth is at a convenient distance from the Sun for living organisms. The Sun's rays are strong enough, but not too strong. The Earth's average temperature is about 15 °C, which is suitable for life. If we were closer to the Sun, we might be intolerably hot like Venus, where the average surface temperature is over 400 °C. Further out, things are colder. Saturn is roughly ten times as far from the Sun, so the Sun in the sky looks one-tenth of the diameter that we see it, and its radiation has only one-hundredth of the intensity. Saturn's surface temperature is about −180 °C.

Most of the energy we use comes from the Sun, but only a very little is used directly from the Sun. On a cold but sunny morning, you might sit in the sunshine to warm your body. Your house might be designed to collect warmth from the Sun's rays, perhaps by having larger windows on the sunny side. However, most of the energy we use comes only indirectly from the Sun. It must be converted into a more useful form, such as electricity (Figure 7.1).

Figure 7.1 We use energy from the Sun in many different ways – for example, for producing electricity.

Figure 7.2 is a chart showing the different fuels that contribute to the world's energy supplies. This chart reflects patterns of energy consumption in the early years of the 21st century. Many people today live in industrialised countries and consume large amounts of energy, particularly from fossil fuels (coal, oil and gas). People living in less-developed countries consume far less energy – mostly they use biomass fuels, particularly wood. A thousand years ago, the chart would have looked very different. Fossil fuel consumption was much less important then. Most people relied on burning wood to supply their energy requirements. We will now look at these groups of fuels in turn.

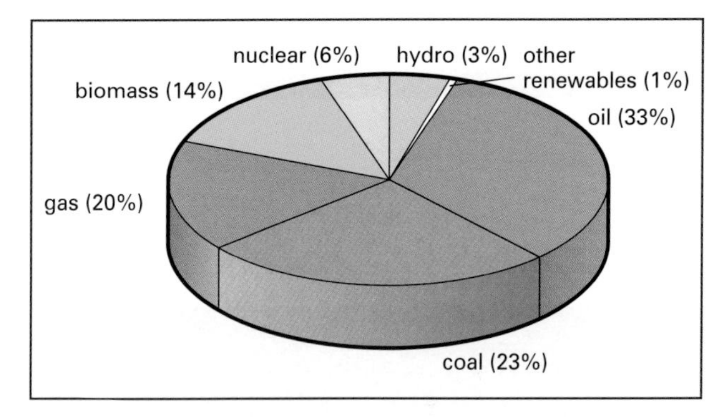

Figure 7.2 World energy use, by fuel. This chart shows the contributions made by different fuels to energy consumption by people in 2006, across the world. Three-quarters of all energy is from fossil fuels.

Energy direct from the Sun

In hot, sunny countries, **solar panels** are used to collect thermal (heat) and light energy from the Sun. The Sun's rays fall on a large solar panel, on the roof of a house, for example. This absorbs the energy of the rays, and water inside the panel heats up. This provides hot water for washing. It can also be pumped round the house, through radiators, to provide a cheap form of central heating.

We can also make electricity directly from sunlight (Figure 7.3). The Sun's rays shine on a large array of **solar cells** (also known as a **photocells**). The energy of the rays is absorbed, and electricity is produced. As this technology becomes cheaper, it is finding more and more uses. It is useful in remote locations – for example, for running a refrigerator that stores medicines in central Africa, or for powering roadside emergency phones in desert regions such as the Australian outback. Solar cells have also been used extensively for powering spacecraft. Ideally, a solar cell is connected to a rechargeable battery, which stores the energy collected, so that it can be available during the hours of darkness.

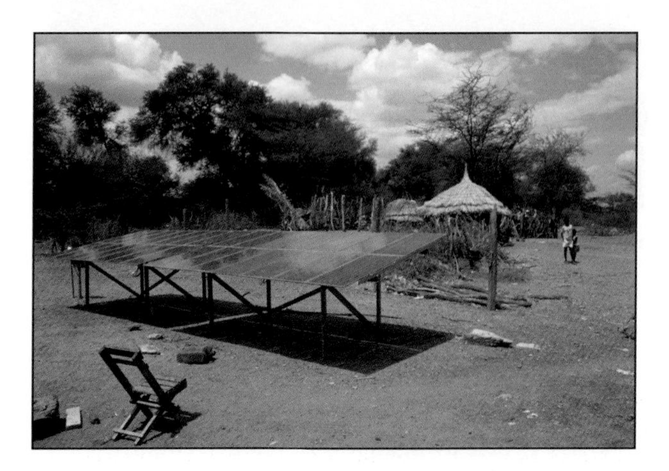

Figure 7.3 This array of solar cells provide electricity for a water pump in a Kenyan village.

Wind and wave power

Wind and waves are also caused by the effects of the Sun. The Sun heats some parts of the atmosphere more than others. Heated air expands and starts to move around – this is a convection current (see Chapter **11**). This is the origin of winds. Most of the energy of winds is given up to the sea as waves are formed by friction between wind and water. There are many technologies for extracting energy from the wind. Windmills for grinding and pumping are traditional, and modern wind turbines can generate electricity (see Figure 7.4).

Figure 7.4 These giant turbines are part of a wind farm at Xinjiang in China. They produce as much electricity as a medium-sized coal-fired power station.

Wave technology is trickier. The up-and-down motion of waves must be used to spin a turbine, which then turns a generator. This is tricky to achieve, and rough seas are a hazardous place to work. On calm days, the system produces no power.

QUESTIONS

1 Explain why wind and wave power could not be relied on to provide a country's entire electricity supply.
2 A photovoltaic cell produces electricity when the Sun shines. What energy conversion is going on here?
3 When a wave travels across the sea, the water moves up and down. What two forms of energy does a wave have?

Biomass fuels

For many people in the world, wood is the most important fuel. It warms their homes and provides the heat necessary for cooking their food. Wood is made by trees and shrubs. It stores energy that the plant has captured from sunlight in the process of photosynthesis. When we burn wood, we are releasing energy that came from the Sun in the recent past, perhaps ten or a hundred years ago.

Wood is just one example of a **biomass fuel**. Others include animal dung and biogas, generated by rotting vegetable matter. These can be very important fuels in societies where most people live by farming. As you can see from Figure 7.2, biomass fuels account for about one-seventh of all energy consumption in the world. This figure can only be a rough estimate, because no-one keeps track of all the wood consumed as fuel. However, we can say that this segment of the chart represents the energy consumption of about three-quarters of the world's population. The remaining one-quarter (who live in developed, industrial nations) consume roughly six times as much.

Fossil fuels

Oil, coal and gas are all examples of **fossil fuels**. These are usually hydrocarbons (compounds of hydrogen and carbon). When they are burned, they combine with oxygen from the air. In this process, the carbon becomes carbon dioxide. The hydrogen becomes 'hydrogen oxide', which we usually call water. Energy is released.

We can write this as an equation:

$$\text{hydrocarbon} + \text{oxygen} \rightarrow \text{carbon dioxide} + \text{water} + \text{energy}$$

Hence, we can think of a fossil fuel as a store of energy. They store energy as chemical energy. Where has this energy come from?

Fossil fuels (Figure 7.5) are the remains of organisms (plants and animals) that lived in the past. Many of the Earth's coal reserves, for example, formed from trees that lived in the Carboniferous era, between 286 and 360 million years ago. ('Carboniferous' means 'coal-producing'.) These trees captured sunlight by photosynthesis, they grew and eventually they died. Their trunks fell into the swampy ground, but they did not rot completely, because there was insufficient oxygen.

Figure 7.5 Coal is a fossil fuel. A fossil is any living material that has been preserved for a long time. Usually, its chemical composition changes during the process. Coal sometimes, as here, shows evidence of the plant material from which it formed. Sometimes you can see fossilised creatures that lived in the swamps of the Carboniferous era. These creatures died along with the trees that eventually became coal.

As material built up on top of these ancient trees, the pressure on them increased. Eventually, millions of years of compression turned them into underground reserves of coal (see Figure 7.5). Today, when we burn coal, the light that we see and the warmth that we feel have their origins in the sunlight trapped by trees hundreds of millions of years ago.

Oil and gas are usually found together. They are formed in a similar way to coal, but from the remains of tiny shrimp-like creatures called microplankton that lived in the oceans. The oilfields of the Persian Gulf, North Africa and the Gulf of Mexico, which contain half of the world's known oil reserves, all formed in the Cretaceous era, 75 to 120 million years ago.

QUESTIONS

4 a Name **three** fossil fuels.
  b Name **three** non-fossil fuels.

5 What energy conversion is happening when charcoal is used as the fuel for a barbecue?

Nuclear fuels

Nuclear power was developed in the second half of the 20th century. It is a very demanding technology, which requires very strict controls, because of the serious damage that can be caused by an accident.

The fuel for a nuclear power station (Figure 7.6) is usually uranium, sometimes plutonium. These are radioactive materials. Inside a nuclear reactor, their radioactive decay is speeded up so that the energy they store is released much more quickly. This is the process of **nuclear fission**.

Uranium is a very concentrated store of energy in the form of nuclear energy. A typical nuclear power station will receive about one truckload of new fuel each week. Coal is less concentrated. A similar coal-fired power station is likely to need a whole trainload of coal every hour. A wind farm capable of generating electricity at the same rate would cover a large area of ground – perhaps 20 square kilometres.

Figure 7.6 This nuclear power station generates electricity. Its fuel is uranium. As the fuel is used up, highly radioactive waste products are produced. These have to be dealt with very carefully to avoid harm to the surroundings. Here, checks are being carried out to ensure that the level of radioactive materials near the power station is safe.

In some countries that have few other resources for generating electricity, nuclear power provides a lot of energy. In France, for example, nuclear power stations generate three-quarters of the country's electricity. Excess production is exported to neighbouring countries, including Spain, Switzerland and the UK.

Nuclear fuel is a relatively cheap, concentrated energy resource. However, nuclear power has proved to be expensive because of the initial cost of building the power stations, and the costs of disposing of the radioactive spent fuel and decommissioning the stations at the end of their working lives.

Water power

One of the smallest contributions to the chart in Figure 7.2 is water or **hydro-power**. For centuries, people have used the kinetic energy of moving water to turn water-wheels, which then drive machinery of all sorts – for example, to grind corn and other crops, pump water and weave textiles. Today, water power's biggest contribution is in the form of hydro-electricity (see Figure 7.7). Water stored behind a dam is released to turn turbines, which make generators spin. This is a very safe, clean and reliable way of producing electricity, but it is not without its problems. A new reservoir floods land that might otherwise have been used for hunting or farming. People may be made homeless, and wildlife habitats destroyed.

Figure 7.7 The giant Itaipú Dam on the Paraná river generates electricity for Brazil and Paraguay.

Most hydro-power comes ultimately from the Sun. The Sun's rays cause water to evaporate from the oceans and land surface. This water vapour in the atmosphere eventually forms clouds at high altitudes. Rain falls on high ground, and can then be trapped behind a dam. This is the familiar **water cycle**. Without energy from the Sun, there would be no water cycle and much less hydro-power.

A small amount of hydro-power does not depend on the Sun's energy. Instead, it is generated from the tides. The Moon and the Sun both contribute to the oceans' tides. Their gravitational pull causes the level of the ocean's surface to rise and fall every twelve-and-a-bit hours. At high tide, water can be trapped behind a dam. Later, at lower tides, it can be released to drive turbines and generators. Because this depends on gravity, and not the Sun's heat and light, we can rely on tidal power even at night and when the Sun is not shining.

Geothermal energy

The interior of the Earth is hot. This would be a useful source of energy – if we could get at it! People do make use of this **geothermal energy** where hot rocks are found at a shallow depth below the Earth's surface. (These rocks are hot because of the presence of radioactive substances inside the Earth.) To make use of this energy, water is pumped down into the rocks, where it boils. High-pressure steam returns to the surface, where it can be used to generate electricity.

Suitable hot underground rocks are usually found in places where there are active volcanoes. Iceland, for example, has several geothermal power stations. These also supply hot water to heat nearby homes and buildings.

QUESTIONS

6 Name **three** energy resources for which the original energy source is not radiation from the Sun.

7 What energy conversion happens when a nuclear power station uses uranium fuel to produce electricity?

Renewables and non-renewables

Figure 7.2 on page 70 shows that most of the energy supplies we use are fossil fuels – coal, oil and gas. There are limited reserves of these, so that, if we continue to use them, they will one day run out. They are described as **non-renewables**. Once used, they are gone for ever.

Other sources of energy, such as wind, solar and biomass, are described as **renewables**. This is because, when we use them, they will soon be replaced. The wind will blow again, the Sun will shine again – and, after harvesting a biomass crop, we can grow another.

Ideally, we should develop an 'energy economy' based on renewables. Then we would not have to worry about supplies that will run out. We would also avoid the problems of global warming and climate change.

Comparing energy sources

We use fossil fuels a lot because they represent concentrated sources of energy. A modern gas-fired power station might occupy the space of a football ground and supply a town of 100 000 people. To replace it with a wind farm might require 50 or more wind turbines spread over an area of several square kilometres – the wind is a dilute source of energy.

This illustrates some of the ideas that we use when comparing different energy sources. If you look back through this chapter, you will find many comments about different energy sources. Each has its advantages and disadvantages. We need to think about the following factors:

- **Cost**. We should separate initial costs from running costs. A solar cell is expensive to buy but there are no costs for fuels – sunlight is free!
- **Reliability**. Is the energy supply constantly available? The wind is variable, so wind power is unreliable. Wars and trade disputes can interrupt fuel supplies.
- **Scale.** As discussed above, a fossil-fuel power station can be compact and still supply a large population. It would take several square metres of solar panels to supply a small household.
- **Environmental impact**. The use of fossil fuels leads to climate change. A hydro-electric dam may flood useful farmland. Every energy source has some effect on the environment.

Activity 7.1 Renewables versus non-renewables

Explain why some energy resources are described as 'renewables'.

Why should we make more use of renewables, and what are their problems?

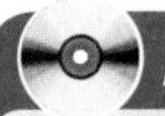

Activity 7.2 Future energy

Make a plan for a world that does not rely on fossil fuels for most of its energy.

QUESTION

8 Explain whether the following energy sources are renewable or non-renewable:
  a uranium-fuelled nuclear power
  b wave power.

7.2 Fuel for the Sun

The Sun releases vast amounts of energy, but it is not burning fuel in the same way as we have seen for fossil fuels. The Sun consists largely of hydrogen, but there is no oxygen to burn this gas. Instead, energy is released in the Sun by the process of **nuclear fusion**. In fusion, two energetic hydrogen atoms collide and fuse (join up) to form an atom of helium.

Nuclear fusion requires very high temperatures and pressures. The temperature inside the Sun is close to 15 million degrees. The pressure is also very high, so that hydrogen atoms are forced very close together, allowing them to fuse.

Scientists and engineers would like to be able to make fusion happen in a similar way here on Earth. Experimental reactors have been built, but it is very tricky to create the necessary conditions for fusion to happen in a controlled way. Perhaps, one day, fusion will prove a safe, clean way of producing a reliable electricity supply.

Summary

Most of our energy comes, directly or indirectly, from the Sun.

Useful energy resources include heat and light from the Sun, biomass and fossil fuels, water, wind, nuclear fuels and geothermal energy.

These energy resources are often used to produce electricity or other useful forms of energy.

A renewable energy resource is replaced after it has been used. When a non-renewable resource is used, it is gone forever.

The Sun releases energy by the fusion of hydrogen to form helium.

End-of-chapter questions

7.1 Explain how the following energy resources rely on energy from the Sun:

a biomass fuel, such as wood [2]

b electricity from a hydro-electric power station. [3]

7.2 Electricity supplied by solar cells is expensive. This is because, although sunlight is free, the cells themselves are expensive to produce.

a Explain why solar cells are a suitable choice for powering a spacecraft but are less likely to be used for providing domestic electricity to consumers in a city such as London, Dubai or Hong Kong. [3]

b Suggest **one** other situation in which solar cells would be a good choice, and justify your suggestion. [2]

c Why are solar cells often used in conjunction with a battery? [2]

7.3 In a hydro-electric power station, water is stored behind a dam. It flows down past a turbine, so that the turbine spins. This causes a generator to turn and produce electricity.

a What form of energy is stored by the water when it is behind the dam? [1]

b What form of energy does the spinning turbine have? [1]

c Write down the **two** energy transformations that occur in a hydro-electric power station. [2]

E **7.4** Fission and fusion are two nuclear processes that release energy.

a i Which is used in a nuclear power station? [1]

ii What is the fuel used for this? [1]

b i Which is the Sun's energy source? [1]

ii What element is the fuel? [1]

iii What element is produced? [1]

Work and power

Core Understanding the ideas of work and power

Extension Calculating work and power

8.1 Doing work

Figure **8.1** shows one way of lifting a heavy object. Pulling on the rope raises the heavy box. As you pull, the force on the box moves upwards.

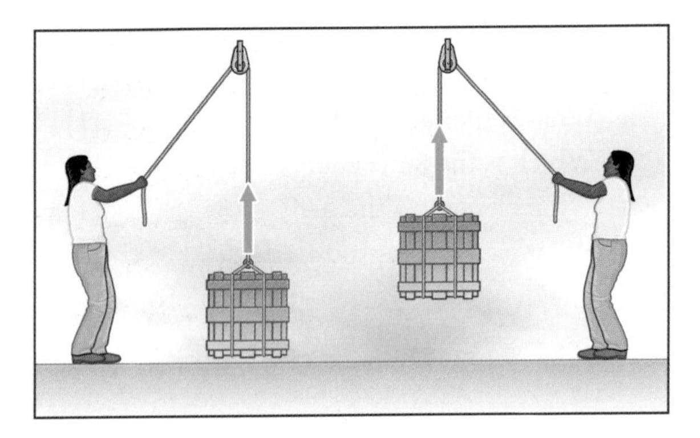

Figure 8.1 Lifting an object requires an upward force, pulling against gravity. As the box rises upwards, the force also moves upwards. Energy is being transferred by the force to the box.

To lift an object, you need a store of energy (as chemical energy, in your muscles). You give the object more gravitational potential energy (g.p.e.). The force is your means of transferring energy from you to the object. The name given to this type of energy transfer by a force is **doing work**.

The more work that a force does, the more energy it transfers. The amount of **work done** is simply the amount of energy transferred:

work done = energy transferred

Three further examples of forces doing work are shown in Figure **8.2**.

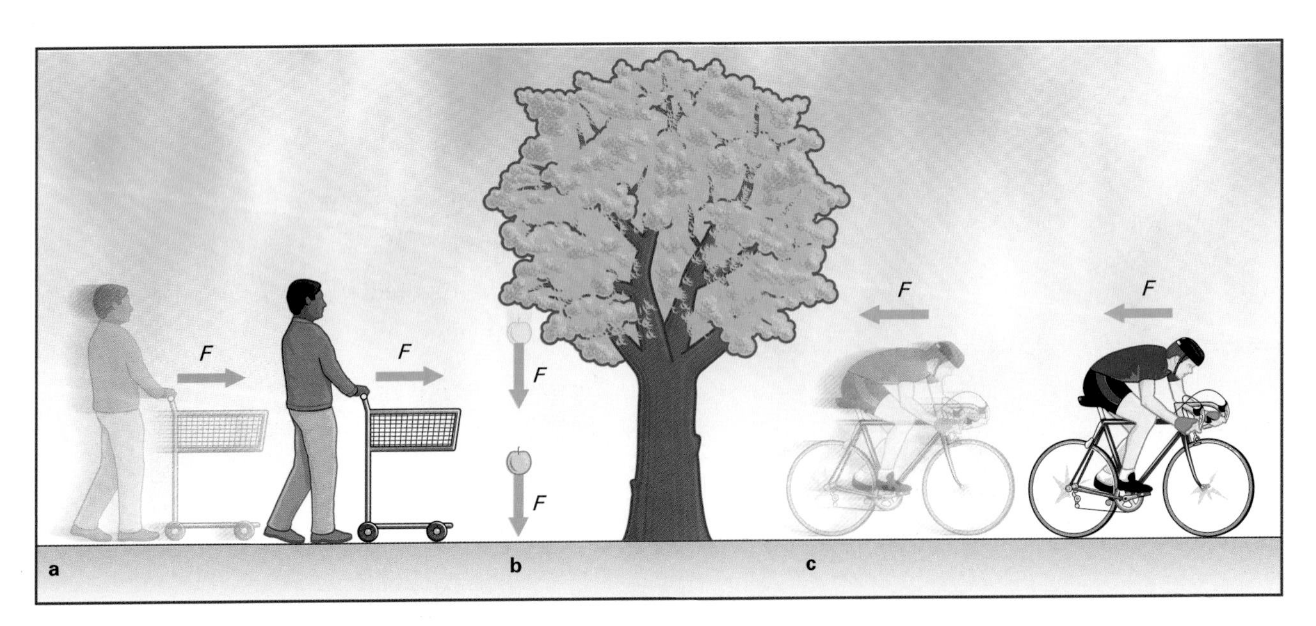

Figure 8.2 Three examples of forces doing work. In each case, the force moves as it transfers energy.

- Pushing a shopping trolley to start it moving. The pushing force does work. It transfers energy to the trolley, and the trolley's kinetic energy (k.e.) increases.
- An apple falling from a tree. Gravity pulls the apple downwards. Gravity does work, and the apple's k.e. increases.
- Braking to stop a bicycle. The brakes produce a backward force of friction, which slows down the bicycle. The friction does work, and reduces the bicycle's k.e. Energy is transferred to the brakes, which get hot.

How much work?

Think about lifting a heavy object, as shown in Figure **8.1**. A heavy object needs a big force to lift it. The heavier the object is, and the higher it is lifted, the more its g.p.e. increases. This suggests that the amount of energy transferred by a force depends on two things:

- the size of the force – the greater the force, the more work it does
- the distance moved in the direction of the force – the further it moves, the more work it does.

So a big force moving through a big distance does more work than a small force moving through a small distance.

Words in physics

You will by now understand that 'work' is a word that has a specialised meaning in physics, different from its meaning in everyday life. When physicists think about the idea of 'work', they think about forces moving.

If you are sitting thinking about your homework, no forces are moving and you are doing no work. It is only when you start to write that you are doing work in the physics sense. To make the ink flow from your pen, you must push against the force of friction, and then you really are working.

Many words have specialised meanings in science.

In earlier chapters, we used these words:

force mass weight velocity moment energy

Each has a carefully defined meaning in physics. This is important because physicists have to agree on the terms they are using. However, if you look these words up in a dictionary, you will find that they have a range of everyday meanings, as well as their specialised scientific meaning. This is not a problem, provided you know whether you are using a particular word in its scientific sense or in a more everyday sense. (Some physicists get very upset if they hear shopkeepers talking about weights in kilograms, for example, but no-one will understand you if you ask for 10 newtons of oranges!)

QUESTIONS

1. Which requires more work, lifting a 10 kg sack of coal or lifting a 15 kg bag of feathers?
2. Which force does work when a ball rolls down a slope?

8.2 Calculating work done

When a force does work, it transfers energy to the object it is acting on. The amount of energy transferred is equal to the amount of work done. We can write this as a simple equation:

$$\Delta W = \Delta E$$

In this equation, we use the symbol Δ (Greek letter delta) to mean 'amount of' or 'change in'. So

ΔW = amount of work done

ΔE = change in energy

How can we calculate the work done by a force? Above, we saw that the work done depends on two things:

- the size of the force F
- the distance d moved by the force.

We can write:

work done = force × distance moved by the force

$$\Delta W = F \times d$$

We use the symbol ΔW to represent the amount of work done. Since this is the same as the amount of energy transferred, it is measured in joules (J), the unit of energy.

Joules and newtons

The equation for the work done by a force ($\Delta W = F \times d$) shows us the relationship between joules and newtons. If we replace each quantity in the equation by its SI unit, we get

$$1\,\text{J} = 1\,\text{N} \times 1\,\text{m} \qquad \text{or} \qquad 1\,\text{J} = 1\,\text{N}\,\text{m}$$

So a joule is a newton-metre. More formally the **joule** (J) is defined as follows:

> One joule (1 J) is the energy transferred (or the work done) by a force of one newton (1 N) when it moves through a distance of one metre (1 m).

Worked example 1

A crane lifts a crate upwards through a height of 20 m. The lifting force provided by the crane is 5 kN (see Figure **8.3**). How much work is done by the force? How much energy is transferred to the crate?

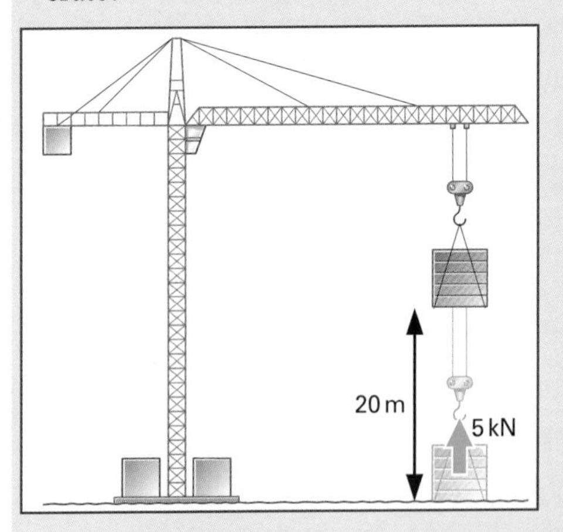

Figure 8.3 A crane provides the upward force needed to lift a crate. The force transfers energy from the crane to the crate. The crate's g.p.e. increases.

Step 1: Write down what you know, and what you want to know.

$$F = 5\,\text{kN} = 5\,000\,\text{N}$$
$$d = 20\,\text{m}$$
$$\Delta W = ?$$

Step 2: Write down the equation for work done, substitute values and solve.

$$\Delta W = F \times d$$
$$= 5000\,\text{N} \times 20\,\text{m}$$
$$= 100\,000\,\text{J}$$

So the work done by the force is 100 000 J, or 100 kJ.

Since work done = energy transferred, this is also the answer to the second part of the question: 100 kJ of energy is transferred to the crate.

Work done and *mgh*

Worked example **1** in which the crane lifts the crate illustrates an important idea. The force provided by the crane to lift the crate must equal the crate's weight mg. It lifts the crate through a height h. Then the work it does is force × distance, or $mg \times h$. Hence the gain in g.p.e. of the crate is mgh. This explains where the equation for g.p.e. comes from.

In Figure **8.4**, the child slides down the slope. Gravity pulls her downwards, and makes her speed up. To calculate the work done by gravity, we need to know the vertical distance h, because this is the distance moved **in the direction of the force**. If we calculated the work done as weight × distance moved down the slope, we would get an answer that was too large. So we have:

work done = force × distance moved in the direction of the force

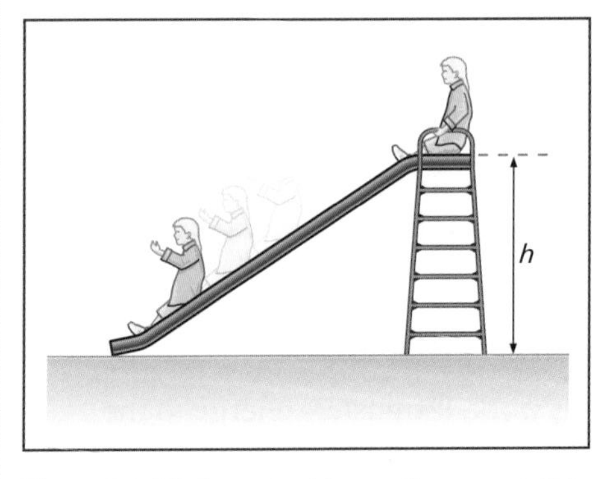

Figure 8.4 It is important to use the correct distance when calculating work done by a force. Gravity makes the child slide down the slope. However, to calculate the energy transferred by gravity, we must use the vertical height moved.

Forces doing no work

If you sit still on a chair (Figure **8.5a**), there are two forces acting on you. These are your weight (mg), acting downwards, and the upward contact force C of the chair, which stops you from falling through the bottom of the chair.

Neither of these forces is doing any work to you. The reason is that neither of the forces is moving, so they do not move through any distance d. Hence, from $\Delta W = F \times d$, the amount of work done by each force is zero. When you sit still on a chair, your energy does not increase or decrease as a result of the forces acting on you.

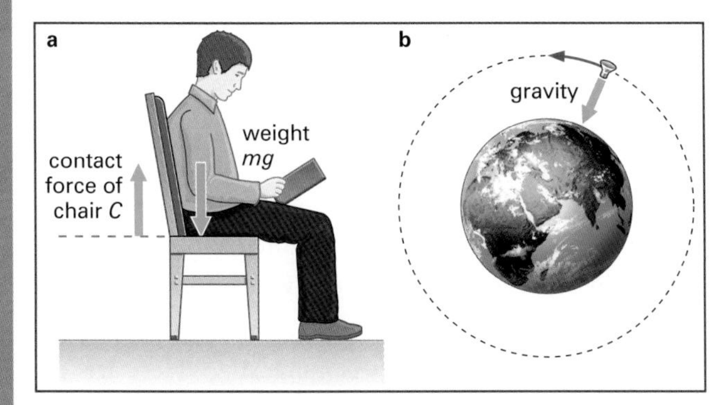

Figure 8.5 a When you sit still in a chair, there are two forces acting on you. Neither transfers energy to you. **b** The spacecraft stays at a constant distance from the Earth. Gravity keeps it in its orbit without transferring any energy to it.

Figure **8.5b** shows another example of a force that is doing no work. A spacecraft is travelling around the Earth in a circular orbit. The Earth's gravity pulls on the spacecraft to keep it in its orbit. The force is directed towards the centre of the Earth. However, since the spacecraft's orbit is circular, it does not get any closer to the centre of the Earth. There is no movement in the direction of the force, and so gravity does no work. The spacecraft continues at a steady speed (its k.e. is constant) and at a constant height above the Earth's surface (its g.p.e. is constant). Of course, although the force is doing no work, this does not mean that it is not having an effect. Without the force, the spacecraft would escape from the Earth and disappear into the depths of space.

Worked example 2

A girl can provide a pushing force of only 200 N. To move a box weighing 400 N onto a platform, she uses a plank as a ramp (Figure **8.6**). How much work does she do in raising the box? How much g.p.e. does the box gain?

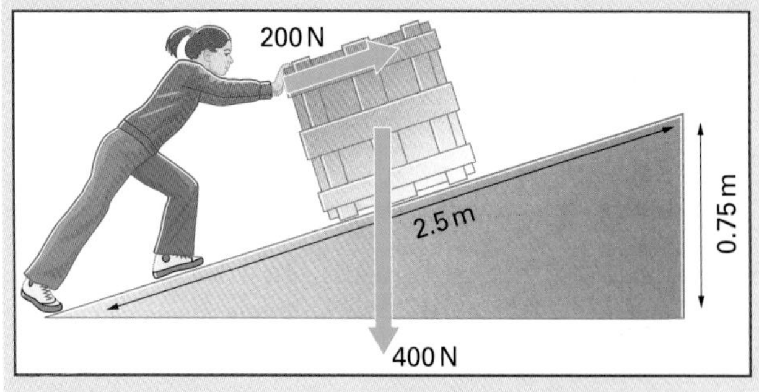

Figure 8.6 A ramp can allow you to lift a heavy load, but you do more work than if you could raise it unaided. From the diagram, you can see that the box is raised 0.75 m vertically, but the girl has to push it 2.5 m along the slope.

Step 1: Write down what you know, and what you want to know.

pushing force along the slope $F = 200\,\text{N}$
distance moved along slope $d = 2.5\,\text{m}$
weight of box downwards $mg = 400\,\text{N}$
vertical distance moved $h = 0.75\,\text{m}$
work done along the slope $\Delta W = ?$
work done against gravity $\Delta W' = ?$

Step 2: Calculate the work done by the pushing force along the slope, ΔW.

$$\Delta W = \text{pushing force along slope} \times \text{distance moved along slope}$$
$$= F \times d$$
$$= 200\,\text{N} \times 2.5\,\text{m}$$
$$= 500\,\text{J}$$

Step 3: Calculate the gain in g.p.e. of the box. This is the same as the work done against gravity, $\Delta W'$.

$$\Delta W' = \text{weight of box} \times \text{vertical distance moved}$$
$$= mg \times h$$
$$= 400\,\text{N} \times 0.75\,\text{m}$$
$$= 300\,\text{J}$$

So the girl does 500 J of work, but only 300 J is transferred to the box. The remaining 200 J is the work done against friction as the box is pushed along the slope.

E

Activity 8.1 Doing work

Push a load up a slope so that you do work. Where does your energy go?

QUESTIONS

3 In what units do we measure the work done by a force?

4 A fast-moving car has 0.5 MJ of kinetic energy. The driver brakes and the car comes to a halt. How much work is done by the force provided by the brakes?

5 **a** How much work is done by a force of 1 N moving through 1 m?

b How much work is done by a force of 5 N moving through 2 m?

6 Which does more work, a force of 500 N moving through 10 m or a force of 100 N moving through 40 m?

7 A steel ball of weight 50 N hangs at a height of 5 m above the ground, on the end of a chain 2 m in length. How much work is done on the ball by gravity, and by the tension in the chain?

8.3 Power

Exercising in the gym (Figure 8.7) can put great demands on your muscles. Speeding up the treadmill means that you have to work harder to keep up. Equally, your trainer might ask you to find out how many times you can lift a set of weights in one minute. These exercises are a test of how powerful you are. The faster you work, the greater your power.

In physics, the word **power** is often used with a special meaning. It means the rate at which you do work (that is, how fast you work). The more work you do, and the shorter the time in which you do it, the greater your power.

Power is the rate at which energy is transferred, or the rate at which work is done.

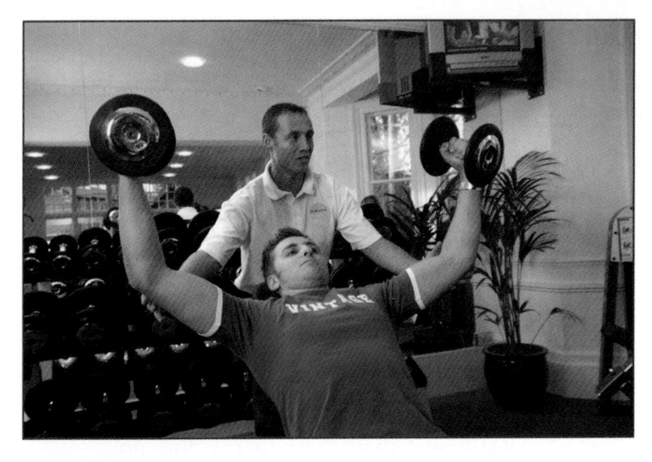

Figure 8.7 It is hard work down at the gym. It is easier to lift small loads, and to lift them slowly. The greater the load you lift and the faster you lift it, the greater the power required. It is the same with running on a treadmill. The faster you have to run, the greater the rate at which you do work.

Fast working

Power tells you about the rate at which a force does work – in other words, the rate at which it transfers energy. When you lift an object up, you are giving it energy. (Its potential energy is increasing.) Here are two ways you can increase your power:

- lift a heavier object in the same time
- lift the object more quickly.

It is not just people who do work. Machines also do work, and we can talk about their power in the same way.

- A crane does work when it lifts a load. The bigger the load and the faster it lifts the load, the greater is the power of the crane.
- A locomotive pulling a train of coaches or wagons does work. The greater the force with which it pulls and the greater the speed at which it pulls, the greater is the power of the locomotive.

QUESTION

8 Your neighbour is lifting bricks and placing them on top of a wall. He lifts them slowly, one at a time. State **two** ways in which he could increase his power (the rate at which he is transferring energy to the bricks).

8.4 Calculating power

We can write these ideas about power as an equation:

$$\text{power} = \frac{\text{work done}}{\text{time taken}} \qquad P = \frac{\Delta W}{t}$$

Since work done = energy transferred, we can also write a similar equation:

$$\text{power} = \frac{\text{energy transferred}}{\text{time taken}} \qquad P = \frac{\Delta E}{t}$$

Units of power

Power is measured in watts (W). One watt (1 W) is the power when one joule (1 J) of work is done in one second (1 s). So one watt is one joule per second:

1 W = 1 J/s
1 000 W = 1 kW (kilowatt)
1 000 000 W = 1 MW (megawatt)

Take care not to confuse (italic) *W* for work done (or energy transferred) with (upright) W for watts. In books, the first of these is shown in *italic* type (as here), but you cannot tell the difference when they are written.

Worked example 3

A car of mass 800 kg accelerates from rest to a speed of 25 m/s in 10 s. What is its power?

Step 1: Calculate the work done. This is the increase in the car's kinetic energy.

$$\text{k.e.} = \tfrac{1}{2}mv^2$$
$$= \tfrac{1}{2} \times 800\,\text{kg} \times (25\,\text{m/s})^2$$
$$= 250\,000\,\text{J}$$

Step 2: Calculate the power.

$$\text{power} = \frac{\text{work done}}{\text{time taken}}$$
$$= \frac{\Delta W}{t}$$
$$= \frac{250\,000\,\text{J}}{10\,\text{s}}$$
$$= 25\,000\,\text{W}$$
$$= 25\,\text{kW}$$

So the energy is being transferred to the car (from its engine) at a rate of 25 kW, or 25 kJ per second.

Car engines are not very efficient. In this example, the car's engine may transfer energy at the rate of 100 kW or so, although most of this is wasted as thermal (heat) energy.

Power in general

We can apply the idea of power to any transfer of energy. For example, electric light bulbs transfer energy supplied to them by electricity. They produce light and heat. Most light bulbs are labelled with their power rating – for example, 40 W, 60 W, 100 W – to tell the user about the rate at which it transfers energy.

There is more about electrical power in Chapter **19**.

QUESTIONS

9 **a** How many watts are there in a kilowatt?
b How many watts are there in a megawatt?

10 It is estimated that the human brain has a power requirement of 40 W. How many joules is that per second?

11 A light bulb transfers 1000 J of energy in 10 s. What is its power?

12 An electric motor transfers 100 J in 8 s. If it then transfers the same amount of energy in 6 s, has its power increased or decreased?

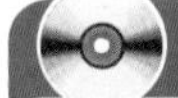

Activity 8.2 Measuring your power

It is hard work running up a flight of stairs. Time yourself and calculate your power.

Summary

When a force moves, it transfers energy. We say that it does work.

The greater the force and the greater the distance it moves, the more work is done.

E

Work done = energy transferred.

Work done = force × distance moved by the force

$$\Delta W = F \times d$$

The distance moved is measured in the direction of the force.

Power is the rate at which energy is transferred, or the rate at which work is done.

The greater the amount of work done and the shorter the time in which it is done, the greater the power.

E

$$\text{Power} = \frac{\text{work done}}{\text{time taken}} \qquad P = \frac{\Delta W}{t}$$

$$\text{Power} = \frac{\text{energy transferred}}{\text{time taken}} \qquad P = \frac{\Delta E}{t}$$

End-of-chapter questions

8.1 Omar and Ahmed are lifting weights in the gym. Each lifts a weight of 200 N. Omar lifts the weight to a height of 2.0 m, whereas Ahmed lifts it to a height of 2.1 m. Who does more work in lifting the weight? Explain how you know. [2]

8.2 Millie and Lily are identical twins who enjoy swimming. Their arms and legs provide the force needed to move them through the water. Millie can swim 25 m in 50 s. Lily can swim 100 m in 250 s.

a Calculate the swimming speed of each twin. [2]

b Which twin has the greater power when swimming? Explain how you can tell. [2]

E

8.3 Jim is pulling a load along a ramp, as shown in Figure **8.8**. The diagram shows the force with which he pulls and the weight of the load.

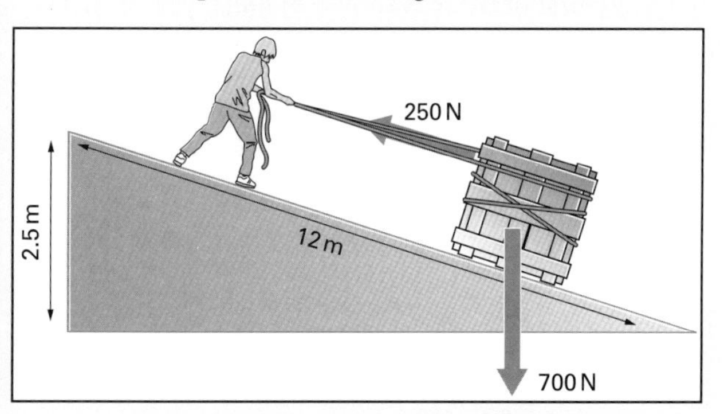

Figure 8.8 Pulling a load up a ramp – for Question **8.3**.

a Calculate the work done by Jim's pulling force. [3]

b What is the gain in potential energy of the load? [3]

E **8.4** Two girls are estimating each other's power. One runs up some steps, and the other times her. Here are their results:

height of one step = 20 cm
number of steps = 36
mass of runner = 45 kg
time taken = 4.2 s

a Calculate the runner's weight. (Acceleration due to gravity $g = 10\,m/s^2$.) [2]

b Calculate the increase in the girl's gravitational potential energy as she runs up the steps. [3]

c Calculate her power. Give your answer in kilowatts (kW). [4]

E **8.5** A car of mass 750 kg accelerates away from traffic lights. At the end of the first 100 m it has reached a speed of 12 m/s. During this time, its engine provides an average forward force of 780 N, and the average force of friction on the car is 240 N.

a Calculate the work done on the car by the force of its engine. [3]

b Calculate the work done on the car by the force of friction. [3]

c Using k.e. = $\frac{1}{2}mv^2$, calculate the increase in the car's kinetic energy at the end of the first 100 m. [2]

d Explain whether your answers are consistent with the principle of conservation of energy. [3]

Block 2

Thermal physics

We are fortunate because we live on the Earth. The Earth is a planet orbiting the Sun at a distance of 150 million kilometres. The Sun keeps the Earth warm, with an average temperature of about 15 °C. At night, the temperature drops because we are facing away from the Sun, outwards into the coldness of space. Because the Earth spins, every part of the globe faces the Sun periodically and has a chance to warm up again after the cooling down of the night.

Things are different out in space. Space is cold and dark. Its average temperature is close to −270 °C. How can we know that? It is simple. Scientists have sent spacecraft like the one shown out into space to measure the temperature. The temperature probe used must be shielded from the Sun's rays and also from any warmth of the spacecraft itself. Measurements show that the average temperature of space is just 2.7 degrees above absolute zero, the coldest possible temperature.

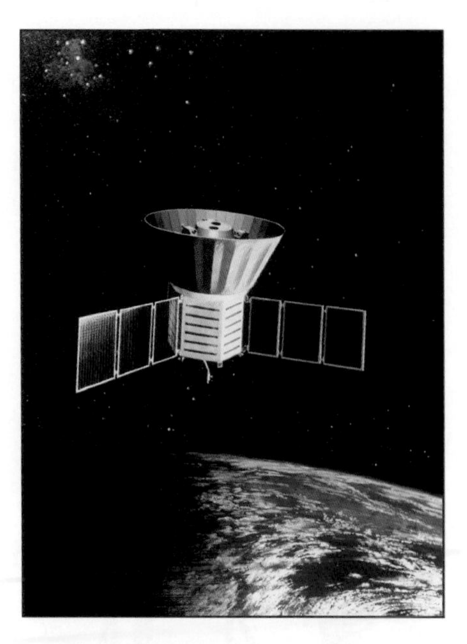

An artist's impression of the Cosmic Background Explorer. This spacecraft measured the temperature of space by detecting the radiation left over from the Big Bang. The gold-coloured 'skirt' shields the temperature sensors from heat from the Sun and Earth.

In fact, the temperature of space had been correctly predicted before anyone had a chance to measure it. The prediction came from the Big Bang theory of the origin of the Universe. The idea is that, roughly 13.7 billion years ago, the Universe 'exploded' into existence. Ever since, it has been expanding and so cooling. When scientists measure the temperature of space, they are detecting the last remnants of the great fireball that was the early Universe.

In this block, we will look at some of these ideas in more detail. In particular, we will look at what we mean by temperature and how thermal (heat) energy travels around.

9 The kinetic model of matter

Core Describing solids, liquids and gases
Core Describing changes of state
Core Using the kinetic model to explain changes of state
E **Extension** Explaining the kinetic model in terms of the forces between particles
Core Explaining the behaviour of gases
E **Extension** Calculating changes in pressure and volume

Snow

Young people usually enjoy snow (Figure **9.1**). You may live in a country where snow is rarely seen. Alternatively, you may be snow-bound for several months of the year. If you do experience snow, you will know the excitement of the first fall of the winter. Everyone rushes out to have snowball fights, or to go tobogganing or skiing.

Snow is remarkable stuff. It is simply frozen water. Yet people such as the Inuit who live among snow have many different words for it, depending on how it is packed down, for instance. This can be vital information if you are interested in winter sports, since it determines the avalanche risk.

We are familiar with the changes that happen when snow or ice melts. A white or glassy solid changes into a transparent, colourless, runny liquid. Heat the liquid and it 'vanishes' into thin air. Although this sounds like a magic trick, it is so familiar that it does not strike us as surprising. The Earth is distinctive among the planets of the solar system in being the only planet on which water is found to exist naturally in all three of its physical states.

In this chapter, we will look at what happens when materials change their state – from solid to liquid to gas, and back again. By thinking about the particles, the atoms and molecules of which the material is made, we can build up a picture or model that describes changes of state and explains some of the things we observe when materials change from one state to another.

Figure 9.1 Dubai is a hot place, but you can still ski there on the artificial snow in this covered ski centre.

9.1 States of matter

We think of matter as existing in three states, **solid**, **liquid** and **gas**. What are the characteristic properties of each of these three states? We need to think about shape and size. Table **9.1** shows how these help us to distinguish between solids, liquids and gases. It may help you to think about ice, water and steam as examples of the three states of matter.

State	Size	Shape
solid	occupies a fixed volume	has a fixed shape
liquid	occupies a fixed volume	takes the shape of its container
gas	expands to fill its container	takes the shape of its container

Table 9.1 The distinguishing properties of the three states of matter.

Here is a trick to try on a small child. Pour a drink into a short, wide glass. Then pour it from that glass into a tall, narrow glass. Ask them which drink they would prefer. Many small children ask for the drink in the tall glass because it appears that there is more. Of course, you will realise that, although the drink changes its shape when you pour it from one container to another, its size (volume) stays the same.

Changes of state

Heat a solid and it melts to become a liquid. Heat the liquid and it boils to become a gas. Cool the gas and it becomes first a liquid and then a solid. These are **changes of state**. The names for these changes are shown in Figure **9.2**. They are:

- **melting** – from solid to liquid
- **boiling** – from liquid to gas
- **condensing** – from gas to liquid
- **freezing** – from liquid to solid.

Another term for a liquid changing to a gas is **evaporation**. We will see the difference between evaporation and boiling shortly.

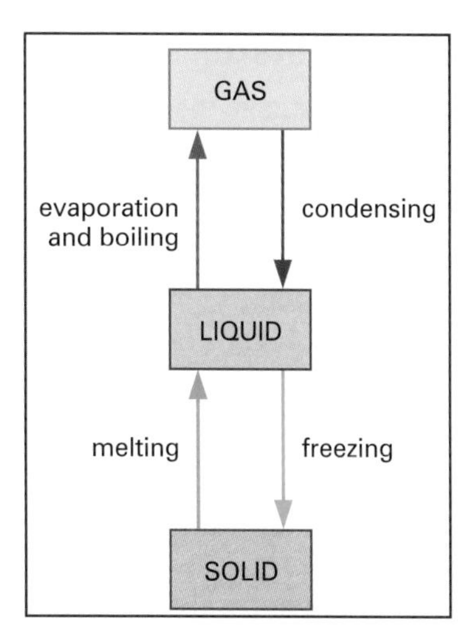

Figure 9.2 Naming changes of state.

Figure **9.3** shows what happens if you take some ice from the deep freeze and heat it at a steady rate. In a deep freeze, ice is at a temperature well below its freezing point (0 °C). From the graph, you can see that the ice warms up to 0 °C, then remains at this temperature while it melts. Lumps of ice float in water; both are at 0 °C. When all of the ice has melted, the water's temperature starts to rise again. At 100 °C, the boiling point of water, the temperature again remains steady. The water is boiling to form steam. This takes longer than melting, which tells us that it takes more energy to boil the water than to melt the ice. Eventually, all of the water has turned to steam. If we can continue to heat it, its temperature will rise again.

Notice that energy must be supplied to change a solid into a liquid. At the same time, its temperature remains constant as it melts. Similarly, when a liquid becomes a gas, its temperature remains constant even though energy is being supplied to it.

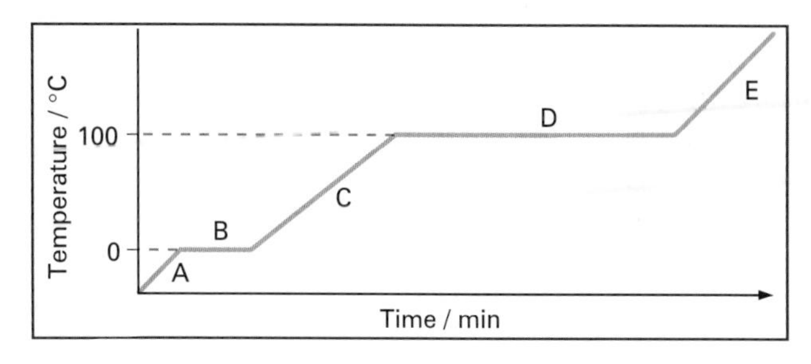

Figure 9.3 A temperature against time graph to show the changes that occur when ice is heated until it eventually becomes steam.

Investigating a change of state

Figure **9.4a** shows one way to investigate the behaviour of a liquid material as it solidifies. The test tube contains a waxy substance called octadecanoic acid. This is warmed up, and it becomes a clear, colourless liquid. It is then left to cool down, and its temperature is monitored using a thermometer (an electronic temperature probe) and recorded using a data-logger. The graph of Figure **9.4b** shows the results. From the graph, you can see that there are three stages in the cooling of the material.

1 The liquid wax cools down. Its temperature drops gradually. The wax is hotter than its surroundings, so it loses heat. Notice that the graph is slightly curved; this is because, as the temperature drops, there is less difference between the temperature of the wax and its surroundings, so it cools more slowly.
2 Now the wax's temperature remains constant for a few minutes. The tube can be seen to contain a mixture of clear liquid and white solid – the wax is solidifying. During this time, the wax is still losing heat, because it is still warmer than its surroundings, but its temperature does not decrease. This is an important observation that needs explaining.
3 The wax's temperature starts to drop again. It is now entirely solid, and it continues cooling until it reaches the temperature of its surroundings.

From the horizontal section of the graph (stage 2) we can draw a horizontal line across to the temperature axis and find the substance's melting point.

From the experiment shown in Figure **9.4**, you can see that a pure substance changes from solid to liquid at a particular temperature, known as the **melting point**. Similarly, a liquid changes to a gas at a fixed temperature, its **boiling point**. Table **9.2** shows the melting and boiling points of some pure substances.

Substance	Melting point / °C	Boiling point / °C
helium	−272	−269
oxygen	−218	−183
nitrogen	−191	−177
mercury	−39	257
water	0	100
iron	2080	3570
diamond (carbon)	4100	5400
tungsten	3920	6500

Table 9.2 The melting and boiling points of some pure substances. Mercury is interesting because it is the only metal that is not solid at room temperature. Tungsten is a metal, and it has the highest boiling point of any substance. Helium has the lowest melting and boiling points of any element. In fact, helium will only solidify if it is compressed as well as cooled.

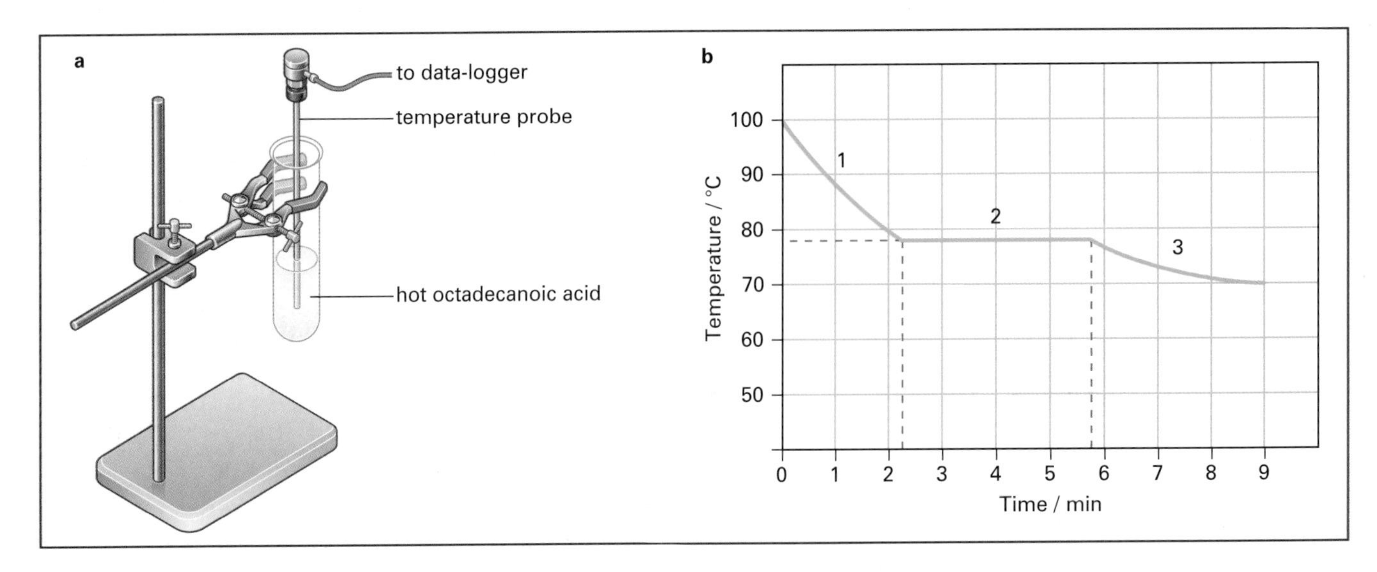

Figure 9.4 **a** As the warm liquid cools, its temperature is monitored by the electronic temperature probe. **b** The graph shows how the temperature of the octadecanoic acid drops as it cools. The temperature remains constant as the liquid solidifies.

Note that we have to be careful here to talk about **pure** substances. The temperature at which a substance melts or boils is different when another substance is dissolved in it. For example, salty water boils at a higher temperature than pure water, and freezes at a lower temperature. A mixture of substances may even melt or boil over a range of temperatures. Candle wax is an example. It is not a single, pure substance, and some of the substances in it melt at lower temperatures than others. Similarly, crude oil is a mixture of different substances, each with its own boiling point. You may have studied the process of fractional distillation, which is used to separate these substances (fractions) at an oil refinery.

There are other ways in which materials can behave when they are heated: some burn, and others decompose (break down) into simpler substances before they have a chance to change state.

Activity 9.1 Measuring melting point

Carry out an experiment to determine the melting point of octadecanoic acid or some other pure substance. Use the method shown in Figure **9.4**.

QUESTIONS

1 To measure the volume of a liquid, you can pour it into a measuring cylinder. Measuring cylinders come in different shapes and sizes – tall, short, wide, narrow. Explain why the **shape** of the cylinder does not affect the measurement of volume.

2 What name is given to the temperature at which a gas condenses to form a liquid?

3 **a** What name is given to the process in which a liquid changes into a solid?
  b What name is given to the temperature at which this happens?

4 **a** Look at Figure **9.3** on page **87**. What is happening in the section marked C?
  b Name the substance or substances present in the section marked D.

5 Look at Figure **9.4b**. From the graph, deduce the melting point of octadecanoic acid.

6 Table **9.2** shows the melting and boiling points of nitrogen and oxygen, the main constituents of air. Why can we not talk about the melting and boiling points of air?

9.2 The kinetic model of matter

Several questions arise from our discussion of changes of state. In this section, we will look at a model for matter that provides one way in which we can answer these questions:

- Why does it take time for a solid to melt? Why does it not change instantly into a liquid?
- Why does it take longer to boil a liquid than to melt a solid?
- Why do different substances melt at different temperatures?
- Why do different substances have different boiling points?

The model we are going to consider is called the **kinetic model of matter**. As we saw in Chapter **6**, the word 'kinetic' means 'related to movement'. In this model, the things that are moving are the particles of which matter is made. The model thus has an alternative name: the **particle model of matter**.

The particles of which matter is made are very tiny. They may be atoms, molecules or ions, but we will simplify things by disregarding these differences and referring only to **particles**. We will also picture a material as consisting of large numbers of identical particles. Thus we are considering a pure substance whose particles are all the same, rather than a mixture that contains two or more types of particle. We will also picture the particles as simple spheres, although in reality they might have more complicated shapes. The molecules of a polymer, for example, may be like long thin strings of spaghetti, rather than like small, round peas.

The idea that matter is made up of spherical particles is a great simplification, but we can still use this idea to find answers to the questions listed above. Later, we will think about whether or not we are justified in using such a simplified model.

Arrangements of particles

Figure **9.5** shows how we picture the particles in a solid, a liquid and a gas. For each picture, we will think about two aspects (see Table **9.3**): how the particles are arranged, and how the particles are moving. (Because these are pictures printed on paper, it is hard to represent the motion of the particles. You may have access to software or video images that can show this more clearly.)

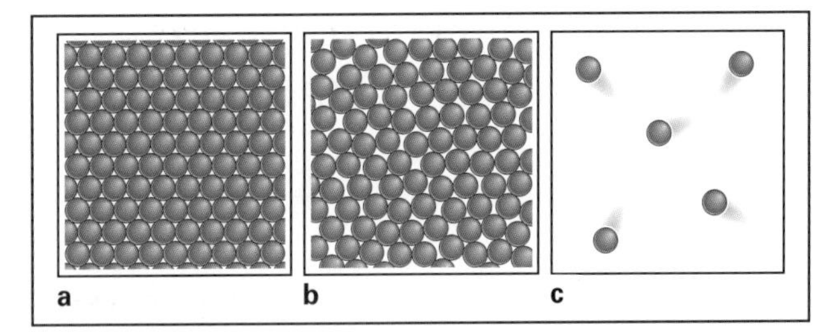

Figure 9.5 Representations of **a** solid, **b** liquid and **c** gas. The arrangement and motion of the particles change as the solid is heated to become first a liquid and then a gas.

Evidence for the kinetic model

We cannot look down a microscope and see the particles that make up matter. We certainly cannot hope to see the particles of a gas as they rush around. However, in the 1820s, the movement of the particles of a gas was investigated by a Scottish botanist, Robert Brown. He was using a microscope to study pollen grains when he noticed tiny particles jiggling about. At first he thought that they might be alive, but when he repeated his experiment with tiny grains of dust, suspended in water, he saw that they also moved around. This motion is now known as **Brownian motion**, and it happens because the moving particles are constantly buffeted by the fast-moving particles of the air.

Today, we can perform a similar experiment using smoke grains. The oxygen and nitrogen molecules that make up the air are far too small to see, so we have to look at something bigger, and look for the effect of the air molecules. We use a smoke cell (Figure **9.6**), which contains air with a small amount of smoke. The cell is lit from the side, and the microscope is used to view the smoke grains.

The smoke grains show up as tiny specks of light, but they are too small to see any detail of their shape.

State	Arrangement of particles	Movement of particles
solid	The particles are packed closely together. Notice that each particle is in close contact with all of its neighbours. In a solid such as a metal, each atom may be in contact with 12 neighbouring atoms.	Because the particles are so tightly packed, they cannot move around. However, they do move a bit. They are able to vibrate about a fixed position. The hotter the solid, the more they vibrate.
liquid	The particles are packed slightly less closely together (compared with a solid). Each particle is still in close contact with most of its neighbours, but fewer than in the case of a solid. The general arrangement is slightly more jumbled and disorderly.	Because the particles are slightly less tightly packed than in a solid, they have the opportunity to move around within the bulk of the liquid. Hence the particles are both vibrating and moving from place to place.
gas	Now the particles are widely separated from one another. They are no longer in contact, unless they collide with each other. In air, the average separation between the particles is about ten times their diameter.	The particles are now moving freely about, bouncing off one another and off the walls of their container. In air at room temperature, their average speed is about 400 m/s.

Table 9.3 The arrangement and movement of particles in the three different states of matter. Compare these statements with the pictures shown in Figure **9.5**.

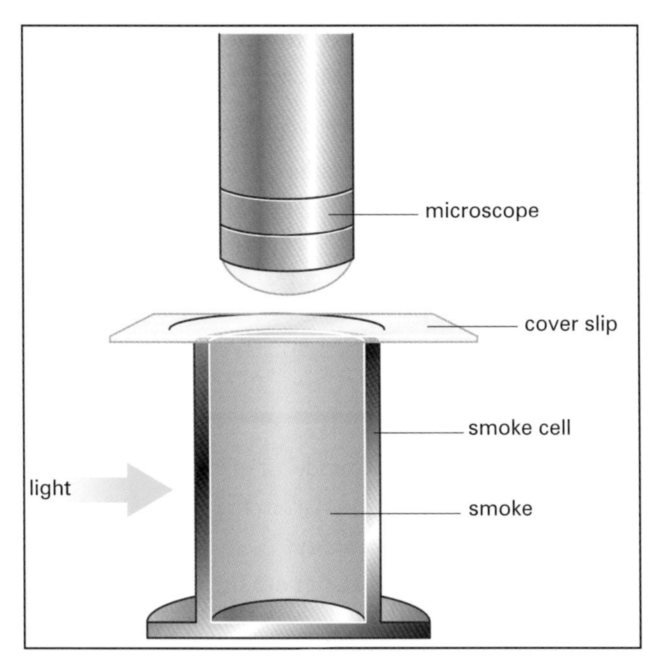

Figure 9.6 An experimental arrangement for observing Brownian motion. The grains of smoke are just large enough to show up under the microscope. The air molecules that collide with them are much too small to see.

What is noticeable is the way they move. If you can concentrate on a single grain, you will see that it follows a somewhat jerky and erratic path. This is a consequence of the grain suffering repeated collisions with air molecules.

Observing Brownian motion of smoke or pollen grains does not mean that we have proved that air and water are made of moving particles. We have not seen the particles themselves. Observing Brownian motion is rather like watching a hockey match from an aircraft high overhead. You may see the players rushing around, but you cannot see the ball. Careful observation over a period of time might lead you to guess that there was a ball moving around among the players, and eventually you might work out the rules of hockey.

However, the kinetic model does give a satisfying explanation of Brownian motion. Much of what scientists have learned since Brown did his first experiments has confirmed his suggestion that he had discovered an effect caused by moving molecules.

E

Observing Brownian motion of smoke particles in air allows us to deduce something important about the motion of air molecules. The air molecules are much smaller than the smoke grains – in other words, they are very light, compared to smoke grains – and yet they can cause the smoke grains to move around. The air molecules can only do this if they are moving around very fast. In fact, the molecules of the air around us move at speeds of the order of 500 m/s – that is a little faster than the speed of sound in air.

Activity 9.2 Observing Brownian motion

Watch brightly lit smoke grains moving in air. You may also be able to watch a video of Brownian motion.

Explanations using the kinetic model

The kinetic model of matter can be used to explain many observations. Here are some:

- Liquids take up the shape of their container, because their particles are free to move about within the bulk of the liquid.
- Gases fill their container, because their particles can move freely about.
- Solids retain their shape, because the particles are packed tightly together.
- Gases diffuse (spread out) from place to place, so that, for example, we can smell perfume across the room. The perfume particles spread about because they are freely mobile.
- Similarly, dissolved substances diffuse throughout a liquid. Sugar crystals in a drink dissolve and molecules spread throughout the liquid, carried by the mobile particles. In a hotter drink, the particles are moving faster and the sugar diffuses more quickly.
- Most solids expand when they melt. The particles are slightly further apart in a liquid than in a solid.
- Liquids expand a lot when they boil. The particles of a gas are much farther apart than in a liquid. We can think about this the other way round. Gases contract a lot when they condense. If all of the air in the room you are now in was cooled enough, it would condense to form a thin layer of liquid, two or three millimetres deep, on the floor.

Evaporation

The boiling point of water is 100 °C, but water does not have to be heated to 100 °C before it will turn into a gas. After a downpour of rain, the puddles eventually dry up even though the temperature is much lower than 100 °C. We say that the water has become water vapour in the air. This is the process of **evaporation**. We can think of a vapour as a gas at a temperature below its boiling point.

A liquid evaporates more quickly as its temperature approaches its boiling point. That is why puddles disappear quickly after a storm in the tropics, where the temperature may be 30 °C, but they may lie around for days in a cold region, where the temperature is close to 0 °C.

How can we use the kinetic model of matter to explain evaporation? Picture a beaker of water. The water will gradually evaporate. Figure **9.7** shows the particles that make up the water. The particles of the water are moving around, and some are moving faster than others. Some may be moving fast enough to escape from the surface of the water. They become particles of water vapour in the air. In this way, all of the water particles may eventually escape from the beaker, and the water has evaporated.

If you get wet, perhaps because you are caught in the rain or you have been swimming, you will notice that you can quickly get cold. The water on your body is evaporating, and this cools you down. Why does evaporation make things cooler?

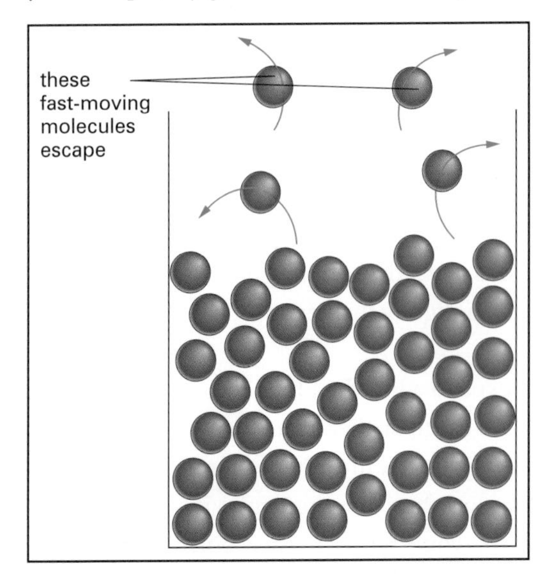

Figure 9.7 Fast-moving particles leave the surface of a liquid – this is how it evaporates.

Look again at Figure **9.7**. The particles that are escaping from the water are the fastest-moving ones. They are the particles with the most energy. This means that the particles that remain are those with less energy, and so the water is colder.

QUESTIONS

7 **a** Why is the kinetic model of matter called **kinetic**?
b In which state of matter do the particles have the most kinetic energy?

8 **a** In which state of matter are the particles most closely packed?
b In which state of matter are they most widely separated?

9 Use the kinetic model of matter to explain why we can walk through air and swim through water but we cannot walk through a solid wall.

10 In an experiment to observe Brownian motion, a student watched a brightly lit grain of dust moving around in water, following a random path.
a Explain why the student could not see the molecules of the water moving around.
b Explain why the grain of dust moved around in the water.

E

9.3 Forces and the kinetic theory

So far, we have seen how the kinetic theory of matter can successfully explain some observations of the ways in which solids, liquids and gases differ. We can explain some other observations if we add another scientific idea to the kinetic theory: we need to consider the forces between the particles that make up matter.

Why do the particles that make up a solid or a liquid stick together? There must be **attractive forces** between them. Without attractive forces to hold together the particles that make up matter, we would live in a very dull world. There would be no solids or liquids, only gases. No matter how much we cooled matter down, it would remain as a gas.

Another way to refer to these forces is to say that there are **bonds** between the particles. Each particle of a solid is strongly bonded to its neighbours. This is because the forces between particles are strongest when the particles are close together. In a liquid, the particles are slightly further apart and so the forces between them are slightly weaker. In a gas, the particles are far apart, so that the particles do not attract each other and can move freely about.

Kinetic theory and changes of state

What happens to these attractive forces as a solid is heated? The particles start to vibrate more and more strongly. Eventually, the particles vibrate sufficiently for some of the bonds to be broken, and a liquid is formed. Heat the material more and eventually the particles have sufficient energy for all of the attractive forces between particles to be overcome. The material becomes a gas.

In a gas, the particles are so far apart and moving so fast that they do not stick together. If you cool down a gas (Figure **9.8**), the particles move more slowly. As they collide with one another, there is more chance that they will stick together. Keep cooling the gas and eventually all of the particles stick together to form a liquid.

More about evaporation

Evaporation is different from boiling. A liquid boils at its boiling point – all of the liquid reaches this temperature and it gradually turns into a gas. Evaporation happens at a lower temperature, below the boiling point.

We have seen (page **92**) that the kinetic model can explain why a liquid cools as it evaporates, because it is the most energetic particles that are leaving its surface. We can use the kinetic model to explain some more observations concerning evaporation (see Table **9.4**).

Observation	Explanation
A liquid evaporates more rapidly when it is hotter.	At a higher temperature, more of the particles of the liquid are moving fast enough to escape from the surface.
A liquid evaporates more quickly when it is spread out, so that it has a greater surface area.	With a greater surface area, more of the particles are close to the surface, and so they can escape more easily.
A liquid evaporates more quickly when a draught blows across its surface.	A draught is moving air. When particles escape from the water, they are blown away so that they cannot fall back in to the water.

Table 9.4 Evaporation – observations and explanations.

Activity 9.3 Using the kinetic model

Discuss how the kinetic model can explain some observations.

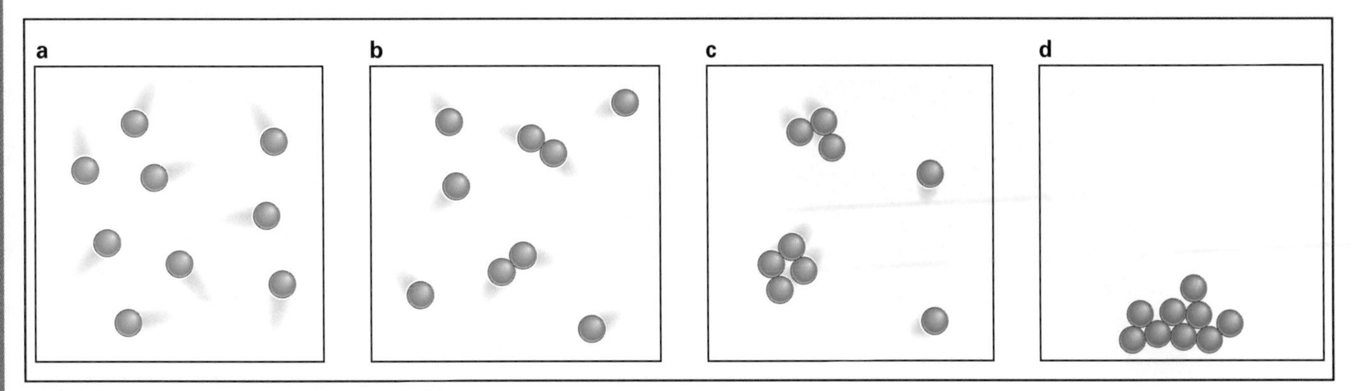

Figure 9.8 **a** As a gas is cooled, it starts to condense. **b** The particles move more slowly and they start to stick together, because of the attractive forces between them. **c** As their energy gets less, they clump together in bigger and bigger groups. **d** Finally, they form a liquid.

QUESTIONS

11 Tungsten melts at a much higher temperature than iron. What can you say about the forces between the tungsten atoms, compared to the forces between the iron atoms?

12 A particular solid material is heated but its temperature does not rise.
  a What is happening to the solid?
  b Where does the energy go that is being supplied to it?

13 If a gas is heated, its molecules move faster. Use the kinetic model to make a prediction: What will happen to the pressure that a gas exerts on the walls of its container when the gas is heated?

9.4 Gases and the kinetic theory

We can understand more about gases if we think about the particles of which they are made. For example:

- Why does a gas exert pressure?
- What happens to a gas when it is heated?
- What happens when a gas is compressed?

If you blow up a balloon (Figure **9.9**), your lungs provide the pressure to push the air into it. Tie up the balloon and the air is trapped. The pressure of the air inside pushes outwards against the rubber, keeping it inflated. The more air you blow into the balloon, the greater its pressure.

Figure 9.9 Inflating a balloon – as you blow, the pressure of the air inside the balloon increases.

Figure **9.10a** shows the particles that make up a gas. The gas is contained in a square box. The volume of the box is the volume of the gas. The gas has mass because each of its particles has mass. If we weighed all the particles individually and added up their masses, we would find the mass of the gas.

Figure **9.10b** shows the same box with twice as many gas particles in it. The mass of the gas is doubled, and so is its density.

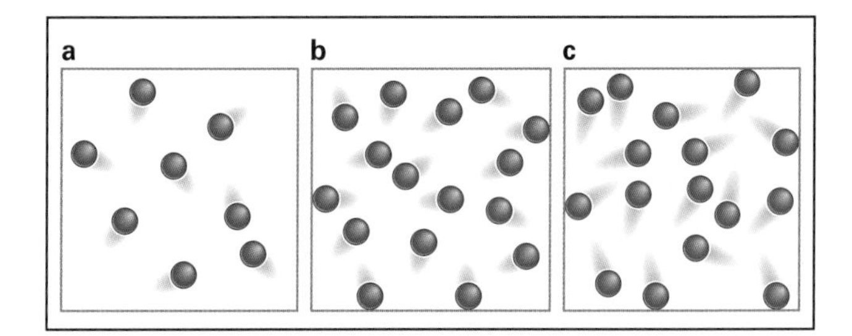

Figure 9.10 **a** The particles of a gas move around inside its container, bumping into the sides. **b** Doubling the number of particles means twice the mass, twice the density and twice the pressure. **c** At a higher temperature, the particles move faster. They have more kinetic energy, and this is what a thermometer records as a higher temperature.

A gas exerts pressure on the walls of its container because its particles are constantly colliding with the walls. They bounce off the walls, exerting a force as they do so. Compare Figures **9.10a** and **9.10b**: with twice as many particles, there are twice as many collisions, so the pressure is doubled.

Figure **9.10c** shows the same gas at a higher temperature. The particles are moving faster, and as a result they have more kinetic energy. So the higher the temperature of a gas, the faster its particles are moving.

Compressing a gas

Figure **9.11** shows some gas trapped in a box. If the box is made smaller, the volume of the gas decreases. At the same time, its pressure increases. From the diagram, you can see why this is. The particles of the gas have been squashed into a smaller volume. So they will collide with the walls of the container more frequently, creating an increased pressure. If the gas is compressed to half its original volume, its pressure will be doubled.

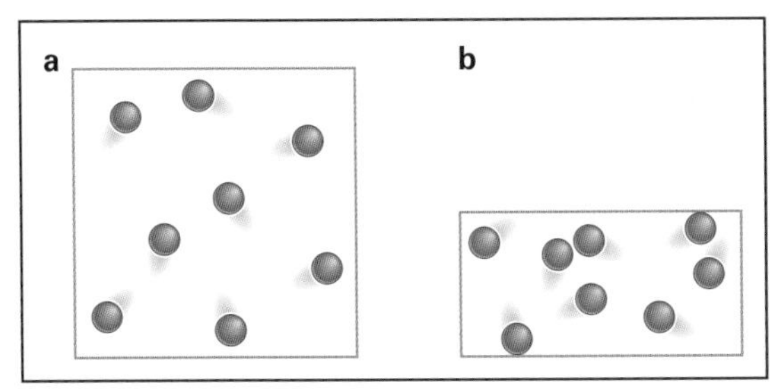

Figure 9.11 With the same number of particles in half the volume, in **b** there are twice as many collisions per second with the walls of the container. The result is twice the pressure in **b** as in **a**.

QUESTIONS

14 Look at Figure **9.10a**. If half of the particles of the gas were removed from the container (and nothing else was changed), how would the following properties of the gas change?
a density b pressure c temperature

15 Draw diagrams of the particles in a gas to explain why, if the volume of the gas is doubled, its pressure is halved.

16 Look at Figure **9.11**. The gas in **b** has twice the pressure as the gas in **a**. How could you change the temperature of the gas in **b** so that its pressure would be the same as that of the gas in **a**? Explain your answer.

Boyle's law

Figure **9.12** shows a method for investigating what happens when the pressure on a fixed mass of gas is increased. In this apparatus, some air is trapped inside the vertical glass tube. The oil in the bottom of the apparatus can be compressed with a pump, so that it pushes up inside the tube, compressing the air. The volume of the air can be read from the scale. The pressure exerted on it by the oil can be read from the dial gauge.

Increasing the pressure on the gas decreases its volume. Table **9.5** shows some typical results. But can we find a mathematical relationship between the pressure p and the volume V of the gas?

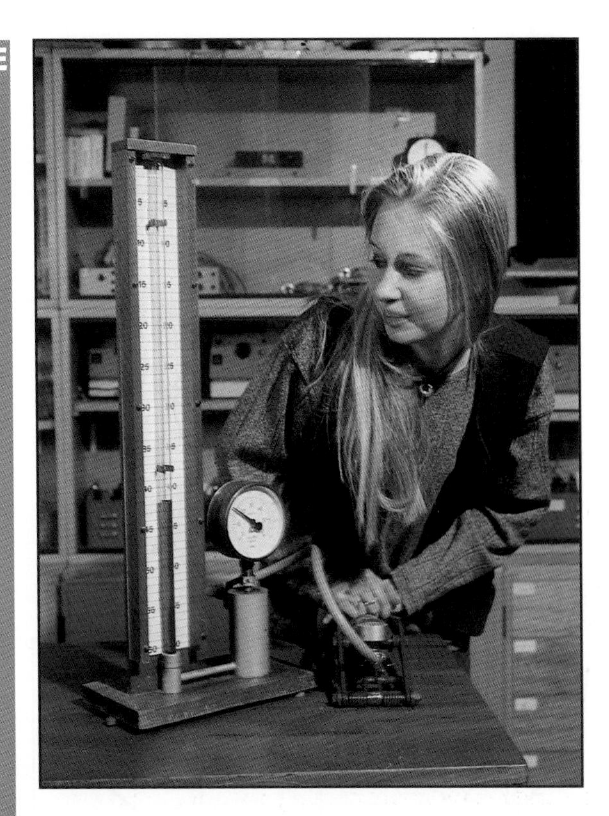

Figure 9.12 Apparatus for increasing the pressure on a gas. A fixed mass of air is trapped inside the tube, and the pressure on it is increased.

Pressure, p / Pa	Volume, V / cm^3	Pressure × volume, pV / Pa cm^3
100	60	6000
125	48	6000
150	40	6000
200	30	6000
250	24	6000
300	20	6000

Table 9.5 Representative results for a Boyle's law experiment, to show their pattern. The temperature of the gas remains constant throughout.

The relationship between p and V was investigated by Robert Boyle, an English physicist and chemist. He published his results in 1662.

The relationship that Boyle found can be stated in a number of different ways:

1 Doubling the pressure has the effect of halving the volume. Three times the pressure gives one-third of the volume, and so on.
2 The graph of Figure **9.13a** shows that increasing pressure leads to decreasing volume.
3 The numbers in Table **9.5** also show this relationship. From the last column in the table, we can see that the quantity pressure × volume is constant, so we can write

$$pV = \text{constant}$$

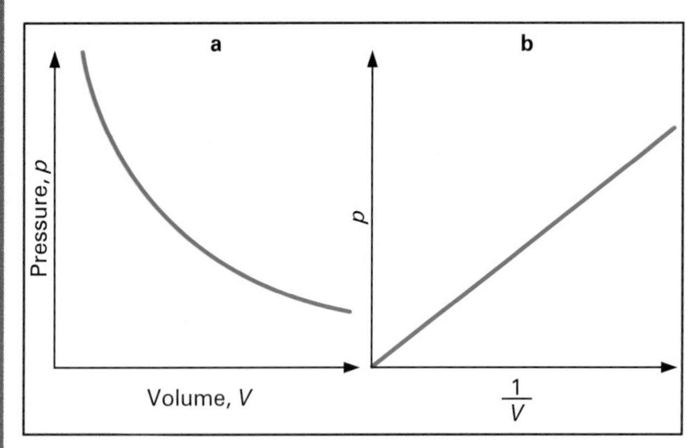

Figure 9.13 Two graphs to represent the results of a Boyle's law experiment. **a** The graph of pressure against volume shows that increasing the pressure causes a decrease in the volume. **b** The mathematical relationship between p and $1/V$ can be seen from this graph. Since it is a straight line through the origin, we can say that pressure is inversely proportional to volume (and vice versa).

4 We can write the same idea in a way that is convenient for doing calculations:

initial pressure × initial volume
= final pressure × final volume

or

$$p_1V_1 = p_2V_2$$

where p_1 and V_1 are one pair of readings of pressure and volume, and p_2 and V_2 are another pair. This equation is easy to memorise, and we shall make use of it in Worked example **1**.

5 If one quantity decreases like this as another increases, we say that one is **inversely proportional** to the other. Using the symbol ∝ ('is proportional to'), we can write

$$p \propto \frac{1}{V} \quad \text{or} \quad V \propto \frac{1}{p}$$

6 The graph of Figure **9.13b** shows that plotting p against $1/V$ gives a straight-line graph, passing through the origin.
7 Finally, we can write the relationship in words:

The volume of a fixed mass of gas is inversely proportional to its pressure, provided its temperature remains constant.

This statement is known as **Boyle's law**.

It is important to understand why Boyle's law includes the phrase 'provided its temperature remains constant'. When a gas is compressed, its temperature rises (because energy is being transferred to the gas), and this tends to make it expand. In the Boyle's law experiment, the trapped air soon loses energy to its surroundings and cools back down to room temperature. While it is hot, its volume is increased. Only when it cools down will we find that it obeys the relationship $pV = \text{constant}$.

Worked example **1** shows how to use the equation $p_1V_1 = p_2V_2$ to find how the volume of a gas changes when the pressure on it is changed. You can use the same equation to work out how the pressure changes when the volume is changed.

Worked example 1

A cylinder contains 50 cm³ of air at a pressure of 120 kPa. What will its volume be if the pressure on it is increased to 400 kPa?

Step 1: Write down the initial and final values of the quantities that we know.

$p_1 = 120\,\text{kPa}$
$V_1 = 50\,\text{cm}^3$
$p_2 = 400\,\text{kPa}$
$V_2 = ?$

Step 2: Write down the Boyle's law equation and substitute values.

$$p_1V_1 = p_2V_2$$

$$120\,\text{kPa} \times 50\,\text{cm}^3 = 400\,\text{kPa} \times V_2$$

Step 3: There is only one unknown quantity in this equation (V_2). Rearrange it and solve.

$$V_2 = \frac{120\,\text{kPa} \times 50\,\text{cm}^3}{400\,\text{kPa}} = 15\,\text{cm}^3$$

So the volume of the air is reduced to $15\,\text{cm}^3$ when it is compressed.

Notice an important feature of the equation $p_1V_1 = p_2V_2$. It does not matter what units we use for p and V, as long as we use the same units for both values of p (for example, Pa or kPa), and the same units for both values of V (for example, m^3, dm^3, or cm^3). In Question **19**, you are asked to use units that you may not be familiar with: atmospheres for pressure, and litres for volume.

Activity 9.4 Pressure and volume of a gas

Solve some problems involving Boyle's law.

QUESTIONS

17 What is the meaning of the subscripts 1 and 2 in the equation $p_1V_1 = p_2V_2$?

18 The pressure on $6\,\text{dm}^3$ of nitrogen gas is doubled at a fixed temperature. What will its volume become?

19 A container holds 600 litres of air at a pressure of 2 atmospheres. If the pressure on the gas is increased to 5 atmospheres, what will its volume become? (Assume that the temperature remains constant.)

20 A gas cylinder has a volume of $0.4\,\text{m}^3$. It contains butane at a pressure of 100 kPa and a temperature of 20 °C. What pressure is needed to compress the gas to a volume of $0.05\,\text{m}^3$ at the same temperature?

Summary

According to the kinetic model, matter is made of moving particles that are close together in solids and liquids, and far apart in gases.

There are attractive forces between particles that act strongly at short distances.

In Brownian motion, the movement of water or air molecules is revealed by their effect on visible grains of pollen or smoke.

As the temperature of a substance increases, the kinetic energy of its particles increases.

During a change of state, energy is supplied but the temperature of a pure substance remains constant.

When a liquid evaporates, the most energetic of its particles escape from the surface so that the liquid cools.

Evaporation occurs at temperatures below the boiling point. It happens faster at higher temperatures, when the surface area is large and when there is a draught.

When the pressure applied to a gas is increased, its volume decreases (at constant temperature).

Pressure and volume are related by pV = constant.

End-of-chapter questions

9.1 For each of the following statements, name the state of matter being described:

a Expands to fill the volume of its container. [1]

b Has a fixed size and shape. [1]

c Has a fixed volume but takes up the shape of its container. [1]

9.2 a 'The particles are packed closely together. They can vibrate about their fixed positions but they cannot move about within the material.' Which state of matter is being described here? [1]

b Write a similar description of the particles that make up a gas. [2]

9.3 A small amount of smoke is blown into a small glass box. A bright light is shone into the box. When observed through a microscope, specks of light are seen to be moving around at random in the box.

a What are these bright specks of light? [1]

b What evidence does this provide for the kinetic model of matter? [2]

9.4 A student pours a small amount of ethanol into a beaker. She places the beaker on an electronic balance to find its mass, and adds a thermometer to measure the temperature of the liquid. Two hours later, she returns to her experiment. She notices that the mass of the beaker and its contents has decreased. She can also see that the temperature of the ethanol has decreased. She guesses that some of the ethanol has evaporated from the beaker.

a Describe how evaporation can explain the decrease in mass. [2]

b Describe how evaporation can explain the decrease in temperature. [3]

9.5 These questions concern the behaviour of gases.

a A rigid container holds a fixed volume of air. The container is heated. How will the pressure of the air change? [1]

b A container is fitted with a piston that allows the pressure on the air in the container to be changed. The piston is pulled outwards so that the volume of the air increases. How will the pressure of the air change? [1]

E

9.6 A small container of water is placed in an oven at 90 °C. The water soon disappears.

a What name is given to process by which a liquid becomes a gas at a temperature below its boiling point? [1]

b Why must energy be supplied to a liquid to turn it into a gas? In your answer, refer to the particles of the liquid and the forces between them. [2]

9.7 A container holds 20 m^3 of air at a pressure of 120 000 Pa. If the pressure is increased to 160 000 Pa, what will the volume of the gas become? Assume that its temperature remains constant. [3]

10 Thermal properties of matter

Core	Measuring temperature
Core	Understanding and using thermometers
E **Extension**	Designing thermometers
Core	Describing the thermal expansion of solids, liquids and gases
Core	Explaining some uses and consequences of thermal expansion
Core	Relating energy supplied to rise in temperature when a body is heated
E **Extension**	Measuring specific heat capacity

Measuring temperature

When someone is about to bath a baby, she or he fills a tub with water and then checks its temperature. This may be done by dipping an elbow into the water (Figure **10.1**). The elbow is sensitive to temperature. If the water feels too hot, adding cold water can cool it to the desired temperature.

The person wants the water to be at body temperature, about 37 °C. She or he is using the fact that there are nerve endings in the skin that are sensitive to temperature. When the water feels neither hot nor cold, it is at the right temperature.

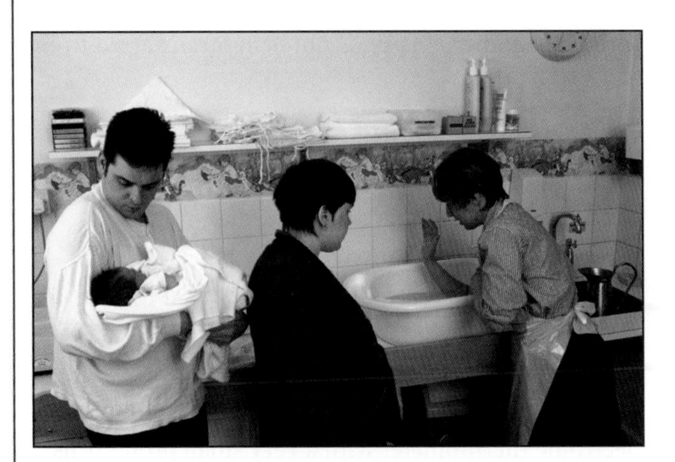

Figure 10.1 The midwife is showing the new-born baby's parents how to check the temperature of the bath water.

If someone is ill, he or she may have a temperature. Another person can test this by touching the ill person's forehead to try to detect a difference in temperature between him- or herself and the ill person. This may seem rather unscientific, but it works!

In science, we use thermometers for measuring temperature. Figure **10.2** shows two thermometers being used to measure human body temperature. One is a liquid-in-glass thermometer (Figure **10.2a**), in which a thin column of mercury expands inside an evacuated glass tube as it gets hotter. The other is a liquid-crystal thermometer (Figure **10.2b**), in which each segment shows up at a particular temperature. This latter type is much safer, particularly for use with children, who might bite and break a glass thermometer.

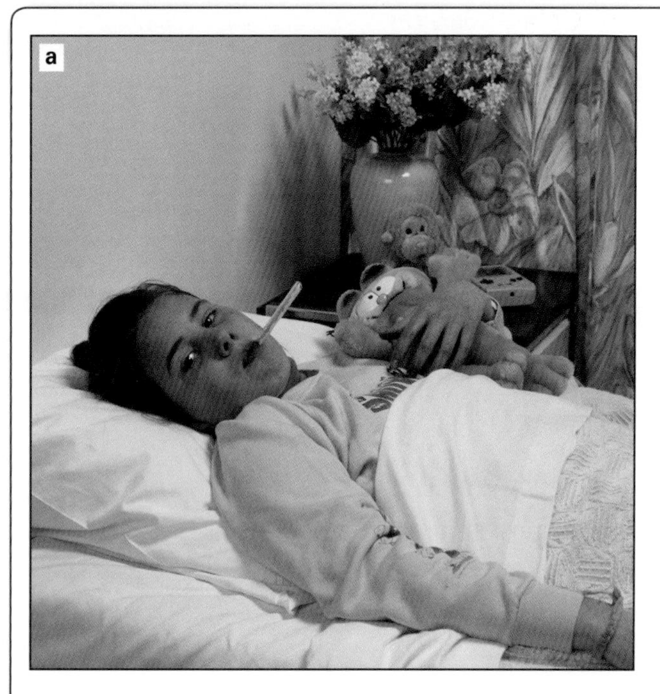

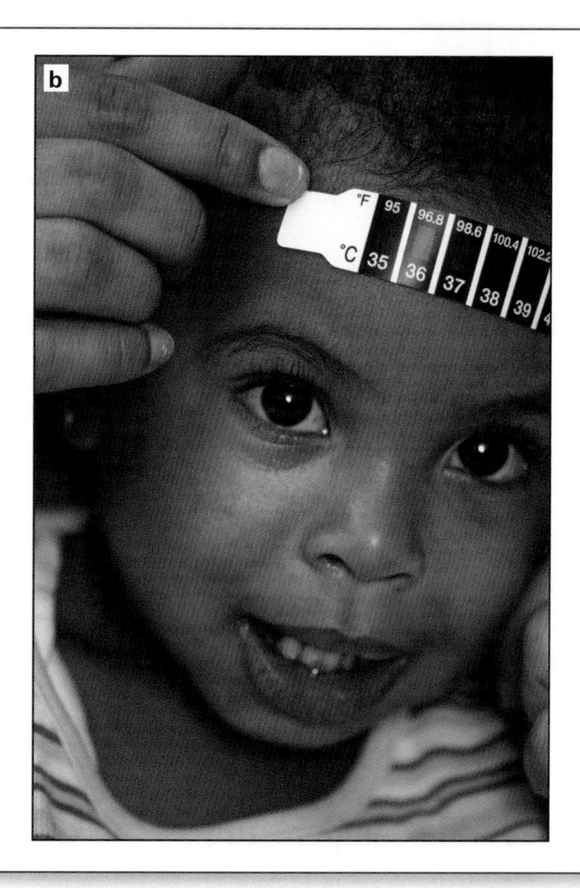

Figure 10.2 Measuring human body temperature using **a** liquid-in-glass and **b** liquid-crystal thermometers.

10.1 Temperature and temperature scales

With both of the thermometers shown in Figure **10.2**, it is important to wait for a minute or two if you want to see the correct reading. This is because the thermometer has probably been stored somewhere relatively cool, perhaps in a drawer at 20 °C. The patient's temperature will be approximately 37 °C, and it takes a short while for the thermometer to reach this temperature.

This gives us an idea of what we mean by **temperature**. The thermometer is placed in contact with the patient's body. It has to warm up until it reaches the same temperature as the patient. Energy from the patient is shared with the thermometer until they are at the same temperature. Then you will get the correct reading. (So the thermometer does not tell you the patient's temperature – it tells its own temperature! However, we know that the patient's temperature is the same as the thermometer's.)

Figure **10.3** shows a thermometer measuring the temperature of some hot water. The molecules of the water are rushing about very rapidly, because the water is hot. They collide with the thermometer and share their energy with it. The bulb of the thermometer gets hotter. Eventually, the thermometer bulb is at the same temperature as the water. (We say that the water and the thermometer bulb are in **thermal equilibrium** with one another. Energy is not being transferred from one to the other.)

You can see from this that it can be important to make a careful choice of thermometer. How could you measure the temperature of a small container containing hot water? If you chose a large, cold thermometer and poked it into the water, it might absorb a lot of energy from the water and thus make it much cooler. You would get the wrong answer for the temperature. A better solution might be to use an electronic thermometer with a very small probe. This would absorb less of the energy of the water.

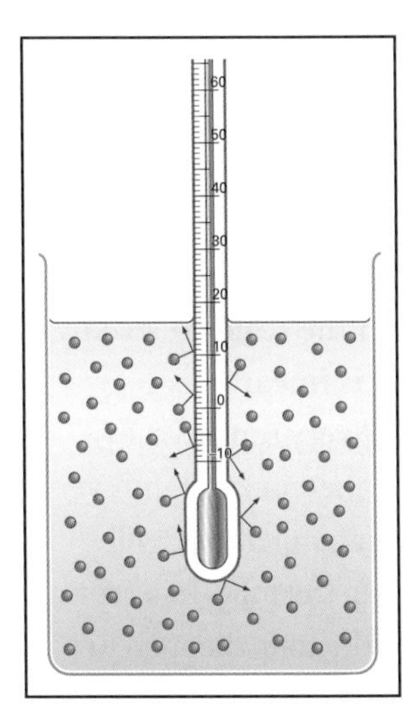

Figure 10.3 A thermometer placed in hot water is bombarded by the fast-moving water molecules. It absorbs some of their energy. Eventually, it reaches the same temperature as the water and gives the correct reading.

Temperature and internal energy

A thermometer thus tells us about the average energy of the particles in the object whose temperature we are measuring. It does this by sharing the energy of the particles. If they are moving rapidly, the thermometer will indicate a higher temperature. Placing a thermometer into an object to measure its temperature is rather like putting your finger into some bath water to detect how hot it is. Your finger does not have a scale from 0 to 100, but it can tell you how hot or cold the water is, from uncomfortably cold to comfortably warm to painfully hot.

Thus the temperature of an object is a measure of the **average kinetic energy** of its particles. Because it is the *average* kinetic energy of a particle, it does not depend on the size of the object. We can compare internal energy and temperature:

- **internal energy** is the **total** energy of **all** of the particles
- **temperature** is a measure of the **average kinetic** energy of the **individual** particles.

So a bath of water at 50 °C has more internal energy than a cup of water at the same temperature, but its individual molecules have the same average kinetic energy as the molecules of the water in the cup.

The Celsius scale

Galileo is credited with devising the first thermometer, in 1593 (Figure **10.4**). The air inside the flask expanded and contracted as the temperature rose and fell. This made the level of the water in the tube change. This could only indicate changes in temperature over a narrow range, and proved unsatisfactory because water evaporated from the reservoir. Galileo knew that air expands as its temperature increases. Modern liquid-in-glass thermometers use mercury or alcohol instead of air. These are also substances that expand when they are heated.

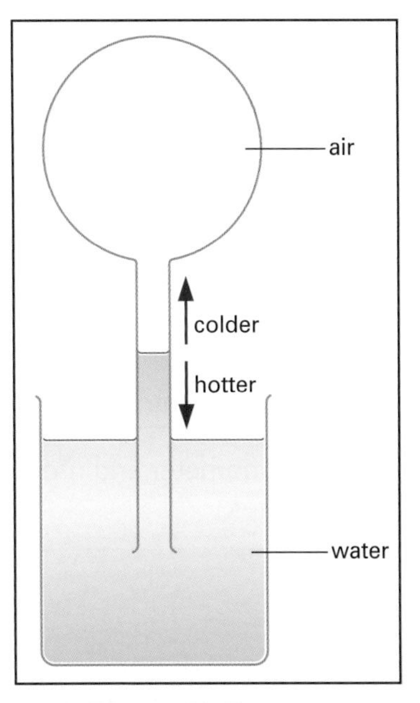

Figure 10.4 The idea behind Galileo's thermometer, the first of all thermometers. It had only a narrow operating range and no scale. As water evaporated and air dissolved in the water, the reading became unreliable.

Anders Celsius, working in Sweden, devised a more successful thermometer than Galileo's. It had a volume of mercury in an enclosed and evacuated tube, with no chance of liquid loss by evaporation. It was like the much more modern Celsius thermometer shown in Figure **10.5**. Celsius also devised a scale of temperature, now known as the Celsius scale. This had two **fixed points**:

- 0 °C – the freezing point of pure water at atmospheric pressure
- 100 °C – the boiling point of pure water at atmospheric pressure.

Each time he made a new thermometer, Celsius could calibrate it quite simply by putting it first into melting ice and then into boiling water, marking the scale each time. Then he could divide the scale into 100 equal divisions. This process is known as **calibration** of the thermometer. (It is interesting to note that, with his first thermometers, Celsius marked the boiling point of water as 0 degrees and the freezing point as 100 degrees. It was a few years later that one of his collaborators decided that it was better to have the scale the other way up.)

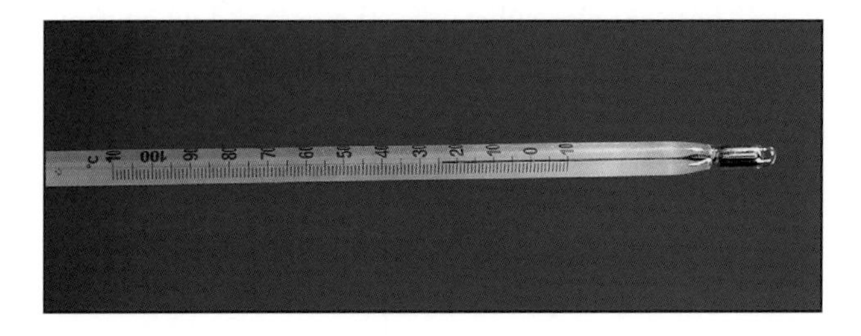

Figure 10.5 A modern Celsius-scale liquid-in-glass thermometer, with a fixed quantity of mercury sealed in a glass tube.

Activity 10.1 Calibrating a thermometer

Mark the scale on a blank thermometer and use it to measure some temperatures.

QUESTIONS

1 Two buckets contain water at 30 °C. One contains 1 kg of water, and the other contains 2 kg of water. State and explain whether the following quantities are the same or different for the water in the two buckets:

a internal energy

b temperature

c average energy of a molecule.

2 What are the **two** fixed points on the Celsius scale?

3 Write step-by-step instructions for the calibration of a thermometer using the Celsius scale.

Designing a thermometer

Mercury-in-glass (and alcohol-in-glass) thermometers are used in many different situations. They are attractive for a number of reasons:

- Mercury expands at a steady rate as it is heated. This means that the marks on the scale are evenly spaced. We say that the scale is **linear**.
- The thermometer can be made very **sensitive**, by making the tube up which the mercury expands very narrow. Then a small change in temperature will push the mercury a long way up the tube. In a typical clinical thermometer, used by doctors, the mercury rises several millimetres for a 1 °C rise in temperature. This makes it possible to measure small changes.
- A mercury thermometer can have a wide **range**, because mercury is liquid between −39 °C and +350 °C. Some domestic ovens have mercury thermometers that read up to 250 °C.

The problem with mercury thermometers is that they have to be read by eye. An alternative is to use an electronic thermometer. Some of these are based on **thermistors**, which are resistors whose resistance changes by a large amount over a narrow temperature range (see Figure **10.6**). These can be very useful, especially as they are robust and can be built into electronic circuits.

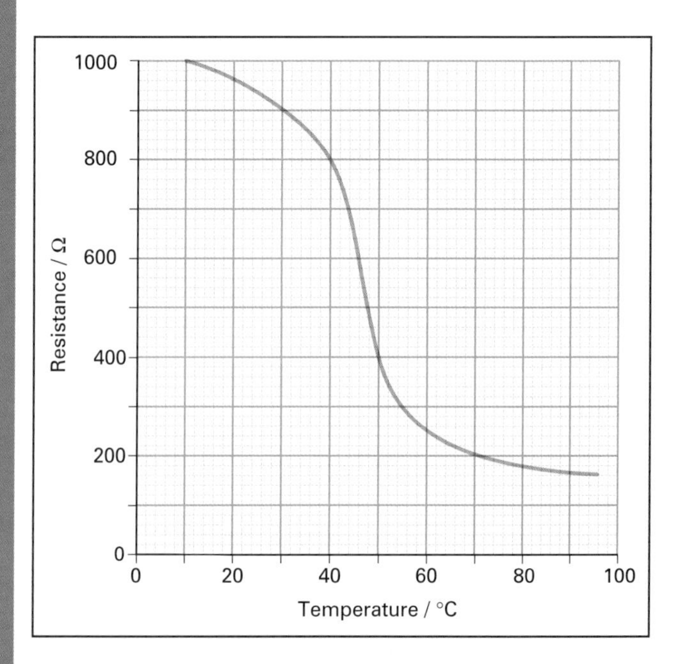

Figure 10.6 The electrical resistance of a thermistor changes over a narrow range of temperatures. This means that it can be used as a temperature probe for an electronic thermometer. However, it will only be sensitive over a narrow range, and its behaviour may be non-linear.

However, from the graph in Figure 10.6, you can see the following:

- The resistance of a thermistor changes in a non-linear way, so that the intervals on a scale will not all be equal in size.
- The range of such a thermometer will be narrow, because the resistance only changes significantly over a narrow range of temperatures. You would need to choose a thermistor whose resistance changes most near the temperature you were trying to measure if you want the thermometer to be sensitive.

A second alternative is to use a **thermocouple**, a device that gives an output voltage that depends on the temperature. Thermocouples are made from pieces of wire made from two different metals. A wire of metal X is joined at each end to wires of metal Y to form two junctions. To use the thermocouple, its ends are connected to a sensitive voltmeter (see Figure 10.7). Then one junction is placed in melting ice at 0 °C while the other is placed in the object whose temperature is to be measured. The voltmeter shows a reading. The greater the voltage produced, the bigger the difference in temperatures between the two junctions. The thermocouple must be calibrated so that the temperature can be deduced from the voltage.

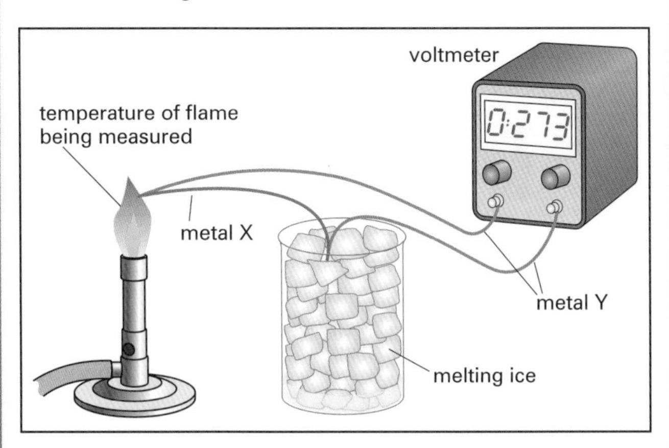

Figure 10.7 Using a thermocouple to measure temperature.

Many electronic thermometers make use of thermocouples (Figure 10.8). The junctions of a thermocouple thermometer can be very small, so that they are robust, and they do not absorb much

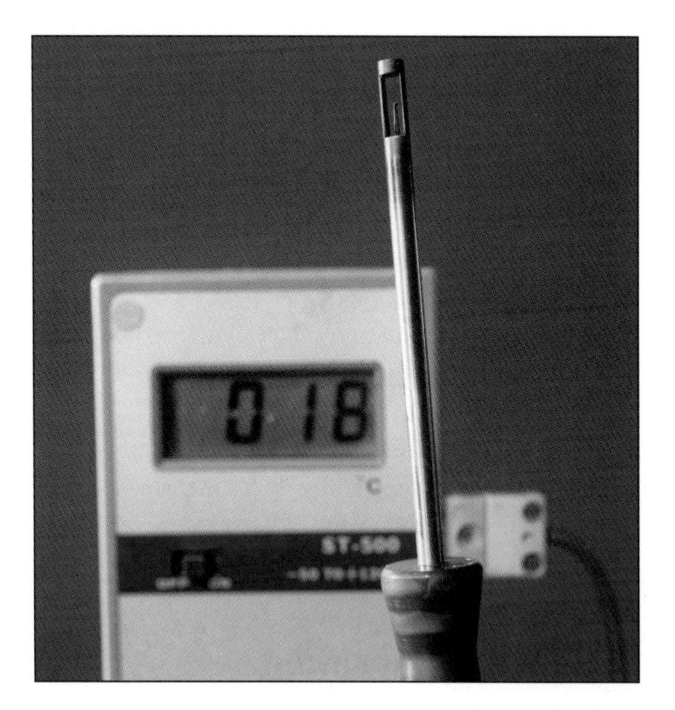

Figure 10.8 This electronic thermometer uses a thermocouple as its probe. You may just about be able to see the thin wires that make up the junction (in the 'eye' of the device). These are connected to a box with electronic circuits that convert the voltage produced to a digital temperature reading.

energy from the material whose temperature they are measuring. Some combinations of metals give bigger voltages than others, so it is important to choose them carefully.

Thermocouples can be used to measure high temperatures (up to the melting point of the metal used). Because they are small, they can heat up and cool down quickly, so they are useful for measuring rapidly varying temperatures.

Thermocouples are used in many gas ovens and heaters that have a pilot flame that burns continuously. One junction is positioned in the flame, giving a voltage of about 20 mV. If the pilot flame goes out, the voltage drops and an electric circuit turns off the gas supply to the burners and the pilot flame.

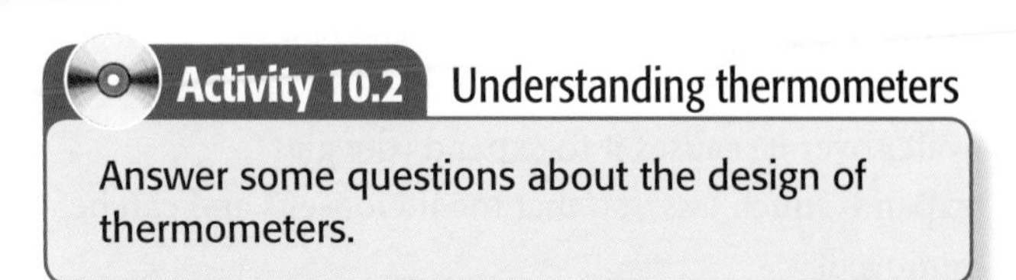

Activity 10.2 Understanding thermometers

Answer some questions about the design of thermometers.

QUESTIONS

4 Look at Figure **10.6**, which shows how the resistance of a thermistor changes with temperature.
   a Over what range of temperatures is the resistance changing most rapidly?
   b Explain why a thermometer that used this thermistor would be less sensitive at 20 °C than at 50 °C.

5 A thermocouple thermometer is better for measuring rapidly varying temperatures than a mercury-in-glass thermometer. Explain why this is so.

10.2 Thermal expansion

Most substances – solids, liquids and gases – expand when they are heated. This is called **thermal expansion** (the word 'thermal' means 'related to heat'). We have already seen that some types of thermometer make use of the thermal expansion of a liquid. Figure **10.9** shows an experiment that demonstrates that a metal bar expands when heated.

- When it is cold, the iron bar will just fit in the gap in the measuring device.
- The bar (but not the measuring device) is heated strongly. Now it is too long to fit in the gap – it has expanded.
- When it cools down, the bar contracts and returns to its original length.

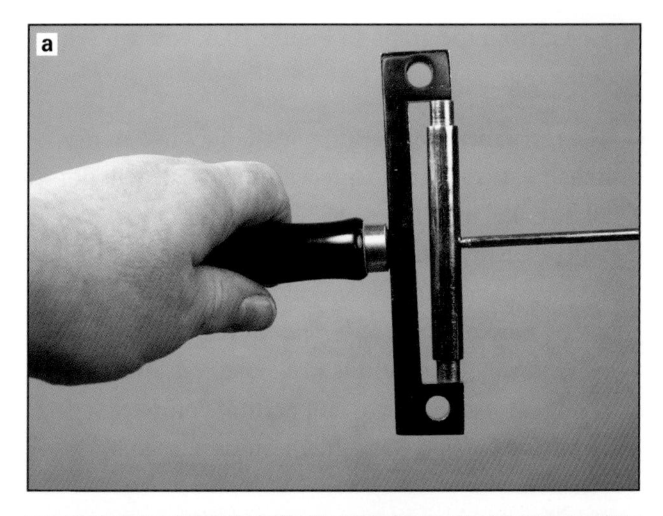

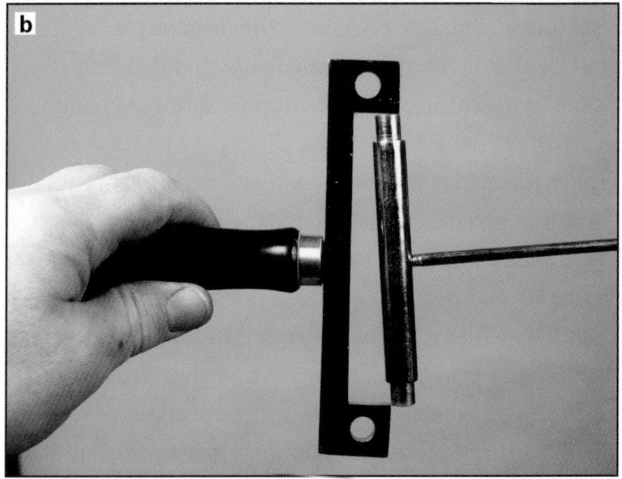

Figure 10.9 In **a**, the metal bar is cold, and fits in the gap in the measuring device. In **b**, it has been heated so that it expands and will no longer fit the gap.

Uses of expansion

Rivets are used in shipbuilding and other industries to join metal plates. A red-hot rivet is passed through holes in two metal plates and then hammered until the ends are rounded (Figure **10.10**). As the rivet cools, it contracts and pulls the two plates tightly together.

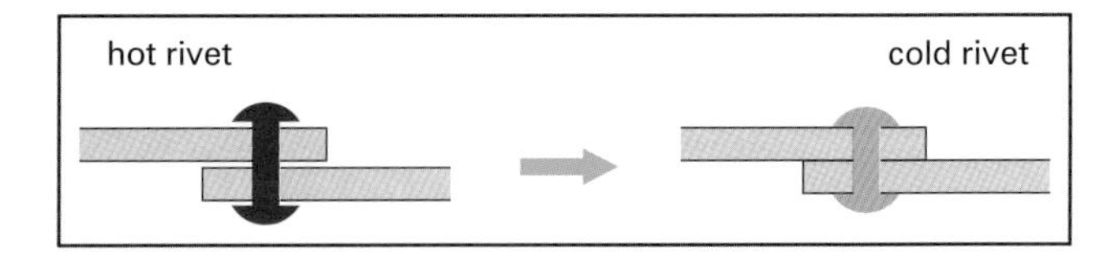

Figure 10.10 Joining two metal plates using a rivet.

A metal lid or cap may stick on a glass jar or bottle. Heating the lid (for example, by running hot water over it) causes it to expand (the glass expands much less), so that the lid loosens and can be removed.

A steel 'tyre' may be fitted on to the wheel of a railway locomotive while the tyre is very hot. It then cools and contracts, so that it fits tightly on to the wheel.

A bimetallic strip (Figure **10.11**) is designed to bend as it gets hot. The strip is made of two metals joined firmly together. One metal expands more rapidly than the other. As the strip is heated, this metal expands rapidly, causing the strip to bend. (The metal that expands more is on the outside of the curve, because the outer curve is longer than the inner one.) These strips are used in devices such as fire alarms and thermostats (which control the temperature of ovens, irons, water heaters, refrigerators, and so on).

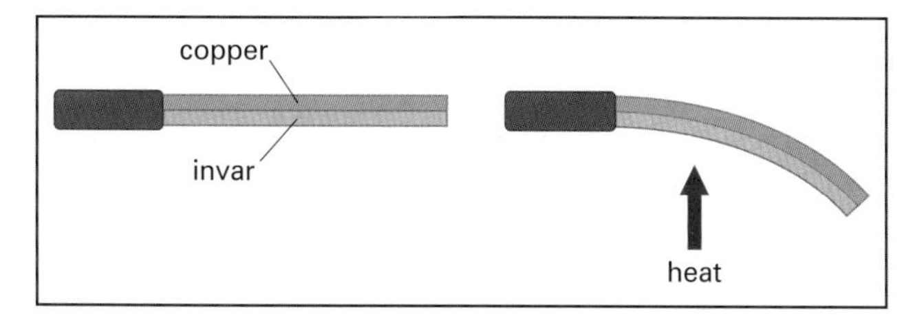

Figure 10.11 A bimetallic strip. 'Invar' is a metal alloy that expands very little when it is heated. Copper expands more readily when it is heated. This difference in expansion forces the strip to bend.

Consequences of expansion

The expansion of materials can cause problems. For example, metal bridges and railway lines expand on hot days, and there is a danger that they might buckle. To avoid this, bridges are made in sections, with expansion joints between the sections (Figure **10.12**). On a hot day, the bridge expands and the gaps between sections decrease. Railway lines are now usually made from a metal alloy that expands very little. On a concrete roadway, you may notice that the road surface is in short sections. The gaps between are filled with soft pitch, which becomes squashed as the road expands.

Glass containers may crack when hot liquid is placed in them. This is because the inner surface of the glass expands rapidly, before the heat has conducted through to the outer surface. The force of expansion cracks the glass. To overcome this, glass such as Pyrex has been developed that expands very little on heating. An alternative is toughened glass, which has been treated with chemicals to reduce the chance of cracking.

Figure 10.12 This truck is about to cross an expansion joint on a motorway bridge. On a hot day, the bridge expands and the interlocking 'teeth' of the joint move closer together.

The expansion of gases

Gases expand when they are heated, just like solids and liquids. We can understand this using the kinetic model of matter (see Chapter **9**). Figure **10.13** shows some gas in a cylinder fitted with a piston. At first, the gas is cold and its particles press weakly on the piston. When the gas is heated, its particles move faster. Now they push with greater force on the piston and push it upwards. The gas has expanded.

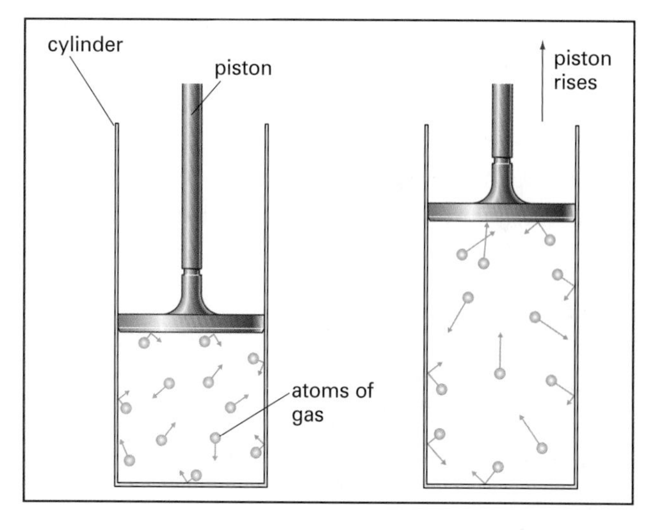

Figure 10.13 A gas expands when it is heated at constant pressure.

The upward force of the gas is balanced by the downward weight of the piston. So, in this situation, the pressure of the gas has remained constant as it has expanded. If the piston did not move, the **volume** of the gas would remain constant when it was heated but its **pressure** would increase.

Activity 10.3 Observing expansion

Try out some experiments to observe the expansion of solids, liquids and gases.

Comparing solids, liquids and gases

Solids, liquids and gases – which expands most for a given rise in temperature?

- Solids expand most slowly when they are heated. Some, such as Pyrex glass and invar metal, have been designed to expand as little as possible.
- Liquids generally expand faster than solids.
- Gases expand faster still.

There are some exceptions to this. For example, liquid paraffin expands very rapidly on heating. Petrol (gasoline) also expands rapidly when it is heated. If, on a hot day, a motorist fills up with petrol from cool underground tanks, the fuel may expand and overflow as it warms up.

QUESTIONS

6 Explain how Galileo's thermometer (see Figure **10.4**) makes use of thermal expansion.

7 Figure **10.14** shows an experiment to demonstrate the expansion of water.

  a Describe and explain what will happen when the flask of cold water is placed in the tank of hot water.

  b How could this experiment be adapted to compare the rates of expansion of water and of liquid paraffin?

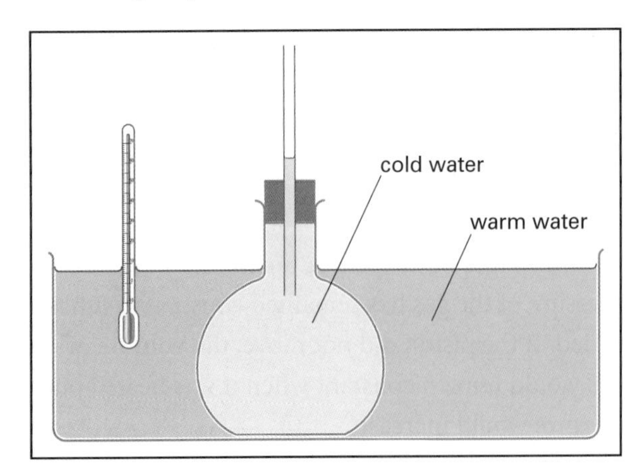

Figure 10.14 Demonstrating the thermal expansion of water – for Question **7**.

10.3 Thermal capacity

Some houses are fitted with night storage heaters. These are electrical heaters that heat up at night, when electricity is cheap. Then, during the day, they remain warm and give out their heat to the room.

Figure **10.15** shows the construction of a night storage heater. The electric heating elements are surrounded by special bricks that store the energy supplied by the electricity. The bricks are made of a material that requires a lot of energy to heat it up. In this way, the bricks store a lot of energy in a small space.

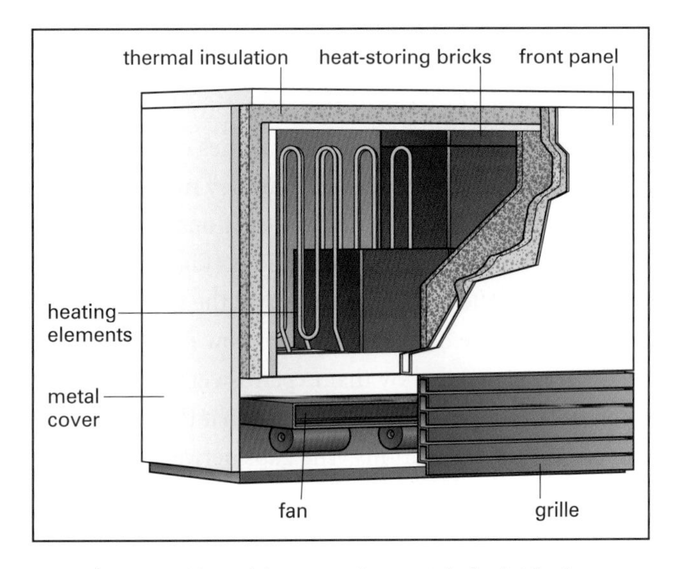

Figure 10.15 Inside a night storage heater. It is the bricks that store energy.

We say that the bricks have a high **thermal capacity**. It takes a lot of energy to raise their temperature by a certain amount. The bricks in a storage heater must be quite big if they are to have sufficient thermal capacity to keep a room warm for several hours. Because of their high thermal capacity, they heat up slowly and they cool down slowly.

The thermal capacity of an object depends on the material it is made of. Metal objects heat up easily – their thermal capacities are low. Objects made of non-metals (such as wood, glass and plastics) and liquids (such as water and oil) have higher thermal capacities.

QUESTIONS

8 A cook places a metal baking tray and a ceramic dish in the oven. She notices that the metal tray is soon too hot to touch while the ceramic dish takes longer to get hot. Which object has the greater thermal capacity? Explain your answer.

9 An electrical storage heater uses bricks to store energy. Explain why brick is a better choice of material than a metal such as steel.

10.4 Specific heat capacity

Suppose that you want to make a hot drink for yourself and some friends. You need to boil some water. You will be wasting energy if you put too much water in the kettle or pan. It is sensible to boil just the right amount. Also, if the water from your tap is really cold, it will take longer, and require more energy, to reach boiling point.

So the amount of energy you need to supply to boil the water will depend on two facts:

1 the mass of the water
2 the increase in temperature.

In order to calculate how much energy must be supplied to boil a certain mass of water, we need to know one other fact:

3 it takes 4200 J to raise the temperature of 1 kg of water by 1 °C.

Let us assume that the cold water from your tap is at 20 °C. You have to provide enough energy to heat it to 100 °C, so its temperature must increase by 80 °C. Let us also assume that you need 2 kg of water for all the drinks. The amount of energy required to heat 2 kg of water by 80 °C is therefore:

$$\text{energy required} = 2 \times 4200 \times 80 = 672\,000\,\text{J} = 672\,\text{kJ}$$

Another way to express the third fact above is to say that the specific heat capacity of water is 4200 J per kg per °C or 4200 J/(kg °C).

In general, the **specific heat capacity (s.h.c.)** of any substance (not just water) is defined as follows:

> The specific heat capacity (s.h.c.) of a substance is the energy required to raise the temperature of 1 kg of the substance by 1 °C.

We can write the equation above as a general formula:

> energy required = mass × specific heat capacity × increase in temperature

Worked example **1** shows how to use this formula in more detail. There is more about the meaning of specific heat capacity (s.h.c.) after Worked example **1**.

Worked example 1

A domestic hot water tank contains 200 kg of water at 20 °C. How much energy must be supplied to heat this water to 70 °C? (Specific heat capacity of water = 4200 J/(kg °C).)

Step 1: Calculate the required increase in temperature.

$$\text{increase in temperature} = 70\,°\text{C} - 20\,°\text{C} = 50\,°\text{C}$$

Step 2: Write down the other quantities needed to calculate the energy.

mass of water = 200 kg
specific heat capacity of water = 4200 J/(kg °C)

Step 3: Write down the formula for energy required, substitute values, and calculate the result.

$$\begin{aligned}\text{energy required} &= \text{mass} \times \text{specific heat capacity} \times \text{increase in temperature}\\ &= 200\,\text{kg} \times 4200\,\text{J/(kg °C)} \times 50\,°\text{C}\\ &= 42\,000\,000\,\text{J}\\ &= 42\,\text{MJ}\end{aligned}$$

So 42 MJ are required to heat the water to 70 °C.

The meaning of s.h.c.

Energy is needed to raise the temperature of any material. The energy is needed to increase the kinetic energy of the particles of the material. In solids, they vibrate more. In gases, they move about faster. In liquids, it is a bit of both.

We can compare different materials by considering standard amounts (1 kg), and a standard increase in temperature (1 °C). Different materials require different amounts of energy to raise the temperature of 1 kg by 1 °C. In other words, they have different **specific heat capacities** (s.h.c.). Table **10.1** shows the values of s.h.c. for a variety of materials.

From the table, you can see that there is quite a wide range of values. The s.h.c. of steel, for example, is one-tenth that of water. This means that, if you supplied equal amounts of energy to 1 kg of steel and to 1 kg of water, the steel's temperature would rise ten times as much.

Type of material	Material	Specific heat capacity / J/(kg °C)
metals	steel	420
	aluminium	910
	copper	385
	gold	300
	lead	130
non-metals	glass	670
	nylon	1700
	polythene	2300
	ice	2100
liquids	water	4200
	sea water	3900
	ethanol	2500
	olive oil	1970
gases	air	1000
	water vapour	2020 (at 100 °C)
	methane	2200

Table 10.1 Specific heat capacities of a variety of materials.

The s.h.c. of water

Water is an unusual substance. As you can see from Table **10.1**, it has a high value of s.h.c. compared to other materials. This has important consequences:

- It takes a lot of energy to heat up water.
- Hot water takes a long time to cool down.

The consequences of this can be seen in our climates. In the hot months of summer, the land warms up quickly (low s.h.c.) while the sea warms up only slowly. In the winter, the sea cools gradually while the land cools rapidly. People who live a long way from the sea (in the continental interior of North America or Eurasia, for example) experience freezing winters and very hot summers. People who live on islands and in coastal areas (such as western Europe) are protected from climatic extremes because the sea acts as a reservoir of heat in the winter, and stays relatively cool in the summer.

Measuring s.h.c.

One method for measuring the specific heat capacity of a metal is shown in Figure **10.16a**. The block of aluminium has a mass of 1 kg. It is heated by a small electric heater, which supplies 50 J every second (its power is 50 W). The thermometer shows the temperature rise of the block.

One approach is to heat the block for a certain length of time, and find the temperature rise. Knowing the time and the power of the heater, you can work out how much energy has been supplied. Then, knowing the temperature rise and the mass of the block, you can calculate the s.h.c.

A better approach is to record the temperature every ten seconds or so, and then plot a graph (Figure **10.16b**) to show the rate at which it is rising. From the slope of the graph, you can then work out how much the temperature rises every second. This is the temperature rise produced by 50 J of energy, and now you can work out the s.h.c.

It is important to evaluate the procedure being used, to judge how accurate the final result is likely to be. The metal block should be insulated, to prevent energy escaping; but, some will still escape. Also, some energy

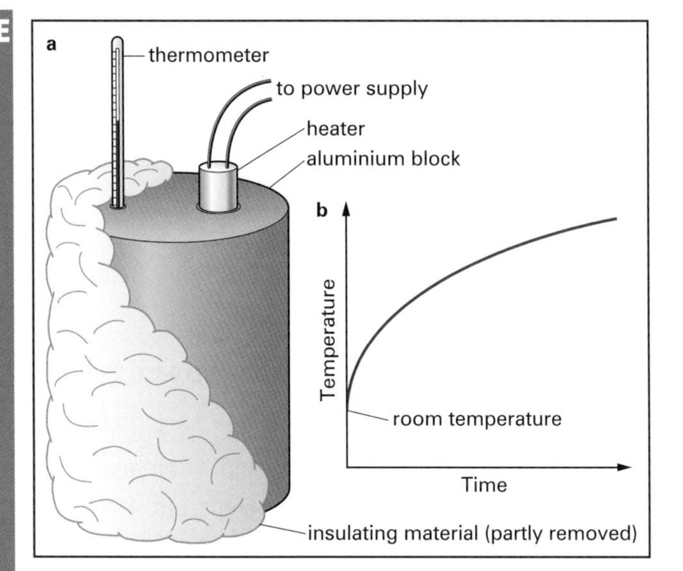

Figure 10.16 Measuring the specific heat capacity of aluminium. **a** The 1 kg aluminium block is heated by an electric heater, and its temperature is recorded at regular intervals. The block is covered in insulating material during the experiment, to reduce heat losses. A small amount of oil fills the gap between the thermometer and the block, ensuring that there is good thermal contact between them. The thermometer then gives reliable readings of the block's temperature. **b** This graph shows how the block's temperature increases. As the temperature rises, more heat escapes to the surroundings and the temperature rises more slowly.

is used in heating the heater itself, rather than the block. Both factors mean that the energy supplied is greater than the energy that heats the block, so the final answer will be too big. Another problem is that the block must not be heated too quickly. It takes time for the heat to conduct through the metal. It is desirable for the whole block to heat up at the same time, otherwise the thermometer will only indicate the temperature of part of the block.

Activity 10.4 Measuring s.h.c.

Measure the specific heat capacity of a metal in the form of a block.

QUESTIONS

10 The specific heat capacity of steel is 420 J/(kg °C).
  a How much energy is required to heat 1 kg of steel by 20 °C?
  b How much energy is required to heat 5 kg of steel by 20 °C?

11 A beaker contains 1 kg of water at 20 °C. A student heats a 1 kg block of aluminium to 100 °C and then drops it into the water. After a short while, the water and the block have both reached a temperature of 38 °C. The student said that this showed that water has a greater specific heat capacity than aluminium. Was he correct? Explain your answer.

12 A thermocouple can be used as a thermometer. Such a thermometer can measure rapidly varying temperatures because of its small thermal capacity.
  a Explain why a thermocouple has a small thermal capacity.
  b Explain why this makes it suitable for measuring rapidly varying temperatures.

10.5 Latent heat

Energy must be supplied to a substance to melt it or to boil it – in other words, to make it change state. This energy does not increase the substance's temperature, and for this reason it is known as **latent heat** (the word 'latent' means 'hidden').

The energy needed to change a liquid into a gas is called the **latent heat of vaporisation**. The energy needed to change a solid into a liquid is called the **latent heat of fusion**. (Here, the word 'fusion' means 'melting'.) To compare different substances fairly, we measure the energy required to change the state of 1 kg of the substances. (Here, as for s.h.c., we use the word 'specific' to mean that it relates to unit mass, that is, 1 kg.). So **specific latent heat** is defined as follows:

The specific latent heat of vaporisation is the energy required to cause 1 kg of a substance to change state from liquid to gas at its boiling point.

The specific latent heat of fusion is the energy required to cause 1 kg of a substance to change state from solid to liquid at its melting point.

As we saw in section **9.3**, this energy is needed to break the bonds between particles.

Measuring latent heat

To determine the specific latent heat of a substance, 1 kg of the substance must be heated at its melting or boiling point until it entirely changes state. The amount of energy supplied must be measured. Then we have:

$$\text{specific latent heat} = \frac{\text{energy supplied}}{\text{mass}}$$

To measure the specific latent heat of fusion of ice, a measured mass of ice at 0 °C is added to warm water in a well-insulated copper container. When the ice has entirely melted, the temperature of the water is measured. Knowing the specific heat capacities of water and copper, the energy they have lost to the ice can be calculated. This is the latent heat, and the energy per kilogram can be calculated.

Similarly, water can be boiled using an electric heater of known power. The mass of water that boils away is measured, and the energy supplied by the heater is calculated.

QUESTIONS

13 Explain why the definition of specific latent heat of fusion includes the phrase '... at its melting point'.

14 It takes 4500 J to turn 2.0 g of water at 100 °C into steam. Calculate the specific latent heat of vaporisation of water.

15 Use the kinetic (particle) model of matter to explain why the specific latent heat of vaporisation of water is much greater than the specific latent heat of fusion of ice.

Summary

Thermometers are used to measure temperature. Any thermometer makes use of a physical property that varies with temperature.

Any temperature scale is based on two fixed points, such as the freezing and boiling points of pure water.

Thermometers can be designed to increase their sensitivity, range and linearity.

Solids, liquids and gases all expand when heated. Thermal expansion can be a problem, but it also has many uses.

In general, gases expand more than liquids, which expand more than solids.

It takes a lot of energy to raise the temperature of an object with a high thermal capacity.

The specific heat capacity of a substance is the energy required to raise the temperature of 1 kg of the substance by 1 °C.

The specific latent heat of vaporisation is the energy required to cause 1 kg of a substance to change state from liquid to gas at its boiling point.

The specific latent heat of fusion is the energy required to cause 1 kg of a substance to change state from solid to liquid at its melting point.

End-of-chapter questions

10.1 A student is using a thermometer to measure temperatures in a laboratory. The thermometer contains mercury. As the temperature increases, the length of the mercury column in the thermometer increases.

a Explain why the mercury column becomes longer. [1]

b The thermometer measures temperatures on the Celsius scale. Table **10.2** gives details of the **two** fixed points of the scale. Copy and complete the table. [2]

	Definition	Value
lower fixed point	melting point of pure ice	……
upper fixed point	……	100 °C

Table 10.2 For Question **10.1b**.

c Give another property of a material that varies with temperature and may be used to measure temperature. [1]

10.2 A student heats an insulated steel block using an electrical heater. The temperature of the block rises.

a The heater supplies energy to the block. In what form does the block store this energy? [1]

b The student then heats a second block, made of copper. The heater supplies energy at the same rate as before. The temperature of this block rises faster than that of the steel block. Which block has the greater thermal capacity? Explain your answer. [2]

E

10.3 A student is investigating two thermometers. She notices that their scales are marked differently.

- Liquid-in-glass thermometer: scale from −10 °C to +110 °C.
- Thermocouple thermometer: scale from −200 °C to +450 °C.

a Which thermometer has the greater range? [1]

The student places both thermometers in pure, melting ice. Each shows that the temperature is 0 °C.

b State another temperature at which you would expect the two thermometers to give the same reading. Explain your answer. [2]

She then places the two thermometers in a beaker of warm water. The liquid-in-glass thermometer shows that the temperature is 45.5 °C. The thermocouple thermometer reads 43 °C.

c Which thermometer is more sensitive? Explain how you know. [2]

d Suggest why the two thermometers do not indicate the same temperature when they are placed in the beaker of water. [2]

10.4 Willem has to measure the specific heat capacity of copper. He has a copper block, which he heats with an electrical heater. The heater supplies energy to the block at a rate of 50 J each second.

Willem records the temperature of the block. Then he switches on the heater for exactly 10 minutes.

a What **two** other measurements will he require in order to calculate the specific heat capacity of steel? [2]

b Explain why the block must be well insulated if he is to obtain an accurate result. [1]

c If the block is poorly insulated, will Willem's result be too high or too low? [1]

11 Thermal (heat) energy transfers

Core Demonstrating conduction, convection and radiation
E **Extension** Explaining conduction
Core Explaining convection and radiation
E **Extension** Comparing good and bad emitters of radiation
Core Discussing applications and consequences of thermal (heat) energy transfer

Warming up, keeping cool

Mammals are warm-blooded creatures. They keep their body temperature at about 35–40 °C. The reason for this is that mammals are active creatures. If they are carnivores, they may have to sprint suddenly to catch their prey. Herbivores may have to graze for most of the day, occasionally running to avoid the carnivores. Muscles work much better at higher temperatures because the reactions that release energy go faster. People are mammals. If you have camped out overnight, you may have experienced the difficulty of getting your muscles to start working when you wake up on a cold morning.

Figure 11.1 The polar bear has thick fur to help it retain heat. It is a very bulky animal, so that its surface area is small compared to its volume. This is another way of retaining heat.

There are problems with being warm-blooded. The polar bear (Figure **11.1**) lives in a very cold climate. It is in constant danger of freezing to death. To avoid this danger, polar bears have thick coats of waterproof fur, so that heat cannot easily escape. They are also very bulky. This means that they have a relatively small surface area, compared to their volume. They have a lot more 'inside' than 'outside', and so they find it easier to retain their body heat. Grizzly bears also live in cold areas, and they too are bulky. Bears that live closer to the equator, such as the European brown bear, tend to be much smaller. They do not have such problems with retaining heat.

African elephants (Figure **11.2**) have the opposite problem. They are large animals living in a hot climate, and they are in danger of over-heating if they are too active. To cool off, they use their ears. On a hot day, more blood flows through the veins in their ear flaps. This warms the air nearby, so that heat escapes by convection. Flapping the ears increases the rate of heat loss. An elephant's ear flaps are thus the equivalent of a car's radiator – a way of getting rid of excess heat. Wallowing in mud can also be cooling. As water evaporates from the elephant's skin, it carries energy away.

All creatures have ways of regulating their body temperatures. They make use of all the different ways in which heat moves around: conduction, convection, radiation and evaporation.

Figure 11.2 An African elephant has large ear flaps, but these are not to improve its hearing. They help it to get rid of excess heat on a hot day, or when they have been very active. Blood flow to the veins in the ears is increased. This warms the nearby air and the heat is carried away by convection.

11.1 Conduction

As we discussed in Chapter 6, thermal (heat) energy is energy transferring from a hotter place to a colder place – in other words, from a higher temperature to a lower temperature. Thermal energy requires a **temperature difference** if it is to be transferred. In this chapter we look at the various ways in which thermal energy is transferred. We start with **conduction**.

Lying on the table are two spoons: one is metal, the other is plastic. You pick up the metal spoon – it feels cold. You pick up the plastic spoon – it feels warm. In fact, both are at the same temperature, room temperature, as a thermometer would prove to you. How can this be? What you are detecting is the fact that metals are good conductors of heat, and plastics are poor conductors of heat. Figure 11.3 shows what is going on.

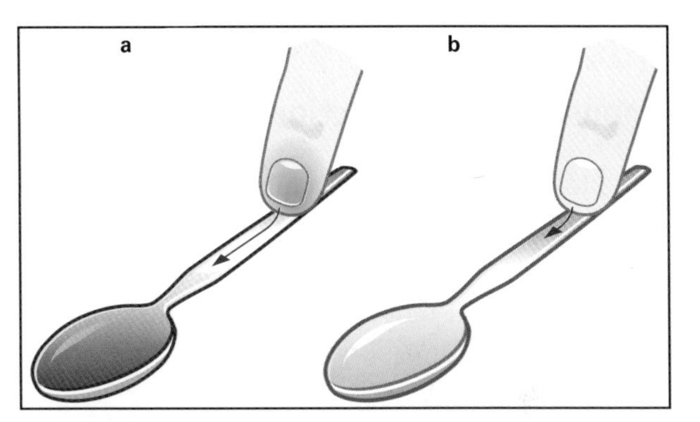

Figure 11.3 Metals feel cold, plastics feel warm. **a** Touching a piece of metal. Heat flows from your finger and into the metal. Because metals are good conductors of heat, heat continues to escape from your finger. Your finger gets colder. **b** Touching a piece of plastic. A small amount of heat conducts into the plastic. But it can go no further, because plastics are good insulators. Your finger stays warm.

a When your finger touches a metal object, heat is conducted out of your finger and into the metal. Because metal is a good **conductor**, heat spreads rapidly through the metal. Heat continues to escape from your finger, leaving it colder than before. The temperature-sensitive nerves in your finger tip tell your brain that your finger is cold. So you think you are touching something cold.

b When you touch a plastic object, heat conducts into the area that your finger is in direct contact with. However, because plastic is a good **insulator**, the heat travels no further. Your finger loses no more heat and remains warm. The message from the nerves in your finger tip is that your finger is warm. So you think you are touching something warm.

(Note that the nerves in your finger tell you how hot your finger is, not how hot the object is that you are touching. This is similar to our discussion of thermometers in Chapter 10. A thermometer in water indicates its own temperature, and we have to assume that the temperature of the water is the same as this.)

Table 11.1 compares conductors and insulators. You can see that, in general, metals are good conductors of heat while non-metals are poor conductors.

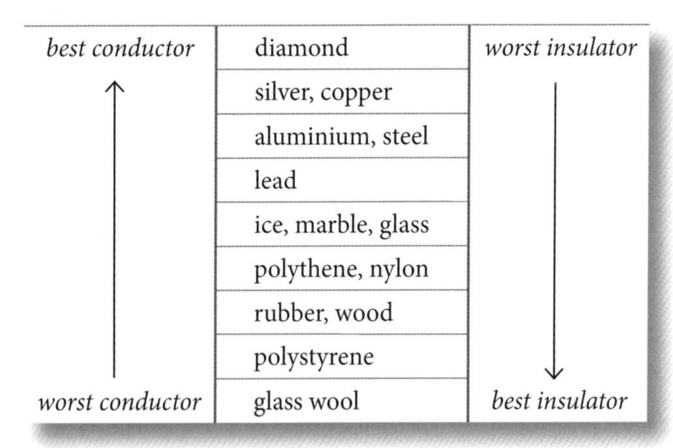

best conductor	diamond	*worst insulator*
↑	silver, copper	↓
	aluminium, steel	
	lead	
	ice, marble, glass	
	polythene, nylon	
	rubber, wood	
	polystyrene	
worst conductor	glass wool	*best insulator*

Table 11.1 Comparing conductors of heat, from the best conductors to the worst. A bad conductor is a good insulator. Almost all good conductors are metals; polymers (plastics) are at the bottom of the list. Glass wool is an excellent insulator because it is mostly air.

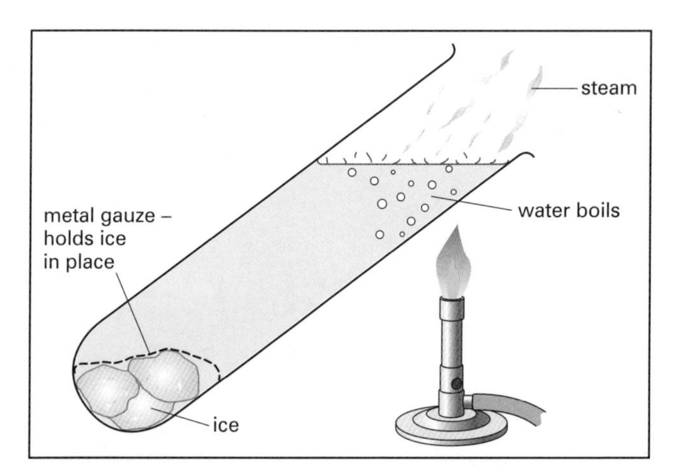

Figure 11.5 Although the water at the top of the tube is boiling, the ice at the bottom remains solid.

Demonstrating conduction

Figure **11.4** shows one way to compare different metals. The metals rods are all the same size. Each has a blob of wax at one end. They are all heated equally at the other end. The best conductor is the metal on which the wax melts first.

Figure **11.5** shows how to demonstrate that water is a poor conductor of heat. A lump of ice is trapped at the bottom of the test tube, held in place by a piece of wire gauze. The water is heated close to the mouth of the tube. The water boils, while the ice remains frozen. Heat has not conducted down to the bottom of the tube. The water there remains cold and the ice does not melt.

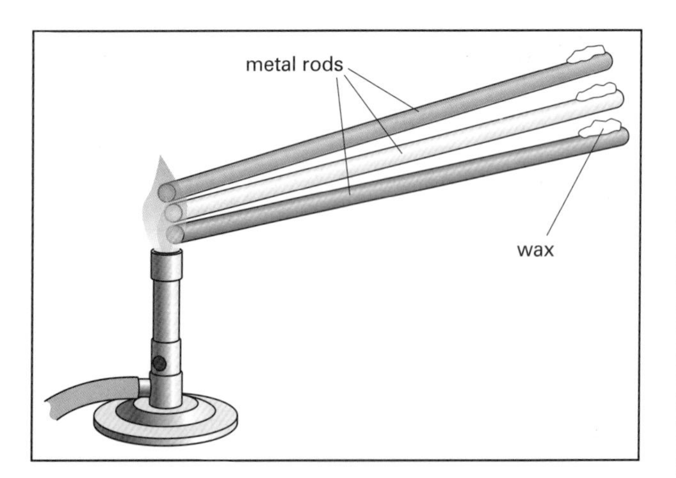

Figure 11.4 An experiment to show which metal is the best conductor of heat.

Activity 11.1 Investigating conduction

Try out some experiments that involve conduction of heat.

QUESTIONS

1 **a** Name a good conductor of heat (a thermal conductor).
  b Name a good thermal insulator.

2 What is needed for heat to flow through a conductor?

3 Look at Table **11.1**. Which will feel colder to the touch, marble or polystyrene?

Explaining conduction in metals and non-metals

Both metals and non-metals conduct heat. Metals are generally much better conductors than non-metals. We need different explanations of conduction for these two types of material.

We will start with **non-metals**. Imagine a long glass rod (Figure **11.6a**). One end is being heated, the other is cold. There is thus a temperature difference between the two ends, and heat flows down the rod. What is going on inside the rod?

We will picture the atoms that make up the glass as shown in Figure **11.6b**. (They are shown as being identical, and regularly arranged, although they are not really like this.) At the hot end of the rod, the atoms are vibrating a lot. At the cold end, they are vibrating much less. As they vibrate, the atoms jostle their neighbours. This process results in each atom sharing its energy with its neighbouring atoms. Atoms with a lot of energy end up with less, those with a little end up with more. The jostling gradually transfers energy from the atoms at the hot end to those at the cold end. Energy is steadily transferred down the rod, from hot to cold.

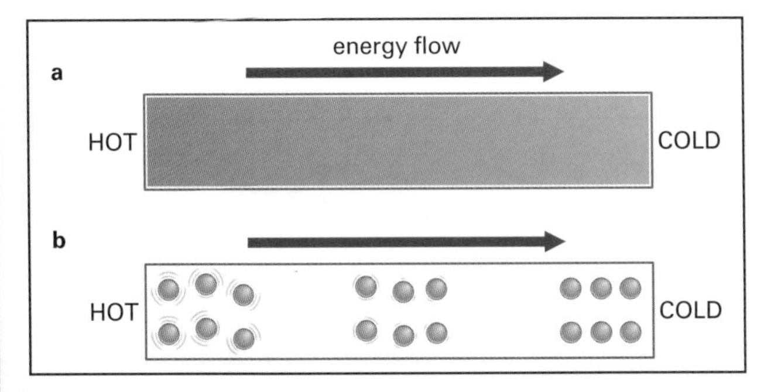

Figure 11.6 Conduction of heat in non-metals. **a** A glass rod, heated at one end and cooled at the other. Heat travels from the hot end to the cold end. **b** Energy is transferred because the vibrating atoms jostle one another. This shares energy between neighbouring atoms. The result is a flow of energy from the hot end to the cold end.

This is the mechanism by which poor conductors (such as glass, ice and plastic) conduct heat. It is also the mechanism in diamond, where the carbon atoms are tightly bonded to their neighbours. Any slight vibration of one atom is rapidly shared with its neighbours, and soon spreads through the whole piece of material.

However, **metals** are good conductors for another reason. In a metal there are particles called electrons that can move about freely. Electrons are smaller than atoms, and they are the particles that carry energy when an electric current flows through a metal. They also carry energy when heat is transferred through a metal.

Finally, **liquids** can also conduct heat, because the particles of which they are made are in close contact with one another. However, convection (see section **11.2** below) is often more important than conduction in the transfer of heat through a liquid.

11.2 Convection

'Hot air rises.' This is a popular saying. It is one of the few ideas from physics that almost everyone who has studied a little science can remember. Figure **11.7** is a photograph made using a technique that shows up currents in the air. You can see hot air rising from the heater, from the computer, and even from the man.

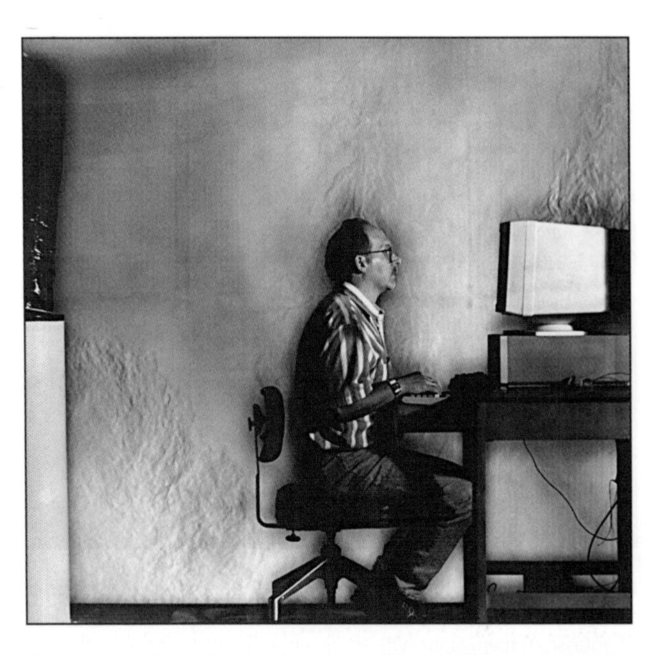

Figure 11.7 Warm air rises above any object that is warmer than its surroundings. In this office scene, there is a heater (lower left) that is producing warm air. Currents also rise above the computer and the operator.

When air is heated, its density decreases (it expands). Since it is less dense than its surroundings, it then floats upwards (just as a cork floats upwards if you hold it under water and then release it). Think about a hot air balloon. If it is to 'fly', the hot air in the balloon, plus the balloon fabric itself, plus the basket that hangs below, complete with occupants, must altogether have a density less than that of the surrounding colder air.

The rising of hot air is just one example of **convection**. Hot air can rise because air is a fluid, and convection is a phenomenon that can be observed in any fluid (liquid or gas).

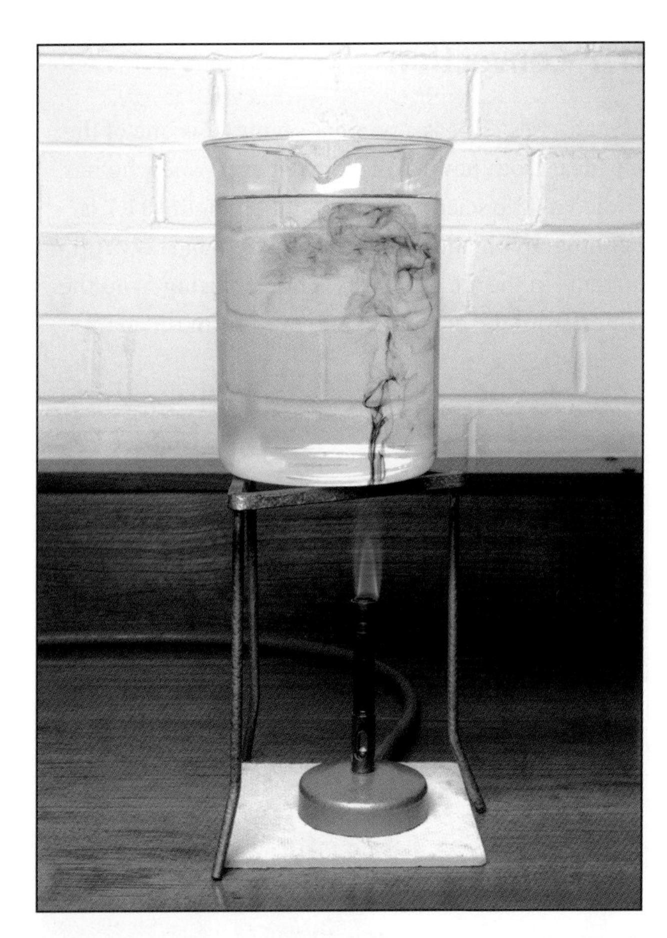

Figure 11.8 Because water is clear and colourless, it can be difficult to see how the water moves to form a convection current. Crystals of potassium manganate(VII) act as a purple dye to show up the movement of the water.

Demonstrating convection

Figure **11.8** shows how a convection current can be observed in water. Above the flame, water is heated and expands. Now its density is less than that of the surrounding water, and it floats upwards. The purple dye shows how it moves. Colder water, which is more dense, flows in to replace it.

A **convection current** is a movement of a fluid that carries energy from a warmer place to a cooler one. This highlights an important difference between convection and conduction.

- In convection, energy is transferred through a material from a warmer place to a cooler place by the movement of the material itself.
- In conduction, energy is transferred through a material from a warmer place to a cooler place without the material itself moving.

Convection currents at work

Convection currents help to share energy between warm and cold places. If you are sitting in a room with an electric heater, energy will be moving around the room from the heater as a result of convection currents, rising from the heater. You are likely to be the source of convection currents yourself, since your body is usually warmer than your surroundings (see Figure **11.9**). Many biting insects make use of this effect. For example, bed bugs crawl across the bedroom ceiling. They can detect a sleeping person below by finding the warmest spot on the ceiling. Then they drop straight down on the sleeper. This is a lot easier than crawling about on top of the bedding.

Figure 11.9 Convection currents rise above the warm objects in a room.

Cold objects also produce convection currents. You may have noticed cold water sinking below an ice cube in a drink. In a refrigerator, the freezing surface is usually positioned at the top and the back, so that cold air will sink to the bottom. Warm air rises to be re-chilled (see Figure **11.10**).

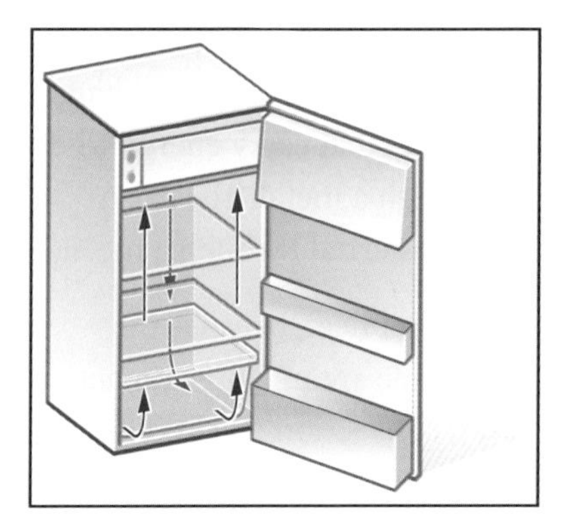

Figure 11.10 In a fridge, cold air sinks from the freezing compartment. If the freezer was at the bottom, cold air would remain there, and the items at the top would not be cooled.

Explaining convection

We have already seen that convection results from the **expansion** of a fluid when it is heated. Expansion means an increase in volume while mass stays constant – hence, density decreases. A less dense material is lighter, and is pushed upwards by the surrounding denser material.

The particles in the hotter fluid have more kinetic energy – they move around faster. As they flow from place to place, they take this energy with them.

Activity 11.2 Convection experiments

Try out some experiments that show convection at work.

QUESTIONS

4 'A thermal (heat) energy transfer by means of the motion of a fluid.' Is this a description of conduction or convection?

5 When a gas is heated, its particles gain energy. Imagine that you could see the particles of a hot gas and of a cold gas (at the same pressure).
   a What difference would you see in their movement?
   b What difference would you see in their separation?

6 What part does convection play in the spreading of energy around a room from an electric heater?

7 Write a brief explanation of convection, using the terms **expansion**, **density** and **gravity**.

8 Why would it not be a good idea to fit an electric heater near the ceiling in a room?

11.3 Radiation

At night, when it is dark, you can see much further than during the day. In the daytime, the most distant object you are likely to be able to see is the Sun, about 150 million kilometres away. At night, you can see much further, to the distant stars. The most distant object visible to the naked eye is the Andromeda galaxy, about 20 million million million kilometres away.

The light that reaches us from the Sun and other stars travels to us through space in the form of **electromagnetic radiation**. This radiation travels as electromagnetic waves. It travels over vast distances, following a straight line through empty space. As well as light, the Earth is bathed in other forms of electromagnetic radiation from the Sun, including infrared and ultraviolet.

The hotter an object, the more **infrared radiation** it gives out. You can use this idea to help you in doing a bit of detective work. Outside the house, a car is parked. How long has it been there? Hold your hands close to the engine compartment to see if you can detect heat radiating from it. Inside the house, the lights are out. Hold your hand close to the light bulb. Can you detect radiation, which will tell you that it was recently lit up?

Our skin detects the infrared radiation produced by a hot object. Nerve cells buried just below the surface respond to heat. You notice this if you are outdoors on a sunny day.

Here are the characteristics of infrared radiation that we have mentioned so far. Infrared radiation:

- is produced by warm or hot objects
- is a form of electromagnetic radiation
- travels through empty space (and through air) in the form of waves
- travels in straight lines
- warms the object that absorbs it
- is invisible to the naked eye
- can be detected by nerve cells in the skin.

Figure **11.11** shows another way of detecting infrared radiation, using a heat-sensitive camera. The photograph shows a boy sitting in front of a camera that detects infrared radiation. It is very sensitive to slight differences in temperature between different parts of the body.

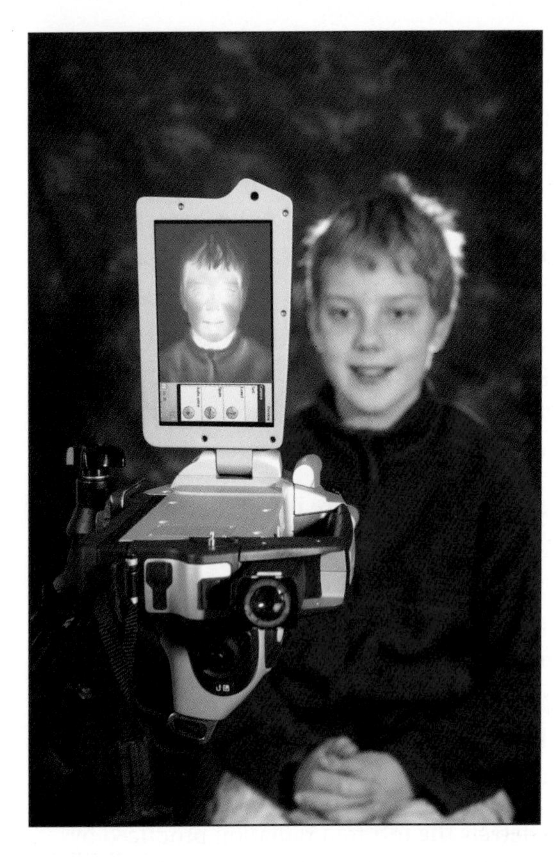

Figure 11.11 Using an infrared-sensitive camera. Slight variations in body temperature show up as different colours. Cameras like this are used in medicine to detect skin disorders and infections.

QUESTIONS

9 How can energy be transferred through the vacuum of space: by conduction, by convection, or by radiation?

10 On Earth, we receive visible light from the Sun. Name **two** other forms of electromagnetic radiation that we receive from the Sun.

11 If an object's temperature is increased, what happens to the amount of infrared radiation it emits?

Good absorbers, good emitters

On a hot, sunny day, car drivers may park their cars with a sunshield behind the windscreen (Figure **11.12**). Such a sunscreen is usually white (or another light colour) or shiny, because this reflects away light and infrared radiation, which would make the car get uncomfortably hot. The black plastic parts of the car

Figure 11.12 A sunscreen reflects away unwanted radiation, which would otherwise make the car unbearably hot.

(such as the steering wheel and dashboard) are very good absorbers of infrared, and they can become too hot to touch.

It is the surface that determines whether an object absorbs or reflects infrared radiation (Figure **11.13**). A surface that is a good reflector is a poor absorber.

On a hot day, you may have noticed how the black surface of a tarred (metalled) road emits heat. Black surfaces readily absorb infrared radiation. They are also good emitters.

- Shiny or white surfaces are the best reflectors (the worst absorbers).
- Matt black surfaces are the best absorbers (the worst reflectors).
- Matt black surfaces are the best emitters.

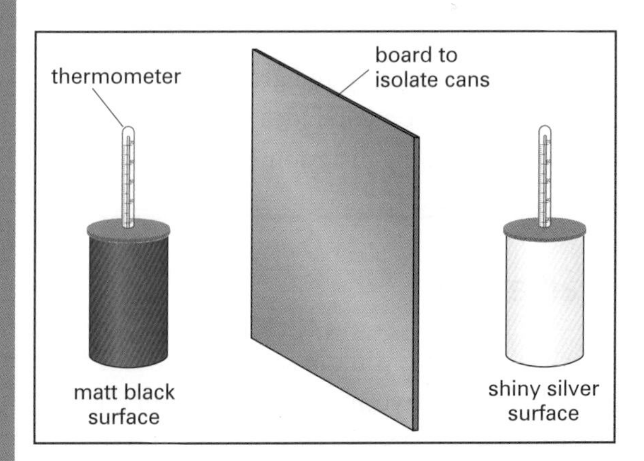

Figure 11.13 Which surface radiates better, black or shiny?

Figure **11.13** shows an experiment to compare the rates at which black and shiny surfaces emit radiation. One can has a matt black surface, and the other is shiny. Both are filled with hot water, and they cool by radiation. The black can cools more rapidly than the shiny one.

QUESTIONS

12 Suppose that you have a matt black surface and a shiny black surface.

- **a** Which is a better absorber of infrared radiation?
- **b** Which is a better emitter of infrared radiation?
- **c** Which is a better reflector of infrared radiation?

13 Look at Figure **11.13**. Use what you know about thermal (heat) energy transfers to explain why the cans must be fitted with lids, and why they should stand on a wooden or plastic surface.

Activity 11.3 Radiation experiments

Carry out some experiments (or watch demonstrations) showing how hot objects radiate.

11.4 Some consequences of thermal (heat) energy transfer

Hot objects have a lot of **internal energy**. As we have seen above, energy tends to escape from a hot object, spreading to its cooler surroundings by conduction, convection and radiation. This can be a great problem. We may use a lot of energy (and money) to heat our homes during cold weather, and the energy simply escapes. We eat food to supply the energy we need to keep our bodies warm, but energy escapes from us at a rate of roughly 100 watts (100 W = 100 J/s).

To keep energy in something that is hotter than its surroundings, we need to **insulate** it. Knowing about conduction, convection and radiation can help us to design effective insulation.

Home insulation

A well-insulated house can avoid a lot of energy wastage during cold weather. Insulation can also help to prevent the house from becoming uncomfortably hot during warm weather. Figure **11.14** shows some ways in which buildings can be insulated. More details of these are listed in Table **11.2**.

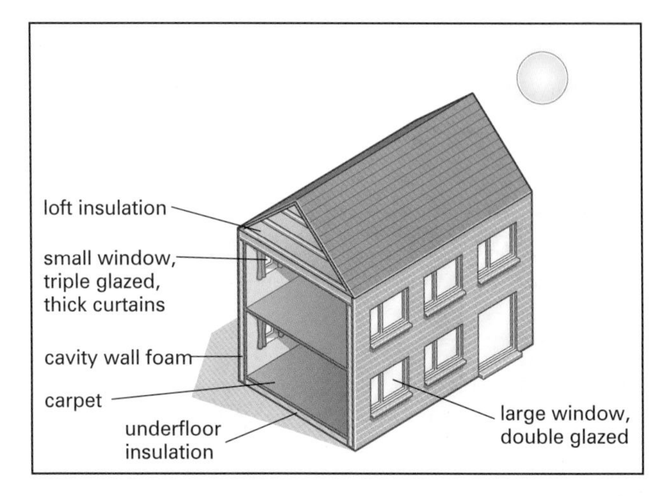

Figure 11.14 This house has been well designed to reduce the amount of fuel needed to keep it warm. The windows on the sunny side are large, so that the rooms benefit from direct radiation from the Sun. The windows on the other side are small, so that little energy escapes through them.

Method	Why it works
thick curtains, draught excluders	stops convection currents, and so prevents cold air from entering and warm air from leaving
loft and underfloor insulating materials	prevents conduction of heat through floors and ceilings
double and triple glazing of windows	vacuum between glass panes cuts out losses by conduction and convection
cavity walls	reduces heat losses by conduction
foam or rockwool in wall cavity	further reduces heat losses by convection

Table 11.2 Ways of retaining energy in a house

Double-glazed windows usually have a vacuum between the two panes of glass. This means that energy can only escape by radiation, since conduction and convection both require a material. Modern houses are often built with cavity walls, with an air gap between the two layers of bricks. It is impossible to have a vacuum in the cavity, and convection currents can transfer energy across the gap (see Figure **11.15a**). Filling the cavity with foam means that a small amount of energy is lost by conduction, although the foam material is a very poor conductor. However, this does stop convection currents from flowing (Figure **11.15b**), so there is an overall benefit.

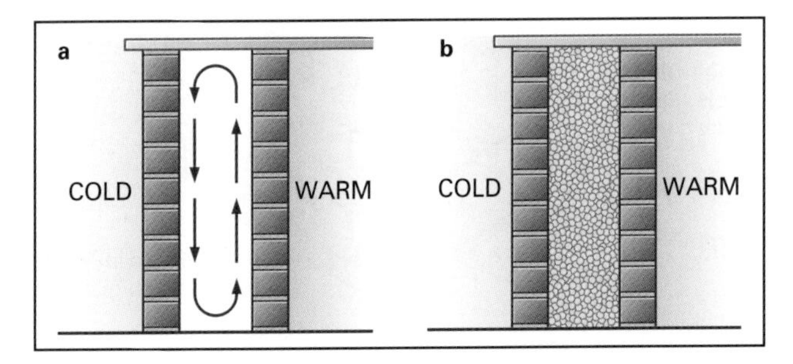

Figure 11.15 a A cavity wall reduces heat loss by conduction, because air is a good insulator. However, a convection current can transfer energy from the inner wall to the outer wall. **b** Filling the cavity with foam or mineral (glass or rock) wool prevents convection currents from forming.

Keeping cool

Vacuum (thermos) flasks are used to keep hot drinks hot. They can also be used to keep cold drinks cold. Giant vacuum flasks are used to store liquid nitrogen and helium at very low temperatures, ready for use in such applications as body scanners in hospitals.

Figure **11.16** shows the construction of a vacuum flask. Glass is generally used, because glass is a good insulator. However, some flasks are made of steel for added strength. The gap between the double walls is evacuated to reduce losses by conduction and convection. Silvering reduces losses by radiation by reflecting back any infrared radiation. A vital part is the stopper, which prevents losses by convection and evaporation.

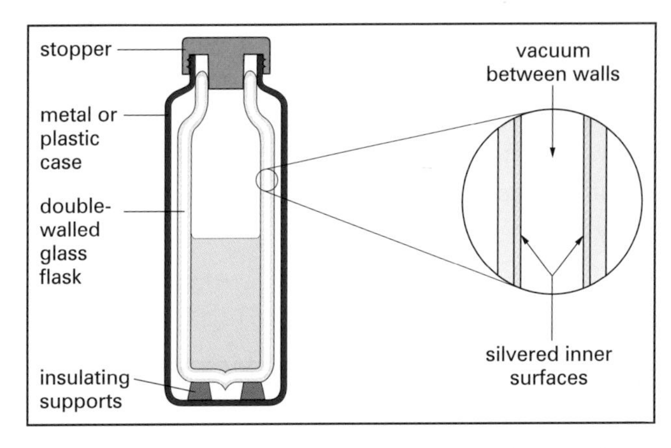

Figure 11.16 A vacuum flask is cleverly designed to keep hot things hot by reducing heat losses. It also keeps cold things cold. Although we might say 'it stops the cold getting out', it is more correct to say that it prevents heat from getting in. The first such flask was designed by James Dewar, a Scottish physicist, in the 1870s. He needed flasks to store liquefied air and other gases at temperatures as low as −200 °C. Soon after, people realised that a flask like this was also useful for taking hot or cold drinks on a picnic.

Convection, climate and weather

Convection currents explain the origins of winds and ocean currents, two of the major factors that control climate patterns around the world. For example, warm air rises above the equator, and colder air sinks in subtropical areas. This creates the pattern of Trade Winds that are experienced in the tropics.

Ocean currents (Figure **11.17**) help to spread warmth from equatorial regions to cooler parts of the Earth's surface. Warm water at the surface of the sea flows towards the poles. In polar regions, colder water sinks and flows back towards the equator. Provided this pattern remains constant, this helps to make temperate regions of the world more habitable. However, there is evidence that the pattern of ocean currents is changing, perhaps as a consequence of global warming.

QUESTIONS

14 List as many features as you can that contribute to the insulation of a house in a cold climate. For each, state whether it reduces heat loss by conduction, by convection or by radiation.

15 Why is it important to wear a hat on a very cold day?

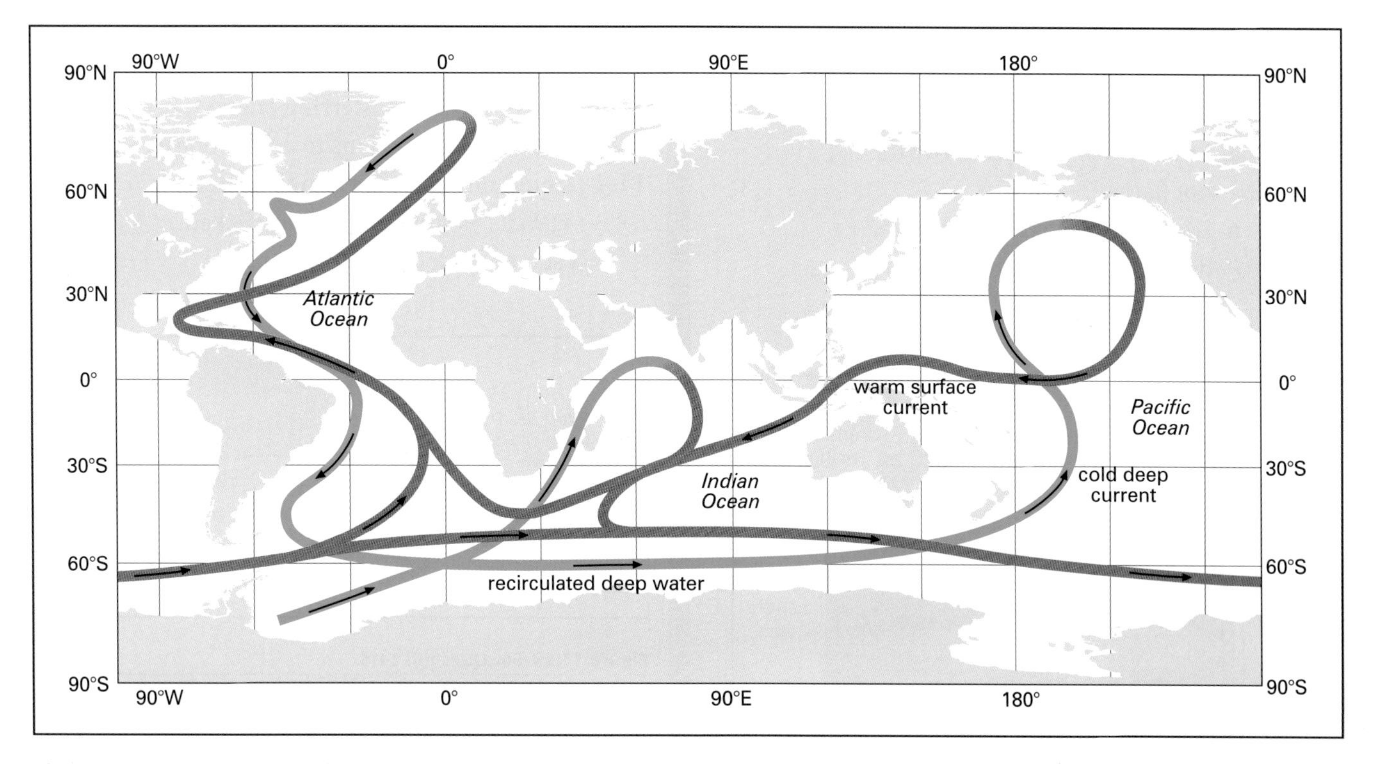

Figure 11.17 Ocean currents help to move energy from the tropics to cooler regions. Colder water from polar regions sinks and flows towards the equator. Warmer water flows closer to the ocean surface.

Summary

In general, metals are better conductors of heat than non-metals.

Energy conducts through a solid when neighbouring particles collide and share energy, or when electrons transfer energy through the material.

Convection currents happen when a fluid expands and rises, because of its lower density.

Infrared radiation is part of the electromagnetic spectrum.

Matt black surfaces are the best emitters and absorbers of radiation. Light, shiny surfaces are the best reflectors of radiation.

End-of-chapter questions

11.1 In cold climates, it is important to keep a house well insulated. Listed below are three ways of insulating a house. For each, explain how it reduces heat loss. In your answers, refer to conduction, convection or radiation, as appropriate.

a Heavy curtains, when closed, trap air next to a window. [2]

b Shiny metal foil is fitted in the loft, covering the inside of the roof. [2]

c Glass wool is used to fill the gap in the cavity walls. [2]

11.2 Figure **11.18** shows a way of demonstrating a convection current in air.

a Explain why air rises above the hot flame. [3]

b Explain why colder air flows downwards through the other 'chimney'. [2]

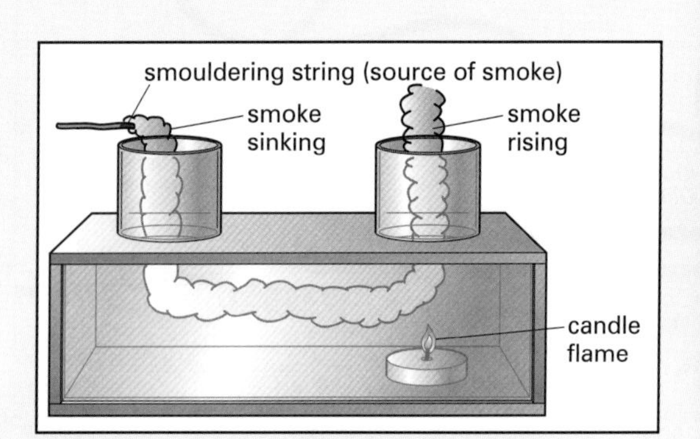

Figure 11.18 For Question **11.2**.

11.3 a One end of a plastic rod is immersed in boiling water. The temperature of the other end gradually increases. Use ideas from the kinetic model of matter to explain how energy travels from one end of the rod to the other. [3]

b If the experiment was repeated using a metal rod of the same dimensions as the plastic rod, what difference would you expect to notice? [2]

c What particles in a metal are involved in transferring energy from hotter regions to colder ones? [1]

11.4 Liquid nitrogen, at a temperature of −196 °C, is stored in a wide-necked vacuum flask (Figure **11.19**).

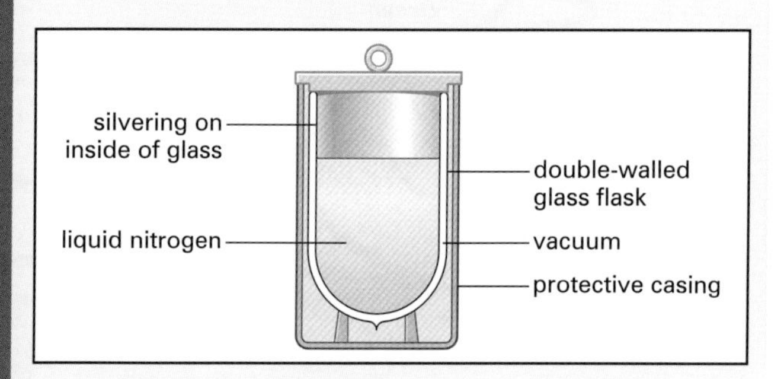

Figure 11.19 For Question **11.4**.

a Explain the features of the design of this flask that help to keep the liquid nitrogen cold. [8]

b When **hot** drinks are stored in a vacuum flask, it is important to keep the stopper in the flask. Why is it less important to have a stopper in a flask that is being used to keep things **cold?** [2]

Block 3
Physics of waves

In December 2004, a giant tsunami caused by an underwater earthquake devastated coastal regions in several countries around the Indian Ocean. Hundreds of thousands of people lost their lives.

Earthquakes are vibrations that carry vast amounts of energy. They travel right through the Earth and can be detected thousands of kilometres away. Surprisingly, earthquakes have also proved useful. Because we can understand how they travel through solids and liquids, geologists have been able to use information from earthquakes to build up a detailed picture of the inner structure of the Earth.

In this block we will look at how the idea of waves can be used to explain sound and light.

The tsunami of 2004 arriving on a beach in Thailand. The woman is running down the beach to warn her children. The family was lucky – they all survived.

12 Sound

Core Describing the production of sounds
Core Measuring the speed of sound
Core Relating pitch and loudness to frequency and amplitude
Core Describing how sound travels

The sound of music

Most musicians have to tune their instruments before they start to play. Guitarists in a band adjust the tension of their strings so that they play the correct notes. If you have heard a symphony orchestra play, you may have noticed that the oboist usually plays a single clear note, and the other instrumentalists tune to this note. If they all played slightly different notes, the effect would be very uncomfortable on our ears.

Most music we hear is played by instruments tuned to a standard scale, like the notes of a piano keyboard. However, not all instruments are tuned in the same way. The Scottish bagpipes, for example, play notes on a slightly different scale. A pipe band playing together can sound very exciting (Figure **12.1**). But when mixed with other instruments, the notes can clash to produce a very unpleasant effect. In a similar way, the instruments of an Indonesian *Gamelan* band (Figure **12.2**) play notes on a different scale.

In this chapter, we will look at musical sounds (and other sounds, too), how they are produced, and how they travel. We will also look at why different instruments sound different to our ears.

Figure 12.1 These pipers play instruments that produce notes on an unusual scale, different from the conventional scale of a piano. Because the scale is different from what we are used to, the music can at first seem off-key. The Scottish pipes were often played before battles, to give the Scottish troops courage and to alarm the enemy.

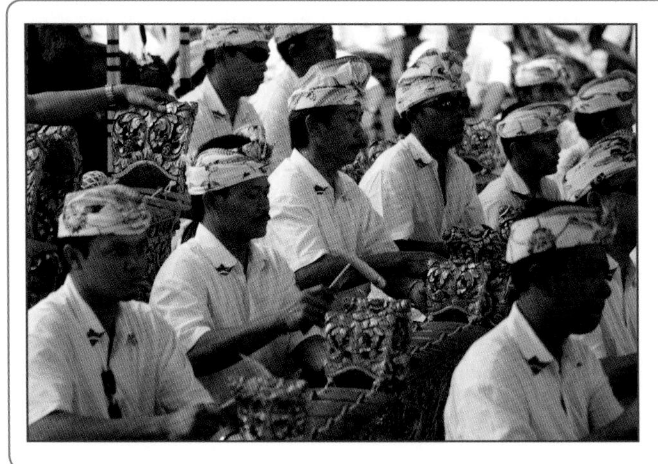

Figure 12.2 *Gamelan* bands can be heard in Indonesia and other countries of the Pacific rim. They include string and woodwind instruments, and are specially noted for their range of percussion instruments – gongs, drums, chimes, marimbas and so on. For people who are used to listening to conventional western music (popular or classical), it can take some time to tune in to the complex rhythms and harmonies produced by a *Gamelan* orchestra.

12.1 Making sounds

Different musical instruments produce sounds in different ways.

- **String instruments.** The strings are plucked or bowed to make them vibrate. In most string instruments, the vibrations are transmitted to the body of the instrument, which also vibrates, along with the air inside it. The vibrations may be too small or too fast to see, but they can be shown up using laser techniques (see Figure 12.3).

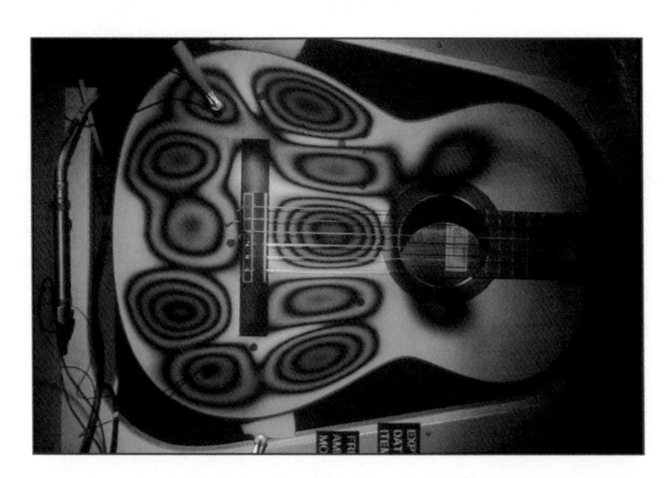

Figure 12.3 Although the player only touches the strings of a guitar, the instrument's whole body vibrates to produce the notes we hear. This is shown up in this image, produced by shining laser light onto the guitar. Different notes produce different patterns of vibration, and this helps to give each note its particular quality.

- **Wind instruments.** The 'air column' inside the instrument is made to vibrate, by blowing across the end of or into the tube (Figure 12.4). The smallest instruments have a straight air column.

Figure 12.4 Two recorders can look very similar, but the lower one is made of wood and the other of plastic. A flute may be of wood or metal. This tells us that it is not the material that the instrument is made of that matters. It is the air inside that vibrates to produce the desired note. Blowing into the instrument causes the air column inside it to vibrate, and the vibrations are transferred to the air outside.

- Bigger instruments capable of playing deeper notes (such as a horn or tuba) have an air column that is bent around so that the instrument is not inconveniently long. Some instruments have a reed in the mouthpiece. This vibrates as the player blows across it, causing the air to vibrate.
- **Percussion instruments.** These instruments are played by striking them (Figure **12.5**). This produces vibrations – of the keys of a xylophone, the skin of a drum, or the metal body of a gong, for example.

In each case, part (or all) of the instrument is made to vibrate. This causes the air nearby to vibrate, and the **vibrations** travel through the air to the audience's ears. Some vibrations also reach us through the ground, so that they make our whole body vibrate (see Figure **12.5**). If you sit close to a loud band or orchestra, you may feel your whole body vibrating in response to the music.

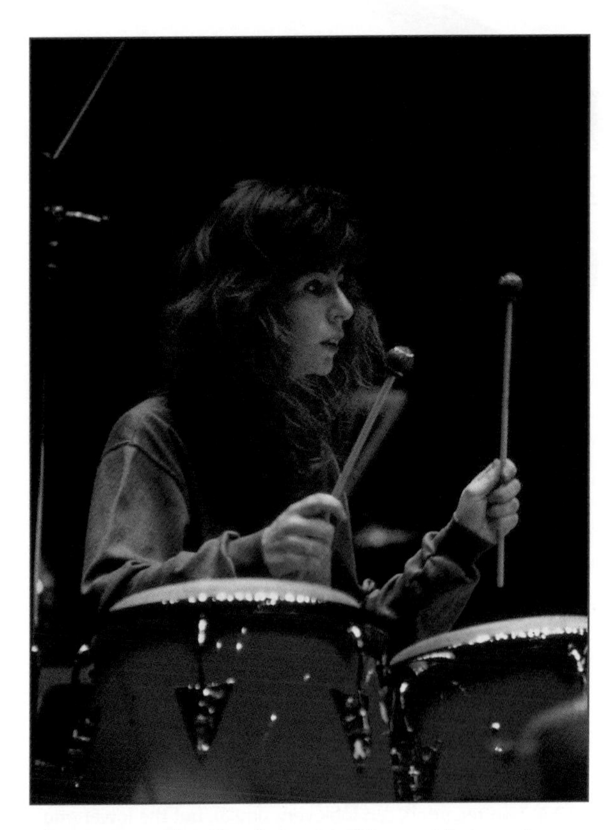

Figure 12.5 Evelyn Glennie is one of the world's top solo percussionists, despite the fact that she is deaf. She has trained herself to be sensitive to vibrations that reach her body through the ground. This allows her to follow the rhythm of a piece of music, as well as to detect the subtle differences in tone between different percussion instruments.

Sounds travel through the air as vibrations. These vibrations can travel through any material – through the solid ground, through the glass panes of a window, through water. If you put a battery-powered radio on the side of the bath and submerge your ears, you will hear the sounds from the radio travelling through the solid bath and the liquid water to your ears.

QUESTIONS

1. Which of the following materials can sound travel through: wood, air, water?
2. When a woodwind instrument such as a flute produces a note, what part of it vibrates?

12.2 At the speed of sound

The speed of sound in air is about 330 m/s, or 1 200 km/h. That is about ten times the speed of cars on a major highway. When someone speaks, it seems to us that we hear the sound they make as soon as they make it. However, it takes a small amount of time to reach our ears. For example, if we are speaking to someone who is just 1 m away, the time for sounds to travel between us is:

$$\frac{1\,\text{m}}{330\,\text{m/s}} = 0.003\,\text{s} = 3\,\text{ms (3 milliseconds)}$$

This is far too short a time for us to notice.

However, there are occasions when we may notice the time it takes for sounds to travel. For example, imagine that you shout at a distance from a long high wall or cliff. After you shout, you may hear an **echo**. The sound has reflected from the hard surface and back to your ears (see Figure **12.6**). Worked example **1** shows how to calculate the time it takes for the sound to travel to a wall and back again.

If you watch people playing a game such as cricket or baseball, you may notice a related effect. You see someone hitting a ball. A split second later you hear the sound of the ball being struck. The time interval between seeing the hit and hearing it occurs because the sound travels relatively slowly to your ears, while

Figure 12.6 An echo is heard when a sound reflects off a hard surface such as a large wall. Sound travels outwards from the source, and bounces off the wall. Some of it will return to the source. If there are several reflecting surfaces, several echoes may be heard.

the light travels very quickly to your eyes. So the light reaches you first, and you see before you hear. When cricket matches are televised, they may use a microphone buried in the pitch to pick up the sounds of the game, so that there is no noticeable gap between what you see and what you hear.

For the same reason, we usually see a flash of lightning before we hear the accompanying roll of thunder. Count the seconds between the flash and the bang. Then divide this by three to find how far away the lightning is, in kilometres. This works because the sound takes roughly 3 s to travel 1 km, whereas the light travels the same distance in a few microseconds.

Worked example 1

A man shouts loudly close to a high wall (see Figure 12.6). He hears one echo. If the man is 40 m from the wall, how long after the shout will the echo be heard? (Speed of sound in air = 330 m/s.)

Step 1: Calculate the distance travelled by the sound. This is twice the distance from the man to the wall (since the sound travels there and back).

$$\text{distance travelled by sound} = 2 \times 40\,\text{m} = 80\,\text{m}$$

Step 2: Calculate the time taken for the sound to travel this distance.

$$\text{time taken} = \frac{\text{distance}}{\text{speed}} = \frac{80\,\text{m}}{330\,\text{m/s}} = 0.24\,\text{s}$$

So the man hears the echo 0.24 s (about a quarter of a second) after his shout.

Measuring the speed of sound

One way to measure the speed of sound in the lab is to find out how long a sound takes to travel a measured distance, just as you might measure the speed of a moving car or cyclist. Since sound travels at a high speed, you need to be able to measure short time intervals. Figure **12.7** shows one method.

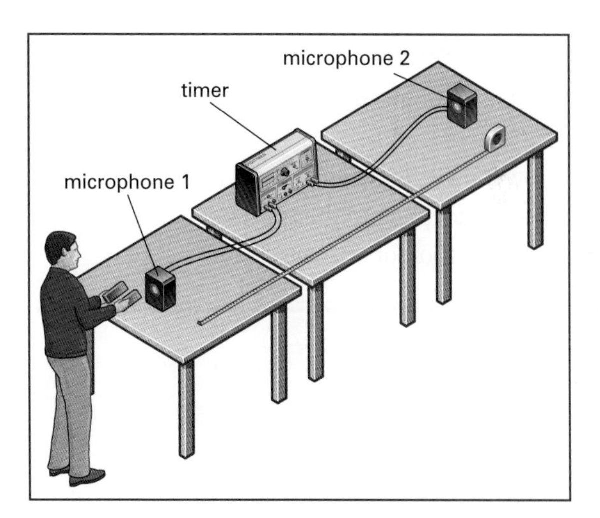

Figure 12.7 A 'time-of-flight' method for measuring the speed of sound. The wooden blocks and the two microphones are arranged in a straight line. The bang from the blocks is picked up first by microphone 1 and then by microphone 2. The first activates the timer, and the second stops it. The speed of sound is calculated from the distance between the two microphones and the time taken by the sound to travel between them.

When the student bangs the two blocks of wood together, it creates a sudden, loud sound. The sound reaches one microphone, and a pulse of electric current travels to the timer. The timer starts running. A fraction of a second later, the sound reaches the second microphone. A second pulse of current stops the timer. Now the timer indicates the time taken for the sound to travel from one microphone to the other.

It is important that the two microphones should be a reasonable distance apart – say, three or four metres. The further apart the better, since this will give a longer 'time of flight' for the sound to travel from one microphone to the other.

Measuring the speed of sound in air

Use echoes to help you to measure the speed of sound in air.

QUESTION

3 Sound takes about 3 ms (3 milliseconds) to travel 1 m.
  a How long will it take to travel from the centre of a cricket pitch to the spectators, 200 m away?
  b What fraction of a second is this?

Different materials, different speeds

We talk about 'the speed of sound' as 330 m/s. In fact, it is more correct to say that this is the speed of sound in air at 0 °C. The speed of sound changes if the temperature of the air changes, if it is more humid, and so on. (Note also that some people talk about 'the velocity of sound', but there is no need to use the word 'velocity' here, since we are not talking about the direction in which the sound is travelling – see Chapter 2.)

Table 12.1 shows the speed of sound in some different materials. You can see that sound travels faster through solids than through gases. Its speed in water (a liquid) is in between its speed in solids and gases.

	Material	Speed of sound / m/s
gases	air	330
	hydrogen	1280
	oxygen	316
	carbon dioxide	268
liquids	water	1500
	sea water	1530
	mercury	1450
solids	glass	5000
	iron, steel	5100
	lead	1200
	copper	3800
	wood (oak)	3800

Table 12.1 The speed of sound in different materials (measured at standard temperature and pressure).

QUESTIONS

4 Look at the experiment to measure the speed of sound shown in Figure 12.7. Explain why the wooden blocks and the two microphones must be in a straight line.

5 Which travels faster, light or sound? Describe **one** observation that supports your answer.

12.3 Seeing sounds

When a flautist plays her flute, she sets the air inside it vibrating. A trumpeter does the same thing. Why do the two instruments sound so different? The flute and the trumpet each contain an 'air column', which vibrates to produce a musical note. Because the instruments are shaped differently, the notes produced sound different to our ears.

An image of the notes can be produced by playing the instrument next to a microphone connected to an oscilloscope (Figure 12.8). The microphone receives the vibrations from the instrument and converts them to an electrical signal, which is displayed on the oscilloscope screen. The trace on the screen shows the regular up-and-down pattern of the vibrations that make up the sound.

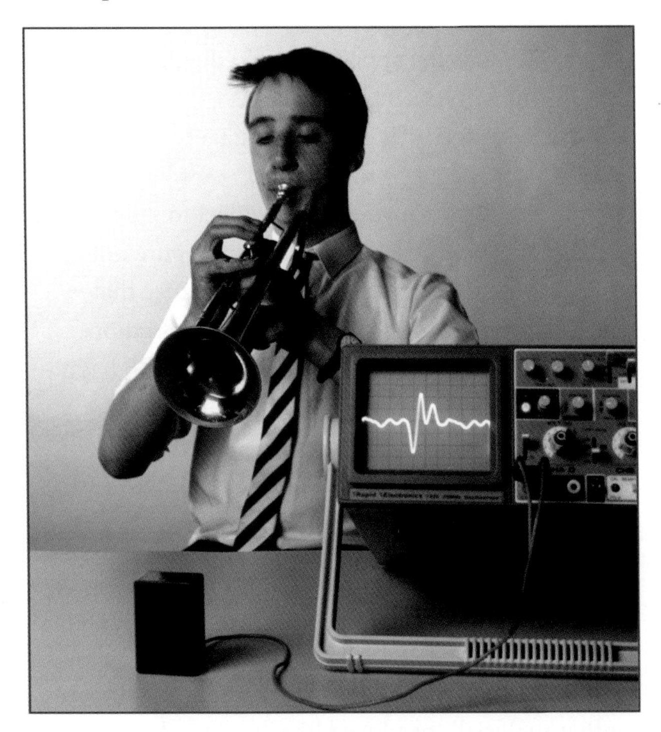

Figure 12.8 To display the vibrations of a musical note, it is converted to an electrical signal by a microphone and displayed on the screen of an oscilloscope. The trace on the screen shows the regular pattern of vibration of the sounds.

Pure notes

A signal generator can produce pure notes that have a very simple shape when displayed on an oscilloscope screen, as shown in Figure **12.9**. As shown in the diagram, we can make an important measurement from this graph. This is the time for one complete vibration, known as the **period** T of the vibration. This is related to the **frequency** f of the sound:

period T = number of seconds for one vibration

frequency f = number of vibrations per second

Hence we can write the following equation:

$$f = \frac{1}{T}$$

Frequency is measured in hertz (Hz). A frequency of 1 Hz is one vibration per second.

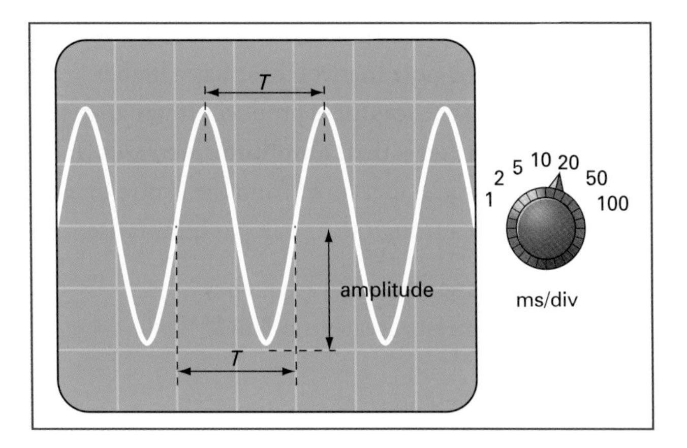

Figure 12.9 A pure note has the shape shown in this oscilloscope trace. The setting of the oscilloscope timebase is indicated on the right. This tells you how much time is represented by the divisions on the horizontal scale.

High and low, loud and soft

You can understand how an oscilloscope works by connecting it up to a signal generator. With a low-frequency note (say, 0.1 Hz), you will see that there is a single dot, which moves steadily across the oscilloscope screen. The electrical signal from the signal generator makes it move up and down in a regular way. Increasing the frequency makes the dot go up and down faster, until it blurs into a continuous line.

Changing the settings on the signal generator allows you to see the traces for notes of different frequencies and loudnesses. A loudspeaker will let you hear them as well. As shown in Figure **12.10**, increasing the

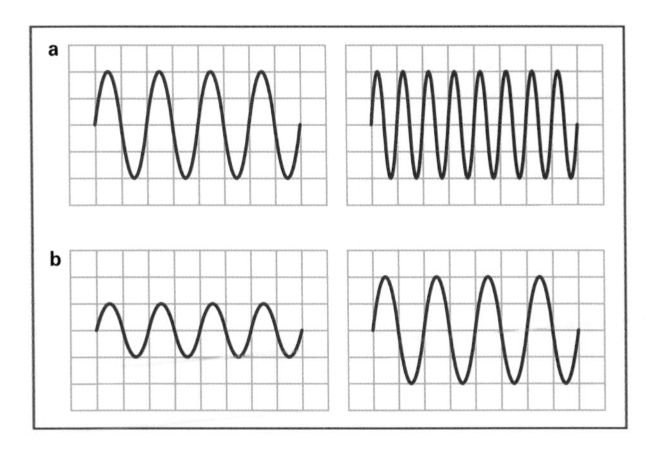

Figure 12.10 **a** Two notes with the same amplitude, and hence the same loudness. The second has more waves squashed into the same space, so its frequency is higher. Its pitch is higher too (it sounds higher). **b** Two notes with the same frequency. The second has a greater amplitude, so that it sounds louder.

frequency of the note squashes the vibrations together on the screen. The note that you hear has a higher **pitch**. Increasing the **loudness** produces traces that go up and down further – their **amplitude** increases. Take care: the amplitude is measured from the centre line to a peak, not from a trough to a peak. To summarise:

- higher pitch means higher frequency
- louder note means greater amplitude.

Range of hearing

A piano keyboard covers a wide range of notes, with frequencies ranging from about 30 Hz at the bottom end to about 3 500 Hz at the top end. Most other instruments cover a narrower range than this. For example, a violin ranges from about 200 Hz to 2 500 Hz. The range of human hearing is greater than this. Typically, we can hear notes ranging from about 20 Hz up to about 20 000 Hz (20 kHz, or 20 kilohertz). However, older people gradually lose the ability to hear high-pitched sounds. Their **upper limit of hearing** decreases by about 2 kHz every decade of their age.

Sounds that are more high-pitched than the upper limit of hearing (above 20 kHz) are too high to hear, and are known as **ultrasound**. Sounds below 20 Hz are too low to hear, and are known as **infrasound**.

Activity 12.2 Seeing sounds

Use a signal generator and an oscilloscope to show traces for different sounds, and test your range of hearing.

QUESTIONS

6 What happens to the pitch of a sound if its frequency increases?

7 What happens to the loudness of a sound if its amplitude decreases?

8 **a** What is the approximate frequency range of human hearing?
  b How does this change with age?

9 What is meant by **ultrasound**?

10 Sketch the trace you would expect to see on an oscilloscope screen, produced by a pure note. On your diagram, indicate the distance that corresponds to the period T of the vibration.

11 Sound A has a period of 0.010 s; sound B has a period of 0.020 s.
  a Which has the greater frequency?
  b Which will sound more high-pitched?

12.4 How sounds travel

Sounds are vibrations that travel through the air (or another material), produced by vibrating objects. How can we picture the movement of the molecules of the air as a sound travels through? Figure **12.11** shows how the vibrations of a tuning fork are transmitted through the air. As the prong of the fork moves to the right, it pushes on the air molecules on that side, squashing them together. These molecules push on their neighbours, which become compressed, and which in turn push on their neighbours, and so on.

It is important to note that the individual air molecules do not travel outwards from the vibrating fork. The air

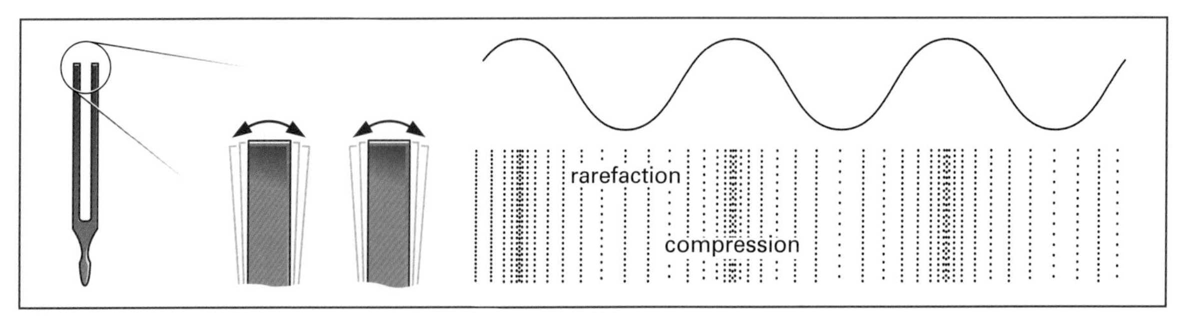

Figure 12.11 A vibrating tuning fork produces a series of compressions and rarefactions as it pushes the air molecules back and forth. This is how a sound travels through the air (or any other material). We can relate this to the wavy trace on an oscilloscope screen.

molecules are merely pushed back and forth. It is the vibrations that travel through the air to our ears.

This picture of how a sound travels also explains why sound cannot travel through a vacuum. There are no molecules or other particles in a vacuum to vibrate back and forth.

Figure **12.11** also shows another way of representing a sound, as a wavy line rather like the trace on an oscilloscope screen. The crests on the wave match the compressions, and the troughs match the rarefactions. It is much easier to represent a sound as an up-and-down wave like this, rather than drawing lots of air molecules pushing each other back and forth.

Here we have used two different **models** to represent sound:

1 vibrations travelling through a material – the particles of the material are alternately compressed together and then rarefied as the sound passes through
2 sound as a wave – a smoothly varying up-and-down line, like the trace on an oscilloscope screen.

The first of these models gives a better picture of what we could see if we could observe the particles of the material through which the sound is passing. The second model is easier to draw. It also explains why we talk about **sound waves**. The wavy line is rather like the shape of waves on the sea. There is much more about sound waves (and other waves) in Chapter **14**.

QUESTIONS

12 Why is it impossible for sounds to travel through a vacuum?

13 How could you convince a small child that, when you speak, it is not necessary for air to travel from your mouth to the ear of a listener?

Compression, rarefaction

Look back to Figure **12.11**. The areas of the sound wave where the air molecules are close together are called **compressions**. As the tuning fork vibrates back and forth, compressions are sent out into the air all around it. In between the compressions are **rarefactions**, areas in which the air molecules are less closely packed together, or rarefied.

The sound wave has been drawn so that the crests on the wave match the compressions, and the troughs match the rarefactions. Thus the wave represents the changes in air pressure as the sound travels from its source.

QUESTION

14 What is the difference between a compression and a rarefaction in a sound wave? Illustrate your answer with a sketch.

Summary

Sounds are vibrations that travel through a material, produced by a vibrating source.

An echo is produced when a sound is reflected off a hard surface.

Sound travels through solids, liquids and gases at speeds of hundreds or thousands of metres per second.

The frequency of a sound is the number of vibrations per second, measured in hertz (Hz).

The greater the frequency of a sound, the higher the pitch.

The greater the amplitude of a sound, the louder it is.

The audible range of sounds is from about 20 Hz to about 20 kHz.

Sounds cannot travel through a vacuum.

The vibrations of a sound travel through a material in the form of compressions and rarefactions of the particles that make up the material.

End-of-chapter questions

12.1 Sounds are produced by vibrating objects.

a When a wind instrument such as a trumpet produces a sound, what is it that is made to vibrate by the player? [1]

b When a stringed instrument such as a violin is played, what is it that is made to vibrate by the player? [1]

c Describe how the sound from the instrument travels through the air to the listener's ears. [3]

12.2 The vibrations of a sound can be detected using a microphone and then displayed on an oscilloscope screen. Figure **12.12** shows three such traces.

a Which trace shows the loudest sound? Explain your answer. [2]

b Which trace shows the sound with the highest pitch? Explain your answer. [2]

12.3 Describe a method for measuring the speed of sound in air, in the laboratory. What measurements are made, and how is the speed of sound calculated from them? [5]

E

12.4 a In which material does a sound travel faster, a solid or a gas? [1]

b Give **one** piece of evidence that shows that sound can travel through solid materials. [2]

To measure the length of a long metal rod, engineers send a pulse of sound into one end of it. The sound travels to the other end and is reflected back. The engineers detect this echo, and determine the time taken for the sound to travel from one end of the rod to the other.

c When making measurements on a steel rod of length 400 m, they find that the echo returns 0.16 s after the initial pulse. What is the speed of sound in steel? [4]

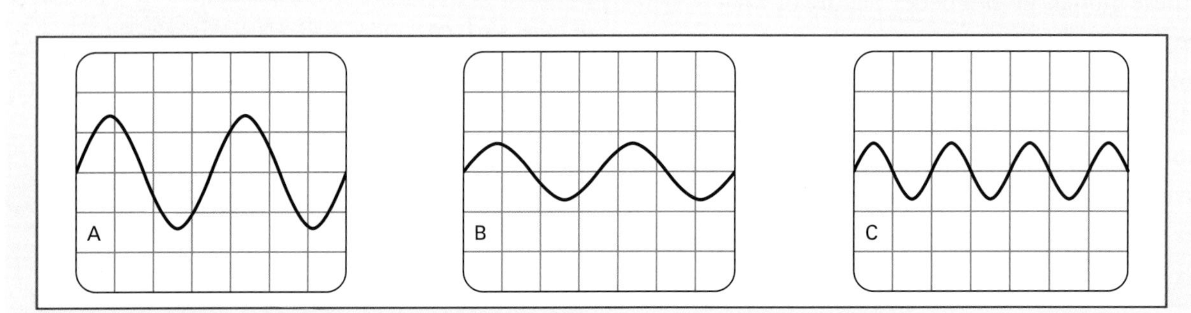

Figure 12.12 For Question **12.2**.

13 Light

Core	Using the law of reflection of light
Core	Describing how a plane mirror forms an image
E Extension	Constructing ray diagrams for reflection
Core	Describing the refraction of light
E Extension	Calculating refractive index and using Snell's law
Core	Describing total internal reflection
Core	Using ray diagrams to explain how a lens forms a real image
E Extension	Explaining how a magnifying glass works

How far to the Moon?

When Apollo astronauts visited the Moon, they left behind reflectors on its surface. These are used to measure the distance from the Earth to the Moon. A laser beam is directed from an observatory on Earth (Figure **13.1**) so that it reflects back from these reflectors left on the lunar surface. The time taken by the light to travel there and back is measured and, because the speed of light is known, the distance can be calculated.

The Moon travels along a slightly elliptical orbit around the Earth, so that its distance varies between 356 500 km and 406 800 km. The laser measurements of its distance are incredibly accurate – to within 30 cm. This means that they are accurate to within one part in a billion. The Moon is gradually slowing down and drifting away from the Earth. With the help of such precise measurements, it is possible to work out just how quickly it is drifting away.

This experiment makes use of two ideas that we will look at in this chapter: the way that light travels in straight lines, and how light is reflected by mirrors.

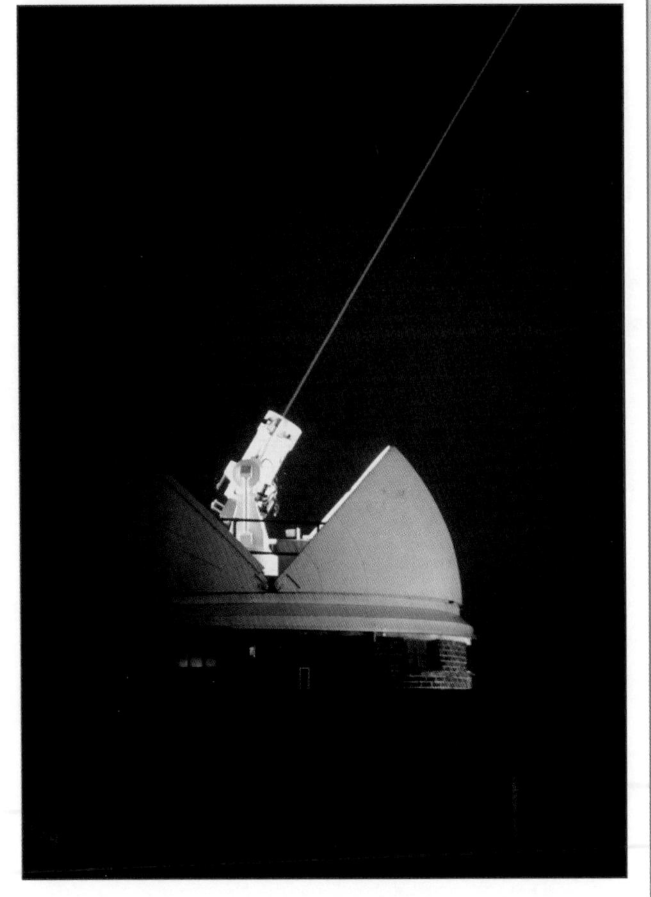

Figure 13.1 A laser beam is directed into space from the Royal Greenwich Observatory (Sussex, UK). The beam reflects off the Moon or a satellite in space. The reflected beam is detected, and the exact distance to the Moon or the satellite can be calculated.

13.1 Reflecting light

Light usually travels in straight lines. It changes direction if it hits a shiny surface, or if it travels from one material into another. This change in direction at a shiny surface such as a mirror is called **reflection**. We look at reflection in this section.

You can see that light travels in a straight line using a **ray box**, as shown in Figure 13.2. A light bulb produces light, which spreads out in all directions. A ray box produces a broad beam. By placing a narrow slit in the path of the beam, you can see a single narrow beam or **ray** of light. The ray shines across a piece of paper. You can record its position by making dots along its length. Laying a ruler along the dots shows that they lie in a straight line.

Figure 13.2 A ray box produces a broad beam of light, which can be narrowed down using a metal plate with a slit in it. Marking the line of the ray with dots allows you to record its position.

You may see demonstrations using a different source of light, a **laser**. A laser (Figure 13.3) has the great advantage that all of the light it produces comes out in a narrow beam. All of the energy is concentrated in this beam, rather than spreading out in all directions (as with a light bulb). The total amount of energy coming from the laser is probably much less than that from a bulb, but it is much more concentrated. That is why it is dangerous if a laser beam gets into your eye.

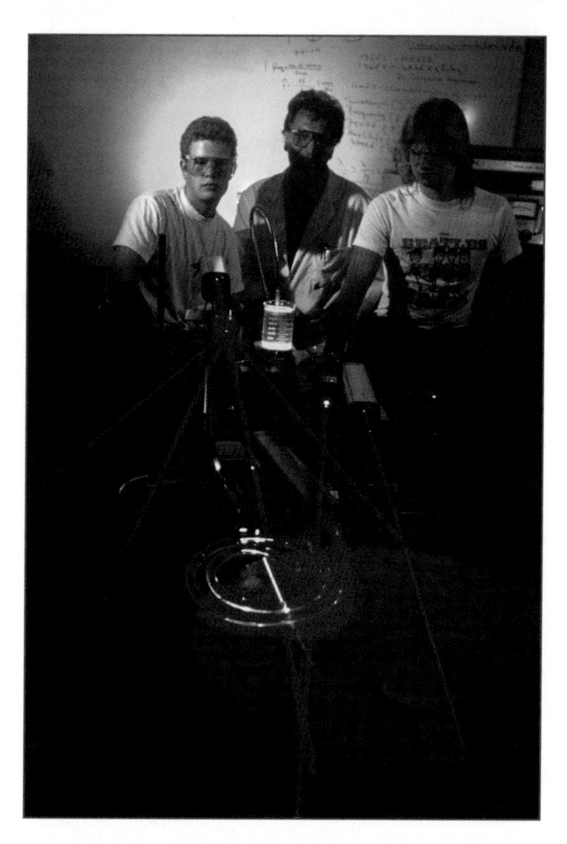

Figure 13.3 Students working with laser beams. They are wearing safety glasses to protect their eyes from stray reflections of the beams.

Looking in the mirror

Most of us look in a mirror at least once a day, to check on our appearance (Figure 13.4). It is important to us to

Figure 13.4 Psychologists use mirrors to test the intelligence of animals. Do they recognise that they are looking at themselves? Apes clearly understand that what they see in the mirror is an image of themselves – they make silly faces at themselves. Other animals, such as cats and dogs, do not – they may even try to attack their own reflection.

know that we are presenting ourselves to the rest of the world in the way we want. Archaeologists have found bronze mirrors over 2000 years old, so the desire to see ourselves clearly has been around for a long time.

Modern mirrors give a very clear image. When you look in a mirror, rays of light from your face reflect off the shiny surface and back to your eyes. You seem to see an image of yourself behind the mirror. To understand why this is, we need to use the law of reflection of light.

When a ray of light reflects off a mirror or other reflecting surface, it follows a path as shown in Figure 13.5. The ray bounces off, rather like a ball bouncing off a wall. The two rays are known as the **incident ray** and the **reflected ray**. The **angle of incidence** i and the **angle of reflection** r are found to be equal to each other. This is the **law of reflection**, which can be written as follows:

angle of incidence = angle of reflection

$$i = r$$

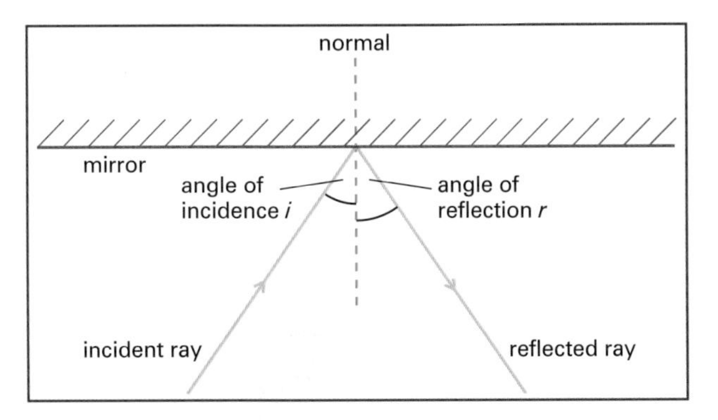

Figure 13.5 The law of reflection of light. The normal is drawn perpendicular to the surface of the mirror. Then the angles are measured relative to the normal. The angle of incidence and the angle of reflection are then equal: $i = r$.

Note that, to find the angles i and r, we have to draw the **normal** to the reflecting surface. This is a line drawn perpendicular (at 90°) to the surface, at the point where the ray strikes it. Of course, the other two angles (between the rays and the flat surface) are also equal. However, we would have trouble measuring these angles if the surface was curved, so we measure the angles relative to the normal. The law of reflection thus also works for curved surfaces, such as concave and convex mirrors.

Activity 13.1 The law of reflection

Check the law of reflection using a ray box and a plane mirror.

The image in a plane mirror

Why do we see such a clear **image** when we look in a plane (flat) mirror? And why does it appear to be behind the mirror?

Figure **13.6** shows how an observer can see an image of a candle in a plane mirror. Light rays from the flame are reflected by the mirror. Some of them enter the

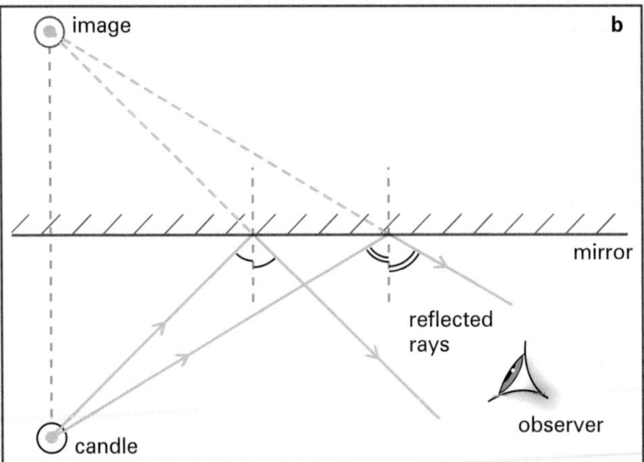

Figure 13.6 **a** Looking in the mirror, the observer sees an image of the candle. The image appears to be behind the mirror. **b** The ray diagram shows how the image is formed. Rays from the candle flame are reflected according to the law of reflection. The dashed lines show that, to the observer, the rays appear to be coming from a point behind the mirror.

observer's eye. In the diagram, the observer has to look forward and slightly to the left to see the image of the candle. Their brain assumes that the image of the candle is in that direction, as shown by the dashed lines behind the mirror. (Our brains assume that light travels in straight lines, even though we know that light is reflected by mirrors.) The dashed lines appear to be coming from a point behind the mirror, at the same distance behind the mirror as the candle is in front of it. You can see this from the symmetry of the diagram.

The image looks as though it is the same size as the candle. Also, it is (of course) a mirror image, that is, it is left–right reversed. You will know this from seeing writing reflected in a mirror.

The image of the candle in the mirror is not a real image. A **real image** is an image that can be formed on a screen. If you place a piece of paper at the position of the image, you will not see a picture of the candle on it, because no rays of light from the candle reach that spot. That is why we drew dashed lines, to show where the rays appear to be coming from. We say that it is a **virtual image**.

To summarise, when an object is reflected in a plane mirror, its image is:

- the **same size** as the object
- the **same distance** behind the mirror as the object is in front of it
- **left–right reversed**
- **virtual**.

QUESTIONS

1 a Write the word AMBULANCE as it would appear when reflected in a plane mirror.
  b Why is it sometimes written in this way on the front of an ambulance?
2 a Draw a diagram to illustrate the law of reflection.
  b Which **two** angles are equal, according to the law?
3 A ray of light strikes a flat, reflective surface such that its angle of incidence is 30°. What angle does the reflected ray make with the surface?
4 What does it mean to say that a plane mirror produces a **virtual** image?

E

Ray diagrams

Figure **13.6b** on page **135** is an example of a **ray diagram**. Such diagrams are used to predict the position of images in mirrors, or when lenses or other optical devices are being used. The idea is first to draw the positions of things that are known (for example, the candle and the mirror). Then rays of light are drawn. These must be carefully chosen if they are to show up what we want to see. The position of the observer is marked, and then the rays are **extrapolated** back, to show where they appear to be coming from. These are the dashed lines shown in the diagram. This is known as a **construction**, and it allows us to mark the position of the image. Worked example **1** shows the steps in constructing a ray diagram.

E

Worked example 1

A small lamp is placed 5 cm in front of a plane mirror. Draw an accurate scale diagram, and use it to show that the image of the lamp is 5 cm behind the mirror.

The steps needed to draw the ray diagram are listed below and shown in Figure **13.7**. (It helps to work on squared paper or graph paper.)

1 Draw a line to represent the mirror, and indicate its reflecting surface, by drawing short lines on the back. Mark the position of the object O.
2 Draw two rays from O to the mirror. Where they strike the mirror, draw in the normal lines.
3 Using a protractor, measure the angle of incidence for each ray. Mark the equal angle of reflection.
4 Draw in the reflected rays, and extend them back behind the mirror. The point where they cross is where the image is formed. Label this point I.

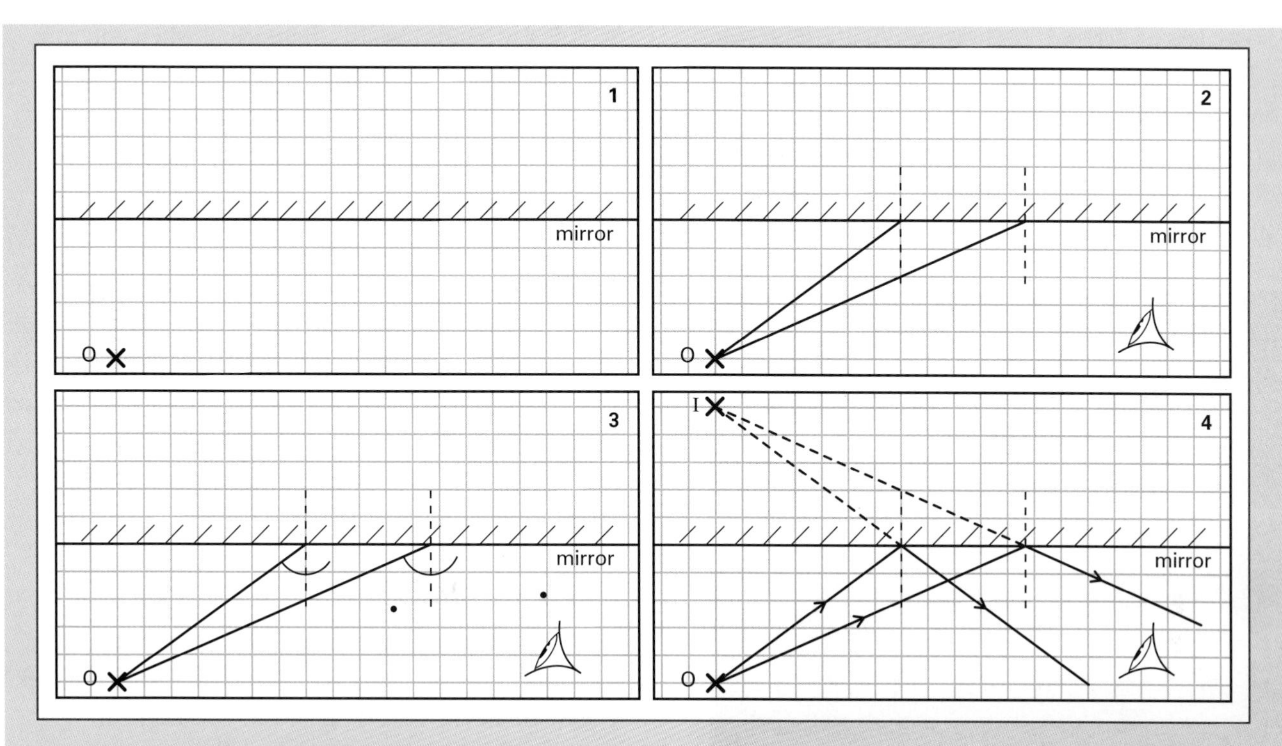

Figure 13.7 The steps in drawing a ray diagram for a plane mirror.

From the diagram for Step 4, it is clear that the image is 5 cm from the mirror, directly opposite the object. The line joining O to I is perpendicular to the mirror.

13.2 Refraction of light

If you look down at the bottom of a swimming pool, you may see patterns of shadowy ripples. The surface of the water is irregular. There are always small disturbances on the water, and these cause the rays of sunlight to change direction. Where the pattern is darker, rays of light have been deflected away, producing a sort of shadow. This bending of rays of light when they travel from one material to another is called **refraction**.

There are many effects caused by the refraction of light. Some examples are the sparkling of diamonds, the way the lens in your eye produces an image of the world around you, and the twinkling of the stars in the night sky. The 'broken stick' effect (Figure 13.8) is another consequence of refraction. The word 'refraction' is related to the word 'fractured', meaning broken.

Refraction occurs when a ray of light travels from one material into another. The ray of light may change direction. You can investigate this using a ray box and a

Figure 13.8 The pencil is partly immersed in water. Because of refraction of the light coming from the part of the pencil that is underwater, the pencil appears broken.

block of glass or Perspex, as shown in Figure **13.9**. Note that the ray travels in a straight line when it is in the air outside the block, and when it is inside the block. It only bends at the point where it enters or leaves the block, so it is the **change of material** that causes the bending.

Figure 13.9 Demonstrating the refraction of a ray of light when it passes through a rectangular block of glass or Perspex. The ray bends as it enters the block. As it leaves, it bends back to its original direction.

From Figure **13.9**, you will notice that the direction in which the ray bends depends on whether it is entering or leaving the glass.

- The ray bends towards the normal when entering the glass.
- The ray bends away from the normal when leaving the glass.

One consequence of this is that, when a ray passes through a parallel-sided block of glass or Perspex, it returns to its original direction of travel, although it is shifted to one side. When we look at the world through a window, we are looking through a parallel-sided sheet of glass. We do not see a distorted image because, although the rays of light are shifted slightly as they pass through the glass, they all reach us travelling in their original direction.

Changing direction

Figure **13.10a** shows the terms used for refraction. As with reflection, we define angles relative to the **normal**. The **incident ray** strikes the block. The **angle of incidence** *i* is measured from the ray to the normal. The **refracted ray** travels on at the **angle of refraction** *r*, measured relative to the normal. (Note that, when we discussed reflection, we used *r* for the angle of reflection; here it stands for the angle of refraction.)

A ray of light may strike a surface head-on, so that its angle of incidence is 0°, as shown in Figure **13.10b**. In this case, it does not bend – it simply passes

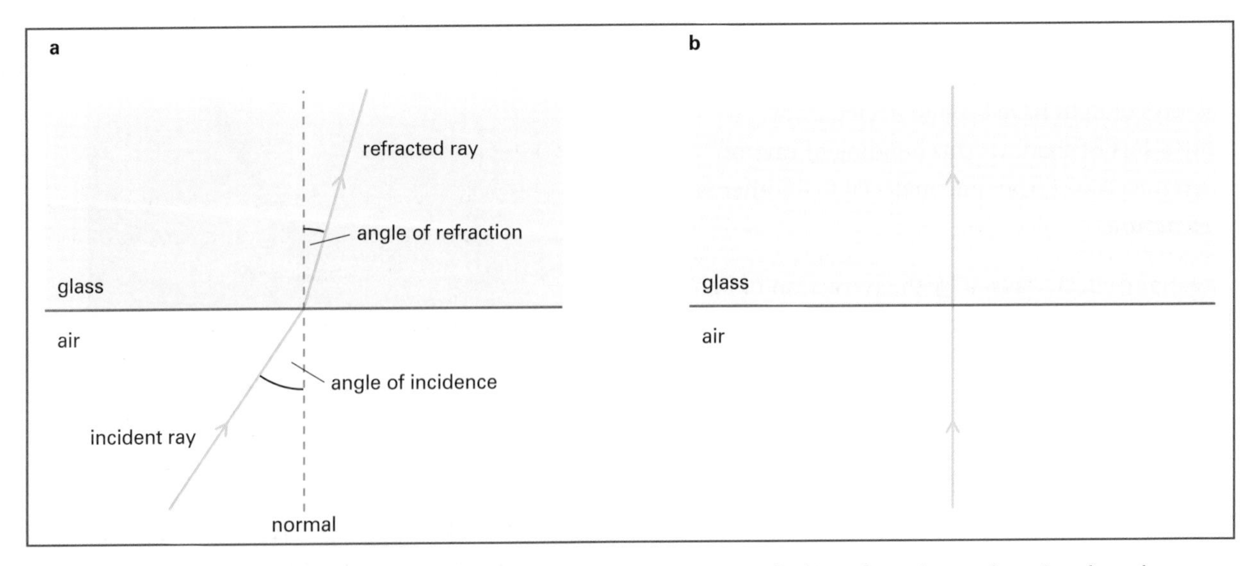

Figure 13.10 **a** Defining the terms used for refraction. The normal is drawn perpendicular to the surface at the point where the ray passes from one material to another. The angles of incidence and refraction are measured relative to the normal. **b** When a ray strikes the glass at 90°, it carries straight on without being deflected.

straight through and carries on in the same direction. Usually we say that refraction is the bending of light when it passes from one material to another. However, we should bear in mind that, when the light is perpendicular to the boundary between the two materials, there is no bending.

Explaining refraction

Why does light change direction when it passes from one material to another? The answer lies in the way its speed changes. Light travels fastest in a vacuum (empty space) and almost as fast in air. It travels more slowly in glass, water and other transparent substances.

One way to explain why a change in speed leads to a change in direction is shown in Figure 13.11. A truck is driving along a road across the desert. The driver is careless, and allows the wheels on the left to drift off the road onto the sand. Here, they spin around, so that the left-hand side of the truck moves more slowly. The right-hand side is still in contact with the road and keeps moving quickly, so that the truck starts to turn to the left.

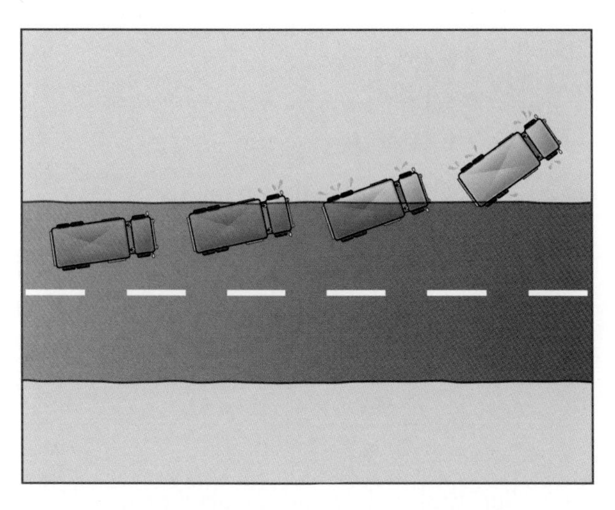

Figure 13.11 To explain why a change in speed explains the bending caused by refraction, we picture a truck whose wheels slip off the road into the sand. The truck veers to the side because it cannot move so quickly through sand.

The boundary between the two materials is the edge of the road. The normal is at right angles to the road. The truck has veered to the left, so its direction has moved towards the normal. Thus we would expect a ray of light to move towards the normal when it enters a material where it moves more slowly. This is indeed what we saw with glass (Figure 13.9). Light travels more slowly in glass than in air, so it bends towards the normal as it enters glass.

Activity 13.2 Investigating refraction

Use a ray box to investigate the refraction of light by a glass or plastic block.

QUESTIONS

5 Draw a diagram to show what we mean by the **angle of incidence** and the **angle of refraction** for a refracted ray of light.

6 A ray of light passes from air into a block of glass. Does it bend **towards** or **away from** the normal?

7 **a** Draw a diagram to show how a ray of light passes through a parallel-sided block of glass or Perspex.

b What can you say about its final direction of travel?

8 A vertical ray of light strikes the horizontal surface of some water.

a What is its angle of incidence?

b What is its angle of refraction?

9 When a ray of light passes from air to glass, is the angle of refraction greater than, or less than, the angle of incidence?

10 Why do we see a distorted view when we look through a window that is covered with raindrops?

E

Refractive index

Light travels very fast – as far as we know, nothing can travel any faster than light. The **speed of light** as it travels though empty space is exactly:

speed of light = 299 792 458 m/s

This fundamental quantity is given its own symbol, c. For most purposes we can round off the value to:

c = 300 000 000 m/s or 3×10^8 m/s

When a ray of light passes from air into glass, it slows down and bends towards the normal. The quantity that describes how much light is slowed down is the **refractive index**. If the speed of light is halved when it enters a material, the refractive index is 2, and so on. Hence we can write an equation for the refractive index n of a material:

$$\text{refractive index } n = \frac{\text{speed of light in a vacuum}}{\text{speed of light in the material}}$$

Water has a refractive index $n = 1.33$. This means that light travels 1.33 times as fast in a vacuum, compared to its speed in water.

Table **13.1** shows the speed of light in different materials. The third column shows the factor by which the light is slowed down – in other words, the refractive index of the material.

Material	Speed of light / m/s	$\frac{\text{speed in vacuum}}{\text{speed in material}}$
vacuum	2.998×10^8	1 exactly
air	2.997×10^8	1.0003
water	2.3×10^8	1.33
Perspex	2.0×10^8	1.5
glass	$(1.8–2.0) \times 10^8$	1.5–1.7
diamond	1.25×10^8	2.4

Table 13.1 The speed of light in some transparent materials. (The value for a vacuum is shown, for comparison.) Note that the values are only approximate.

Snell's law

There is a law that relates the size of the angle of refraction r to the angle of incidence i. This is **Snell's law**. It also involves the refractive index, since the greater the refractive index, the more a ray is bent. The law is written in the form of an equation:

$$n = \frac{\sin i}{\sin r}$$

Worked example **2** shows how to use this equation to find the angle through which a ray is refracted. The equation can also be used to find the value of the refractive index of a material: simply measure values of i and r and substitute them in the equation.

Worked example 2

A ray of light strikes a glass block with an angle of incidence of 45°. The refractive index of the glass is 1.6. What will be the angle of refraction?

The situation is shown in Figure **13.12**.

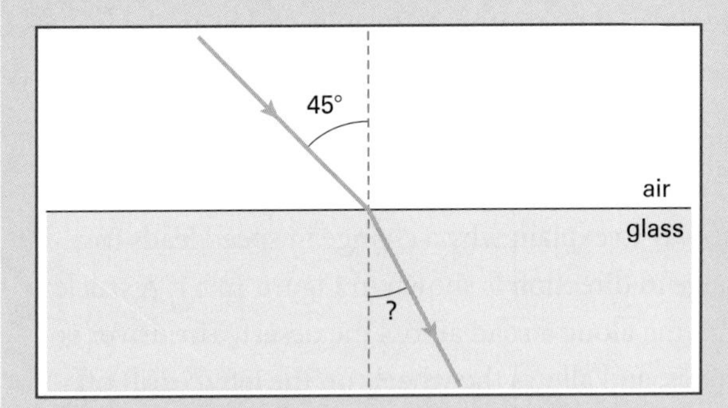

Figure 13.12 See Worked example **2**, on Snell's law.

Step 1: Write down what you know and what you want to know.

$$i = 45°$$
$$n = 1.6$$
$$r = ?$$

Step 2: Write down the equation for Snell's law. Since we want to know r, rearrange it to make $\sin r$ the subject.

$$n = \frac{\sin i}{\sin r}$$

$$\sin r = \frac{\sin i}{n}$$

Step 3: Substitute values and calculate $\sin r$.

$$\sin r = \frac{\sin 45°}{1.6} = 0.442$$

Step 4: Use the $\sin^{-1}$ function on your calculator to find r. (This will tell you the angle whose sine is 0.442.)

$$r = \sin^{-1} 0.442 = 26.2°$$

You can see that Snell's law correctly predicts that the ray will be deflected towards the normal.

QUESTIONS

In these questions you will need to use the fact that the speed of light in a vacuum is 3.0×10^8 m/s.

11 Look back at Table **13.1**. What is the value of the refractive index of diamond?

12 Figure **13.13** shows what happens when a ray of light enters blocks of two different materials, A and B.

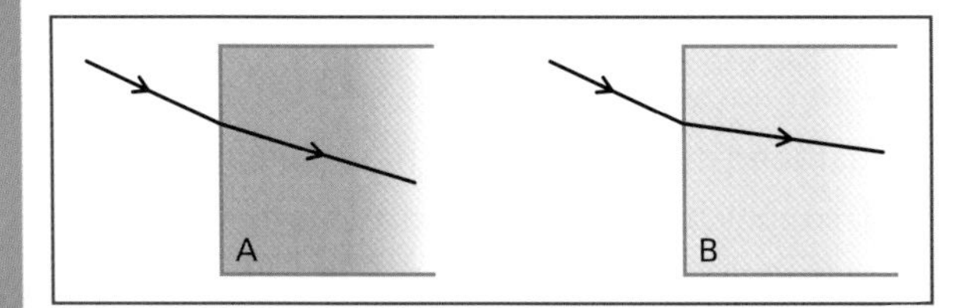

Figure 13.13 For Question **13**.

a In which material does the light travel more slowly, A or B? Explain how you can tell from the diagrams.

b Which material, A or B, has the greater refractive index?

13 Light travels more quickly through water than through glass.

a Which has the greater refractive index, water or glass?

b If a ray passes from glass into water, which way will it bend: towards or away from the normal?

14 The speed of light in a block of glass is found to be 1.9×10^8 m/s. Calculate the refractive index of the glass.

15 A solution of sugar in water is found to have a refractive index of 1.38. Calculate the speed of light in the solution.

16 Perspex is a form of transparent plastic. It has a refractive index $n = 1.5$. A ray of light strikes the flat surface of a Perspex block with an angle of incidence of 40°. What will be the angle of refraction?

13.3 Total internal reflection

If you have carried out a careful investigation of refraction using a ray box and a transparent block, you may have noticed something extra that happens when a ray strikes a block. A reflected ray also appears, in addition to the ray that is refracted. You can see this in Figure **13.9**, but it was ignored in Figure **13.10**. When the ray strikes the block, some of the light passes into the block and is refracted, and some is reflected. When it leaves the block, again some leaves the block and is refracted, and some is reflected. These reflected rays obey the law of reflection:

angle of incidence = angle of reflection

These reflected rays can be a nuisance. If you try to look downwards into a pond or river to see if there are any fish there, your view may be spoilt by light reflected from the surface of the water. You see a reflected image of the sky, or of yourself, rather than what is in the water. On a sunny day, reflected light from windows or water can be a hazard to drivers.

To see how we can make use of reflected rays, you can use the apparatus shown in Figure **13.14**. A ray box shines a ray of light at a semicircular glass block. The ray is always directed at the curved edge of the block, along the radius. This means that it enters the block along the normal, so that it is not bent by refraction. Inside the glass, the ray strikes the midpoint of the flat side, which we call point X.

Figure 13.14 Using a ray box to investigate reflection when a ray of light strikes a glass or Perspex block. The ray enters the block without bending, because it is directed along the radius of the block.

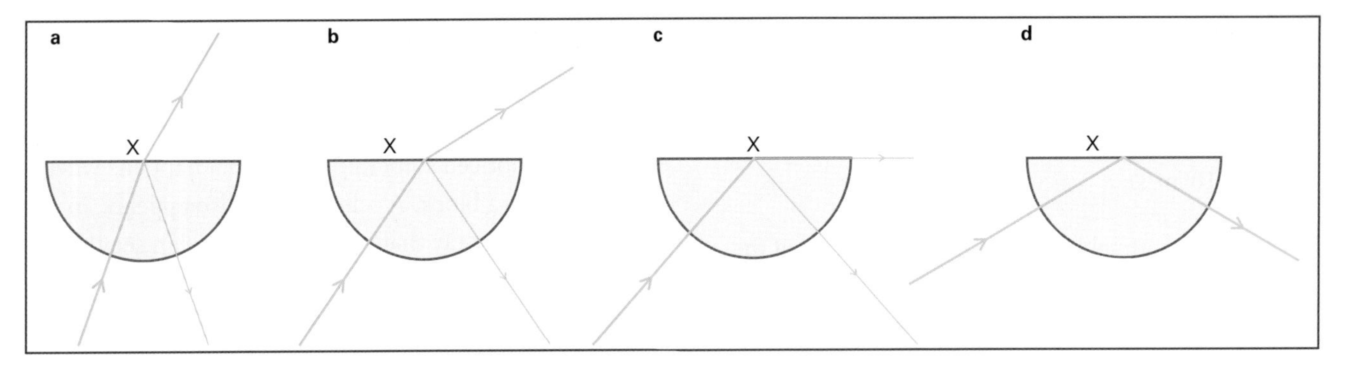

Figure 13.15 How a ray of light is reflected or refracted inside a glass block depends on the angle of incidence. **a, b** For angles less than a certain angle, called the critical angle, some of the light is reflected and some is refracted. **c** At the critical angle, the angle of refraction is 90°. **d** At angles of incidence greater than the critical angle, the light is totally internally reflected – there is no refracted ray.

What happens next? This depends on the angle of incidence of the ray at point X. The various possibilities are listed below and shown above in Figure **13.15**.

a If the angle of incidence is small, most of the light emerges from the block. There is a faint reflected ray inside the glass block. The refracted ray bends away from the normal.

b If the angle of incidence is increased, more light is reflected inside the block. The refracted ray bends even further away from the normal.

c Eventually, at one particular angle, the refracted ray emerges along and parallel to the surface of the block. Most of the light is reflected inside the block.

d Now, at an even greater angle of incidence, all of the light is reflected inside the block. No refracted ray emerges from point X.

We have been looking at how light is reflected inside a glass block. We have seen that, if the angle of incidence is greater than a particular value, known as the **critical angle**, the light is entirely reflected inside the glass. This phenomenon is known as **total internal reflection** (**TIR**):

- **total**, because 100% of the light is reflected
- **internal**, because it happens inside the glass
- **reflection**, because the ray is entirely reflected.

For total internal reflection to happen, the angle of incidence of the ray must be greater than the critical angle. The critical angle depends on the material being used. For glass, it is about 42° (though this depends on the composition of the glass). For water, the critical angle is greater, about 49°. For diamond, the critical angle is small, about 25°. Hence rays of light that enter a diamond are very likely to be totally internally reflected, so they bounce around inside, eventually emerging from one of the diamond's cut faces. That explains why diamonds are such sparkly jewels.

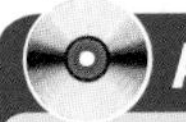

Activity 13.3 Total internal reflection

Use a ray box and a semicircular block to observe total internal reflection.

QUESTIONS

17 Explain the meaning of the words **total** and **internal** in the expression 'total internal reflection'.

18 The critical angle for water is 49°. If a ray of light strikes the upper surface of a pond at an angle of incidence of 45°, will it be totally internally reflected? Explain your answer.

E

Optical fibres

A revolution in telecommunications has been made possible by the invention of fibre optics. Telephone messages and other electronic signals such as Internet computer messages or cable TV signals are passed along fine glass fibres in the form of flashing laser light – a digital signal. Figure **13.16a** shows just how fine these fibres can be. Each of these fibres is capable of carrying thousands of telephone calls simultaneously.

Inside a fibre, light travels along by total internal reflection (see Figure **13.16b**). It bounces along inside the fibre because, each time it strikes the inside of the fibre, its angle of incidence is greater than the critical angle. Thus no light is lost as it is reflected. The fibre can follow a curved path and the light bounces along inside it, following the curve. For signals to travel over long distances, the glass used must be of a very high purity, so that it does not absorb the light.

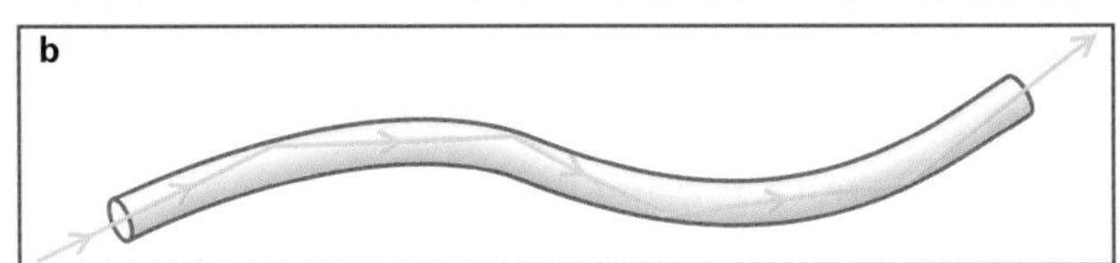

Figure 13.16 The use of fibre optics has greatly increased the capacity and speed of the world's telecommunications networks. Without this technology, cable television and the Internet would not be possible. **a** Each of these very fine fibres of high-purity glass can carry many telephone messages simultaneously. **b** Light travels along a fibre by total internal reflection. Because the reflection is total, and the glass is so pure, the light can travel many kilometres along a single fibre.

Optical fibres are also used in medicine. An endoscope is a device that can be used by doctors to see inside a patient's body – for example, to see inside the stomach. One bundle of fibres carries light down into the body (it is dark in there), while another bundle carries an image back up to the user. The endoscope may also have a small probe or cutting tool built in, so that minor operations can be performed without the need for major surgery.

QUESTIONS

19 Sketch a diagram to show how a ray of light can travel along a curved glass fibre. Indicate the points where total internal reflection occurs.

20 Why must high-purity glass be used for optical fibres used in telecommunications?

13.4 Lenses

We are all familiar with lenses in everyday life – in spectacles and cameras, for example. The development of high-quality lenses has had a profound effect on science. In 1609, using the newly invented telescope, Galileo discovered the moons of Jupiter and triggered a revolution in astronomy. In those days, scientists had to grind their own lenses starting from blocks of glass, and Galileo's skill at this was a major factor in his discovery.

Later in the 17th century, a Dutch merchant called Anton van Leeuwenhoek managed to make microscope lenses that gave a magnification of 200 times. He used these to look at the natural world around him. He was amazed to find a wealth of tiny microorganisms, including bacteria, that were invisible to the naked eye (Figure **13.17**). This provided the clue to how infectious diseases might be spread. Previously people thought infections were carried by smells or by mysterious vapours. A revolution in medicine had begun.

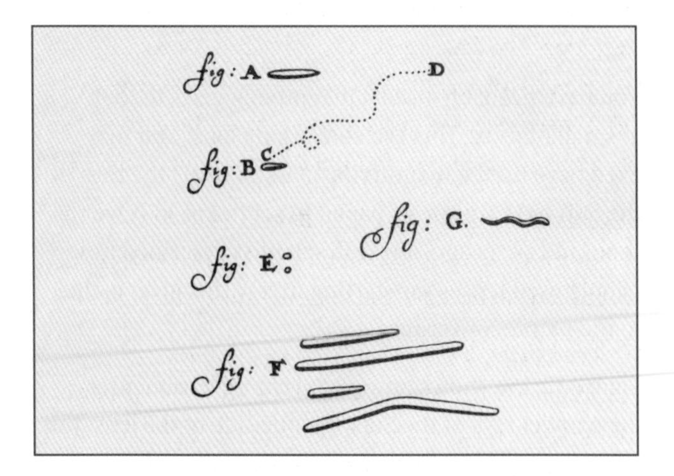

Figure 13.17 Bacteria cannot be seen with the naked eye. These drawings were made by Anton van Leeuwenhoek in 1683 using an early microscope. They show bacteria he obtained by scraping material from between his teeth.

Converging and diverging lenses

Lenses can be divided into two types, according to their shape (Figure **13.18**):

- **converging lenses** are fatter in the middle than at the edges
- **diverging lenses** are thinner in the middle than at the edges.

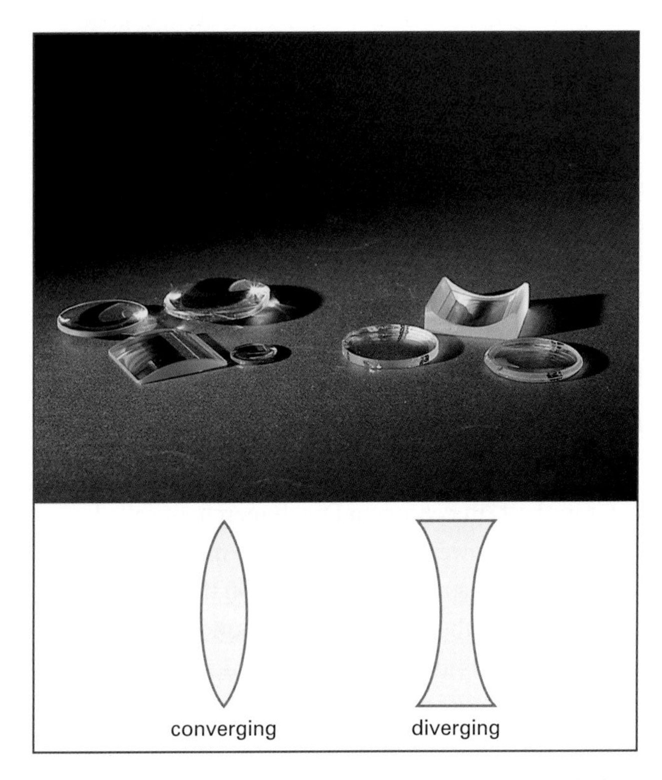

Figure 13.18 The lenses on the left are converging lenses, which are fattest at the middle. On the right are diverging lenses, which are thinnest at the middle. They are given these names because of their effect on parallel rays of light. Usually we simply draw the cross-section of the lens, to indicate which type we are considering.

You have probably used a magnifying glass to look at small objects. This is a converging lens. You may even have used a magnifying glass to focus the rays of the Sun onto a piece of paper, to set fire to it. (Over a thousand years ago, an Arab scientist described how people used lenses for starting fires.) This gives a clue to the name 'converging'.

Figure **13.19a** shows how a converging lens focuses the parallel rays of the Sun. On one side of the lens, the rays are parallel to the **axis** of the lens. After they pass through the lens, they converge on a single point, the **principal focus** or **focal point**. After they have passed through the principal focus, they spread out again.

So a converging lens is so-called because it makes parallel rays of light converge. The principal focus is the point where the rays are concentrated together, and where a piece of paper needs to be placed if it is to be burned. The distance from the centre of the lens to the principal focus is called the **focal length** of the lens. The fatter the lens, the closer the principal focus is to the lens. A fat lens has a shorter focal length than a thin lens.

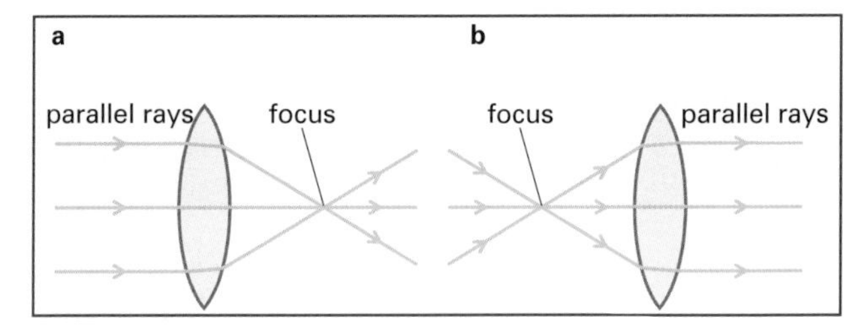

Figure 13.19 The effect of a converging lens on rays of light. **a** A converging lens makes parallel rays converge at the principal focus. **b** Rays from the principal focus of a converging lens are turned into a parallel beam of light.

A converging lens can be used 'in reverse' to produce a beam of parallel rays. A source of light, such as a small light bulb, is placed at the principal focus. As they pass through the lens, the rays are bent so that they become a parallel beam (Figure **13.19b**). This diagram is the same as Figure **13.19a**, but in reverse.

Lenses work by refracting light. When a ray strikes the surface of the lens, it is refracted towards the normal. When it leaves the glass of the lens, it bends away from the normal. The clever thing about the shape of a converging lens is that it bends all rays just enough for them to meet at the principal focus.

Forming a real image

When the Sun's rays are focused onto a piece of paper, a tiny image of the Sun is created. It is easier to see how a converging lens makes an image by focusing an image of a light bulb or a distant window onto a piece of white paper. The paper acts as a screen to catch the image. Figure **13.20** shows an experiment in which an image of a light bulb (the object) is formed by a converging lens.

Figure 13.20 Forming a real image of a light bulb using a converging lens. The image is upside down on the screen at the back right.

There are some things to note. In this experiment, the image is:

- **inverted** (upside down)
- **reduced** (smaller than the object)
- **nearer to** the lens than the object
- **real.**

We say that the image is real, because light really does fall on the screen to make the image. If light only appeared to be coming from the image, we would say that the image was virtual. The size of the image depends on how fat or thin the lens is.

We can explain the formation of this real image using a **ray diagram**. The steps needed to draw an accurate ray diagram are listed below and shown in Figure **13.21**. (It helps to work on squared paper or graph paper.)

1 Draw the lens – a simple outline shape will do – with a horizontal axis through the middle of it.
2 Mark the positions of the principal focuses F on either side, at equal distances from the lens. Mark the position of the object O, an arrow standing on the axis.
3 Draw ray 1, a straight line from the top of the arrow and passing undeflected through the middle of the lens.
4 Draw ray 2, from the top of the arrow parallel to the axis. As it passes through the lens, it is deflected down through the principal focus. Look for the point where the two rays cross. This is the position of the top of the image I.

With an accurately drawn ray diagram, you can see that the image is inverted, reduced and real. Note that we do not bother to draw ray 2 bending twice, at the two surfaces of the lens. It is easier to show it bending once, in the middle of the lens, though this is not a correct representation of what really happens.

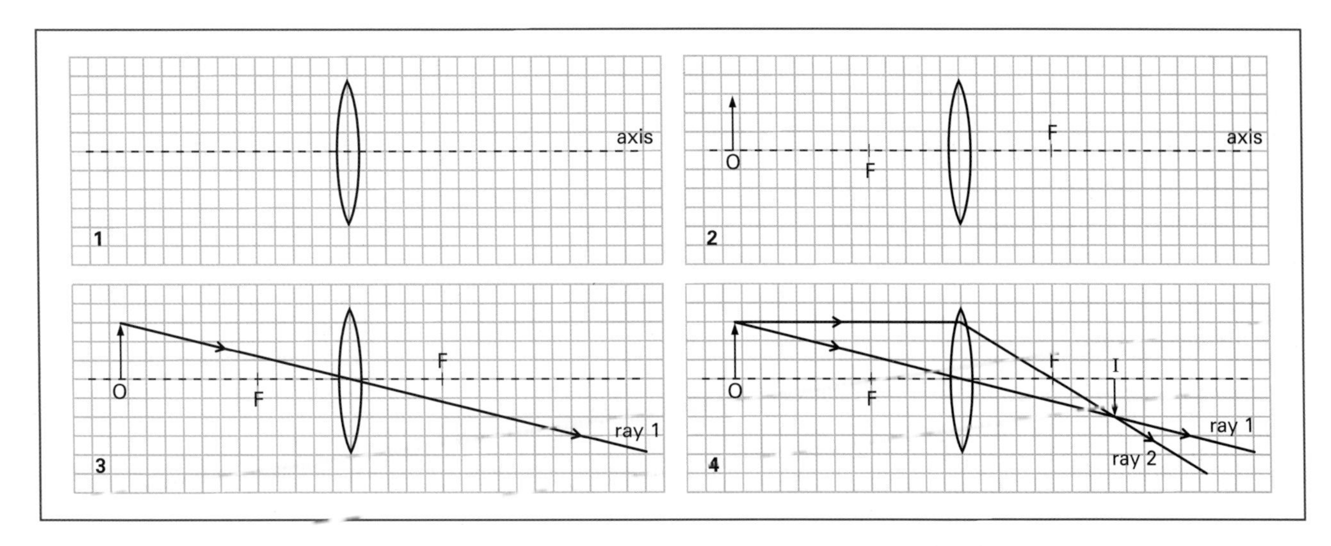

Figure 13.21 A ray diagram can be used to show how an image is formed by a converging lens. The steps are given in the text.

So, to construct a ray diagram like this, draw two rays starting from the top of the object:

- ray 1, undeflected through the centre of the lens
- ray 2, parallel to the axis and then deflected through the principal focus.

Activity 13.4 Investigating converging lenses

Measure the focal length of a lens and draw an accurate ray diagram.

QUESTIONS

21 Draw a diagram to show the difference in shape between a converging lens and a diverging lens.

22 Draw a ray diagram to show how a converging lens focuses parallel rays of light.

23 How would you alter your diagram in question **22** to show how a converging lens can produce a beam of parallel rays of light?

24 What is meant by the principal focus (or focal point) of a converging lens?

25 What is the difference between a real image and a virtual image?

26 Look at the ray diagram shown in Figure **13.21**. How does it show that the image formed by a converging lens is inverted?

Magnifying glasses

A magnifying glass is a converging lens. You hold it close to a small object and peer through it to see a magnified image. Figure **13.22** shows how a magnifying glass can help to magnify print for someone with poor eyesight.

The object viewed by a magnifying glass is closer to the lens than the principal focus. This allows us to draw the ray diagram shown in Figure **13.23**. In the same way as in Figure **13.21**, we draw two rays from the top of the object O, rays 1 and 2:

- ray 1 is undeflected, as it passes through the centre of the lens

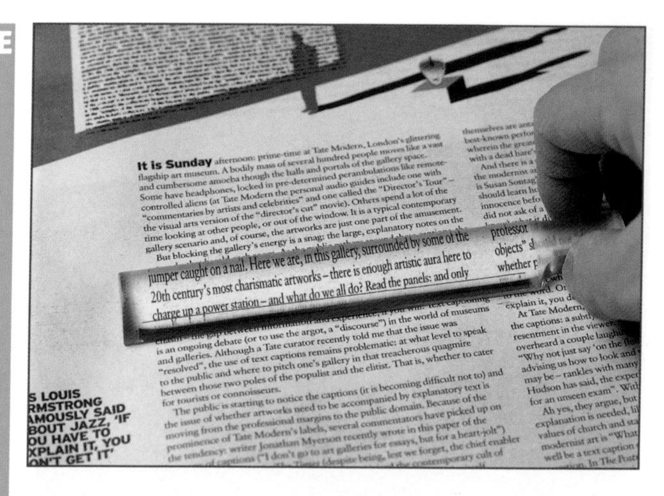

Figure 13.22 This long converging lens is designed to help people to read. It produces a magnified image of a line of print. The user simply slides it down the page.

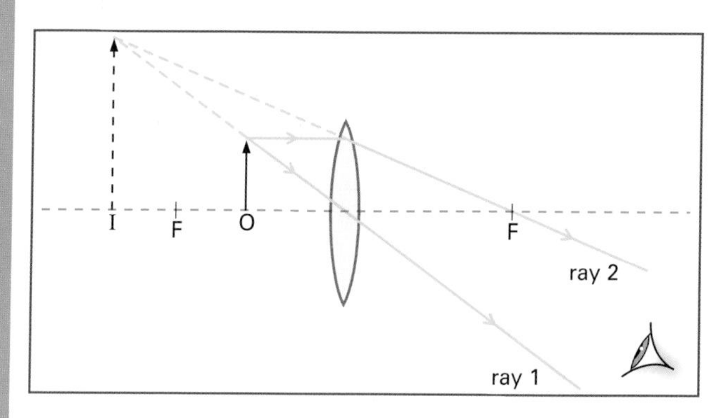

Figure 13.23 A ray diagram to show how a magnifying glass works. The object is between the lens and the focus. The image produced is virtual. To find its position, the rays have to be extended back (dashed lines) to the point where they cross.

- ray 2 starts off parallel to the axis and is deflected by the lens so that it passes through the principal focus.

Rays 1 and 2 do not cross over each other. They are diverging (spreading apart) after they have passed through the lens. However, by extending the rays backwards, as shown by the dashed lines, we can see that they both appear to be coming from a point behind the object. This is the position of the image I. We draw dashed lines because light does not actually travel along these parts of the rays. This tells us that the image formed is virtual. We cannot catch the image on a screen, because there is no light there.

From the ray diagram (Figure **13.23**), we can see the following features of the image produced by a magnifying glass. The image is:

- **upright** (the right way up, not inverted)
- **magnified** (bigger than the object)
- **further from** the lens than the object
- **virtual** (not real).

So, if you read a page of a book using a magnifying glass, the image you are looking at is behind the page that you are reading.

QUESTIONS

27 Look at Figure 13.23. How can you tell from the diagram that the object formed by the magnifying glass is a virtual image?

28 **a** A converging lens has focal length 5 cm. An object is placed 3 cm from the centre of the lens, on the principal axis. Draw an accurate ray diagram to represent this.

b Use your diagram to determine the distance of the virtual image formed from the lens.

Summary

The law of reflection:

angle of incidence = angle of reflection

$$i = r$$

Angles are measured relative to the normal to the surface.

The image formed by a plane mirror is the same size as the object, as far behind the mirror as the object is in front of it, left–right reversed, and virtual.

A light ray changes direction when it meets the boundary between two different materials (unless it meets the boundary at right angles). This bending is known as refraction.

Refractive index:

$$n = \frac{\text{speed of light in a vacuum}}{\text{speed of light in the material}}$$

Snell's law:

$$n = \frac{\sin i}{\sin r}$$

A ray is totally internally reflected when it strikes a boundary at an angle greater than the critical angle.

Total internal reflection is used in optical fibres for telecommunications.

A converging lens makes rays of light parallel to the axis converge at the principal focus.

A magnifying glass is a converging lens, with the object closer than the principal focus. It produces a magnified, virtual image.

End-of-chapter questions

13.1 The law of reflection says that: 'When a ray of light is reflected at a surface, the **angle of incidence** is equal to the **angle of reflection**.'

Draw a diagram to indicate how a ray of light is reflected by a flat mirror, and mark the **two** angles mentioned in the law. [4]

13.2 Windows usually have a flat sheet of glass, so that we can see clearly through them. Frosted glass has an irregular surface, so that we do not see a clear image through it.

a Draw a ray diagram to show how a ray of light passes through a parallel-sided glass block if it hits the glass at 90° (that is, perpendicular to the glass). [2]

b Draw a ray diagram to show how a ray of light passes through a parallel-sided glass block if it hits the glass at an angle other than 90° (that is, obliquely to the glass). [3]

c Explain why we can see clearly through a flat sheet of glass, even though light is refracted as it passes through. [1]

13.3 Figure **13.24** shows two blocks of a material whose critical angle is 40°. In block A, the ray strikes the inner surface with an angle of incidence of 30°. In block B, the ray's angle of incidence is 50°.

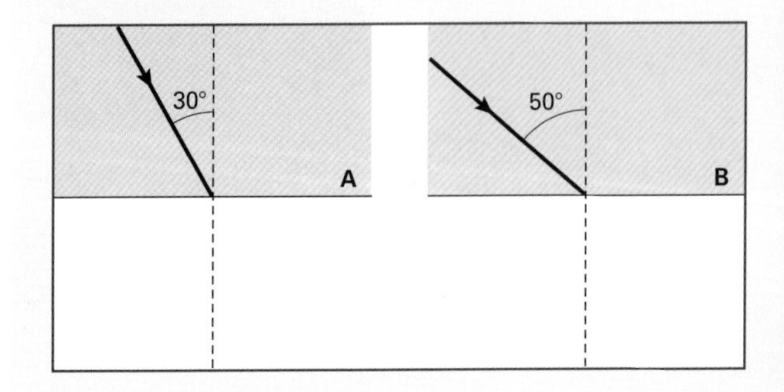

Figure 13.24 For Question **13.3**.

a Copy and complete each diagram to show what happens when the ray strikes the surface. [4]

b Use the diagrams to explain what is meant by **total internal reflection**. [3]

E

13.4 A small lamp is placed at a distance of 4 cm from a plane mirror.

a Draw an accurate ray diagram to show where the image of the lamp in the mirror is formed. [4]

b Explain how you have used the law of reflection in drawing your diagram. [2]

c What does it means to say that the image of the lamp is a **virtual** image? [2]

13.5 Figure **13.25** shows an incomplete ray diagram, which represents the following situation.

A converging lens has a focal length of 4 cm. Its principal focuses are marked F. An object O is placed at a distance of 10 cm from the lens. Ray 1 passes through the centre of the lens. Ray 2 is parallel to the axis of the lens.

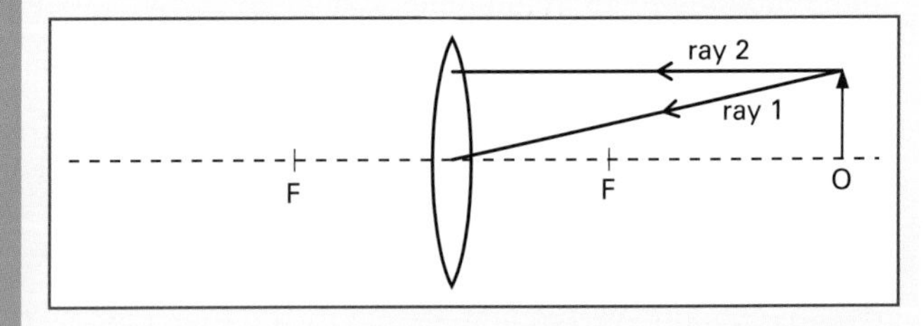

Figure 13.25 For Question **13.5**.

a Copy and complete the ray diagram, on squared paper or graph paper, to find the position of the image formed by the lens. [4]

b Explain whether the diagram shows that the image is real or virtual. [2]

c Explain whether the diagram shows that the image is magnified or diminished (smaller than the object). [2]

d Explain whether the diagram shows that the image is upright or inverted. [1]

14 Properties of waves

Core	Describing transverse and longitudinal waves
Core	Explaining speed, amplitude, frequency and wavelength
E Extension	Calculating wave speed
Core	Describing reflection, refraction and diffraction of waves
E Extension	Explaining reflection, refraction and diffraction of waves

All at sea!

It cannot be much fun to be adrift in a small boat on a rough sea, being tossed up and down. For some birds, this is a regular experience. Many seabirds spend the whole winter on the open sea, at a time when the sea is at its roughest (Figure **14.1**). The waves may be 20 m high, enough to dwarf a two-storey house, but the birds feel safer here than they would on the cliffs, where they nest in the spring. Guillemots, for example, cluster together in 'rafts', carried up and down by the waves. It is this up-and-down motion that is liable to make *you* feel sea-sick if you are on board a ship in stormy weather.

When waves reach the beach, they start to break. The bottom of the wave drags on the seabed and slows down. The top of the wave carries on and gradually tips over to form a breaker. Breaking waves like this are the natural home of the surfer (Figure **14.2**).

Physicists talk about light waves, sound waves, electromagnetic waves, and so on. The idea of a wave is a very useful model in physics. It is not obvious that light and sound are similar to waves on the sea. In this chapter we will see how water waves *can* act as a good model for both light and sound. The water waves that we will be thinking of are more like those on the open sea than breakers on a beach.

Figure 14.1 Many seabirds such as guillemots spend the whole of the winter on the open ocean. They gather together in 'rafts' and spend their days and nights riding up and down on the waves.

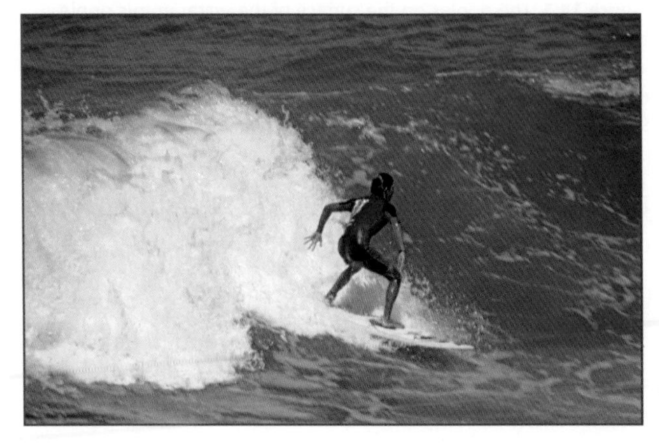

Figure 14.2 Surfers look out for waves that are beginning to break. The top of the wave is tipping over, and this provides the push they need to start them moving along with the crest of the wave.

14.1 Describing waves

Physicists use waves as a **model** to explain the behaviour of light, sound and other phenomena. Waves are what we see on the sea or a lake, but physicists have a more specialised idea of waves. We can begin to understand this model in the laboratory using a **ripple tank** (Figure 14.3). A ripple tank is a shallow glass-bottomed tank containing a small amount of water. A light shining downwards through the water casts a shadow of the **ripples** on the floor below, showing up the pattern that they make.

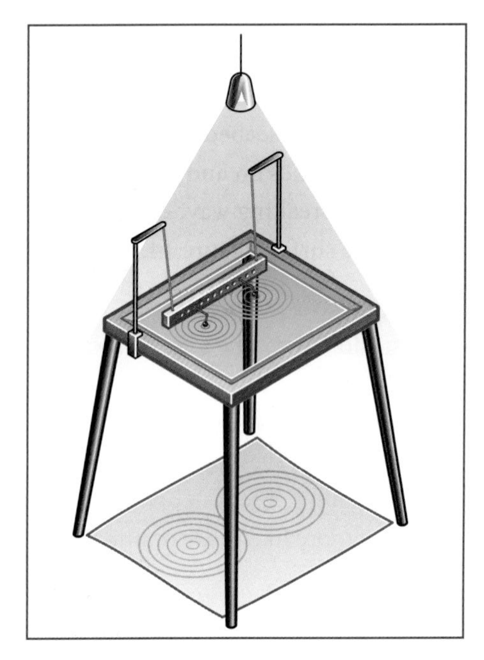

Figure 14.3 The ripples on the surface of the water in this ripple tank are produced by the spherical dippers attached to the bar, which vibrates up and down. The pattern of the ripples is seen easily by shining a light downwards through the water. This casts a shadow of the ripples on the floor beneath the tank.

Figure 14.4 shows two patterns of ripples, straight and circular, which are produced in different ways.

a One way of making ripples on the surface of the water in a ripple tank is to have a wooden bar that just touches the surface of the water. The bar vibrates up and down at a steady rate. This sends equally spaced straight ripples across the surface of the water.

b A spherical dipper can produce a different pattern of ripples. The dipper just touches the surface of the water. As it vibrates up and down, equally spaced circular ripples spread out across the surface of the water.

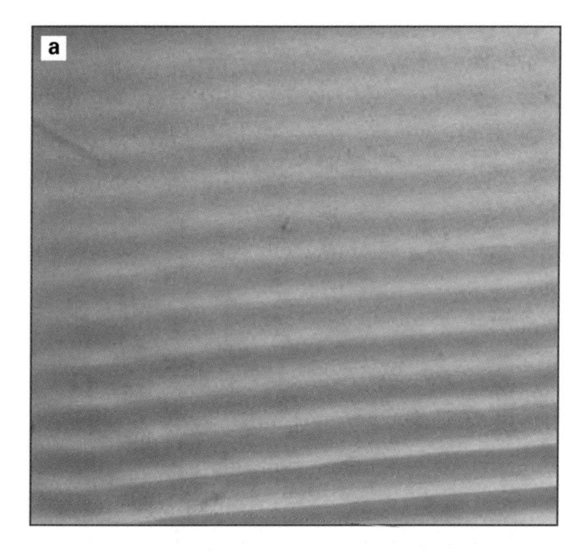

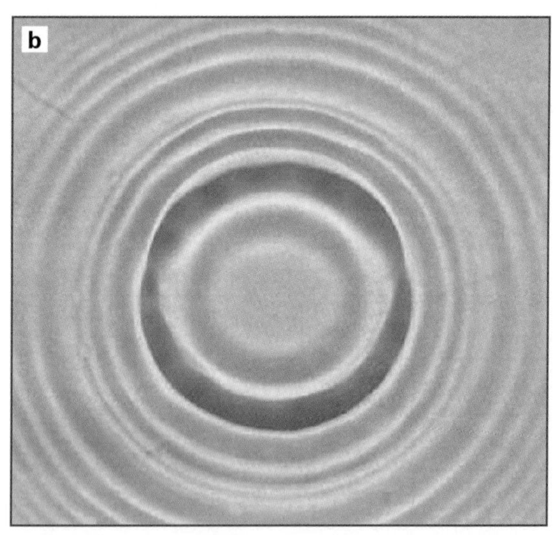

Figure 14.4 Two patterns of ripples on water. **a** Straight ripples are a model for a broad beam of light. **b** Circular ripples are a model for light spreading out from a lamp.

In each case, the ripples are produced by something vibrating up and down **vertically**, but the ripples move out **horizontally**. The vibrating bar or dipper pushes water molecules up and down. Each molecule drags its neighbours up and down. These then start their neighbours moving, and so on. This may make you think of the seabirds we discussed, floating on the rough sea. The waves go past the birds. The birds simply float up and down on the surface of the water.

How can these patterns of ripples be a model for the behaviour of light? The straight ripples are like a beam of light, perhaps coming from the Sun. The ripples move straight across the surface of the water, just as light from the Sun travels in straight lines. The circular ripples spreading out from a vibrating dipper are like light spreading out from a lamp. (The dipper is the lamp.) Throughout this chapter, we will gradually build up the idea of how ripples on the surface of water can be a model for the behaviour of light, other electromagnetic waves and sound.

Wavelength and amplitude

A more familiar way of representing a wave is as a wavy line, as shown in Figure **14.5**. We have already used this idea for sound waves (in Chapter **12**) and we will do so again for electromagnetic waves (in Chapter **15**). This wavy line is like a downward slice though the ripples in the ripple tank. It shows up the succession of **crests** and **troughs** of which the ripples are made.

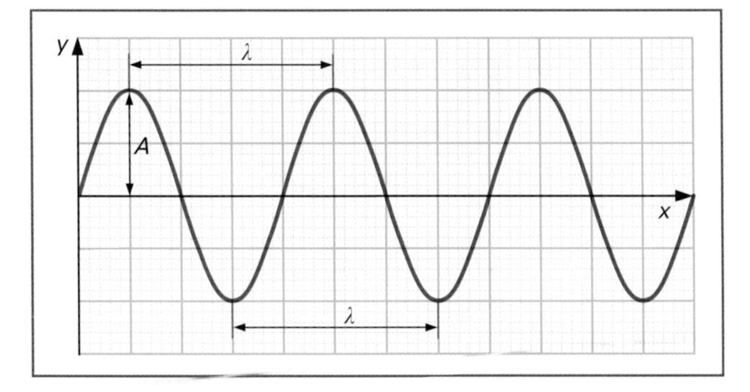

Figure 14.5 Representing a wave as a smoothly varying wavy line. This shape is known as a sine graph. If you have a graphics calculator, you can use it to display a graph of $y = \sin x$, which will look like this graph.

The graph in Figure **14.5** shows a wave travelling from left to right. The horizontal axis (x-axis) shows the distance x travelled horizontally by the wave. The vertical axis (y-axis) shows how far (distance y) the surface of the water has been displaced from its normal level. Hence we can think of the x-axis as the level of the surface of the water when it is undisturbed. The line of the graph shows how far the surface of the water has been displaced from its undisturbed level.

From the representation of the wave in Figure **14.5**, we can define two quantities for waves in general:

- The **wavelength** λ of a wave is the distance from one crest of the wave to the next (or from one trough to the next). Since the wavelength is a distance, it is measured in metres, m. Its symbol is λ, the Greek letter 'lambda'.
- The **amplitude** A of a wave is the maximum distance that the surface of the water is displaced from its undisturbed level – in other words, the height of a crest. For ripples on the surface of water, the amplitude is a distance, measured in metres, m. Its symbol is A.

Note that the amplitude is measured from the undisturbed level up to the crest. It is not measured from trough to crest. For ripples in a ripple tank, the wavelength might be a few millimetres and the amplitude a millimetre or two. Waves on the open sea are much bigger, with wavelengths of tens of metres, and amplitudes varying from a few centimetres up to several metres.

Frequency and period

As the bar in the ripple tank vibrates, it sends out ripples. Each up-and-down movement sends out a single ripple. The more times the bar vibrates each second, the more ripples it sends out. This is shown in the graph of Figure **14.6**. Take care! This looks very similar to the previous wave graph (Figure **14.5**), but here the horizontal axis shows time t, not distance x. This graph shows how the surface of the water at a particular point moves up and down as time passes.

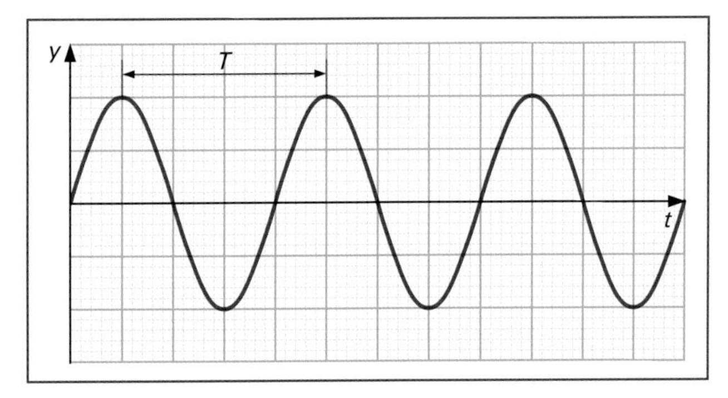

Figure 14.6 A graph to show the period of a wave. Notice that this graph has time t on its horizontal axis.

From the representation of the wave in Figure **14.6**, we can define two quantities for waves in general:

- The **frequency** f of a wave is the number of waves sent out each second. Frequency is measured in hertz, Hz. One hertz (1 Hz) is one complete wave or ripple per second.
- The **period** T of a wave is the time taken for one complete wave to pass a point. The period is measured in seconds, s.

We have already discussed the frequency and period of a sound wave in Chapter **12**. It is important always to check whether a wave graph has time t or distance x on its horizontal axis.

The frequency of a wave is the number of waves sent out or passing a point per second. Its period is the number of seconds for each wave to pass a point. Hence frequency f and period T are obviously related to each other. Waves with a short period have a high frequency.

$$\text{frequency (Hz)} = \frac{1}{\text{period (s)}} \qquad f = \frac{1}{T}$$

$$\text{period (s)} = \frac{1}{\text{frequency (Hz)}} \qquad T = \frac{1}{f}$$

Waves on the sea might have a period of 10 s. Their frequency is therefore about 0.1 Hz. A sound wave might have a frequency of 1 000 Hz. Its period is therefore 1/1 000 s, which means that a wave arrives every 1 ms (one millisecond).

Wave speed

The **wave speed** is the rate at which the crest of a wave travels along. For example, it could be the speed of the crest of a ripple travelling over the surface of the water. Speed is measured in metres per second (m/s).

Waves can have very different speeds. Ripples in a ripple tank travel a few centimetres per second. Sound waves travel at 330 m/s through air. Light waves travel at 300 000 000 m/s through air.

Transverse and longitudinal waves

Ripples in a ripple tank are one way of looking at the behaviour of waves. You can demonstrate waves in other ways. As shown in Figure **14.7a**, a stretched 'slinky' spring can show waves. Fix one end of the spring and move the other end from side to side. You will see that a wave travels along the spring. (You may also notice it reflecting from the fixed end of the spring.) You can demonstrate the same sort of wave using a stretched rope or piece of elastic.

A second type of wave can also be demonstrated with a stretched 'slinky' spring. Instead of moving the free

Figure 14.7 Waves along a stretched spring. **a** A transverse wave, made by moving the free end from side to side. **b** A longitudinal wave, made by pushing the free end back and forth, along the length of the spring.

end from side to side, move it back and forth (Figure **14.7b**). A series of compressions travels along the spring, regions in which the segments of the spring are compressed together. In between are rarefactions, regions where the segments of the spring are further apart. This type of wave cannot be demonstrated on a stretched rope.

These demonstrations in Figure **14.7** show two different types of wave:

- **transverse waves**, in which the particles carrying the wave move from side to side, at right angles to the direction in which the wave is moving
- **longitudinal waves**, in which the particles carrying the wave move back and forth, along the direction in which the wave is moving.

A ripple on the surface of water is an example of a transverse wave. The particles of the water move up and down as the wave travels horizontally.

A sound wave is an example of a longitudinal wave. As a sound travels through air, the air molecules move back and forth as the wave travels. Compare Figure **14.7b** with Figure **12.11** on page **130** to see the similarity. Table **14.1** lists examples of transverse and longitudinal waves.

Transverse waves	Longitudinal waves
ripples on water	sound
light and all other electromagnetic waves	

Table 14.1 Transverse and longitudinal waves.

Activity 14.1 Observing waves

Carry out some experiments to observe transverse and longitudinal waves.

QUESTIONS

1 Describe the motion of molecules of water as a ripple moves across the surface of water in a ripple tank.

2 The two graphs shown in Figures **14.5** and **14.6** are very similar to each other. What is the important difference between them?

3 Draw a diagram to show what is meant by the amplitude of a wave.

4 How could you find the wavelength of the ripples shown in Figure **14.4**?

5 If 10 waves occupy 15 cm, what is their wavelength?

6 **a** If 100 sound waves reach your ear each second, what is their frequency?

b What is their period?

7 Are sound waves transverse or longitudinal?

E

14.2 Speed, frequency and wavelength

How fast do waves travel across the surface of the sea? If you stand on the end of a pier, you may be able to answer this question. Suppose that the pier is 60 m long, and that you notice that exactly five waves fit into this length (Figure **14.8**). From this information, you can deduce that their wavelength is:

$$\text{wavelength} = \frac{60\,\text{m}}{5} = 12\,\text{m}$$

Now you time the waves arriving. The interval between crests as they pass the end of the pier is 4 s. How fast

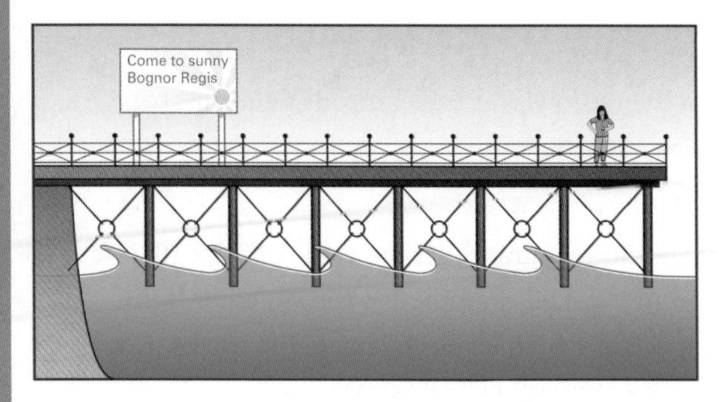

Figure 14.8 By timing waves and measuring their wavelength, you can find the speed of waves.

are the waves moving? One wavelength (12 m) passes in 4 s. So the speed of the waves is:

$$\text{speed} = \frac{12\,\text{m}}{4\,\text{s}} = 3\,\text{m/s}$$

Hence the speed v, frequency f and wavelength λ of a wave are connected. We can write the connection in the form of an equation:

> speed (m/s) = frequency (Hz) × wavelength (m)
>
> $v = f\lambda$

Another way to think of this is to say that the speed is the number of waves passing per second times the length of each wave. If 100 waves pass each second ($f = 100\,\text{Hz}$), and each is 4 m long ($\lambda = 4\,\text{m}$), then 400 m of waves pass each second. The speed of the waves is 400 m/s.

Worked example 1

An FM radio station broadcasts signals of wavelength 3.0 m and frequency 100 MHz. What is their speed?

Step 1: Write down what you know, and what you want to know.

$f = 100\,\text{MHz} = 100\,000\,000\,\text{Hz} = 10^8\,\text{Hz}$
$\lambda = 3.0\,\text{m}$
$v = ?$

Step 2: Write down the equation for wave speed. Substitute values and calculate the answer.

$v = f\lambda$
$v = 10^8\,\text{Hz} \times 3.0\,\text{m}$
$= 3 \times 10^8\,\text{m/s}$

So the radio waves travel through the air at $3.0 \times 10^8\,\text{m/s}$.

You should recognise that the value of $3.0 \times 10^8\,\text{m/s}$ found in Worked example **1** is the **speed of light**, the speed at which all electromagnetic waves travel through empty space (vacuum).

Worked example 2

A pianist plays the note middle C, whose frequency is 264 Hz. What is the wavelength of the sound waves produced? (Speed of sound in air = 330 m/s.)

Step 1: Write down what you know, and what you want to know.

$f = 264\,\text{Hz}$
$\lambda = ?$
$v = 330\,\text{m/s}$

Step 2: Write down the equation for wave speed. Rearrange it to make wavelength λ the subject.

$v = f\lambda$

$$\lambda = \frac{v}{f}$$

Step 3: Substitute values and calculate the answer.

$$\lambda = \frac{330\,\text{m/s}}{264\,\text{Hz}} = 1.25\,\text{m}$$

So the wavelength of the note middle C in air is 1.25 m.

Changing material, changing speed

When waves travel from one material into another, they usually change speed. Light travels more slowly in glass than in air. Sound travels faster in steel than in air. When this happens, the frequency of the waves remains unchanged. As a consequence, their wavelength must change. This is illustrated in Figure **14.9**, which shows light waves travelling quickly through air. They reach some glass and slow down, and their wavelength decreases. When they leave the glass again, they speed up, and their wavelength increases again.

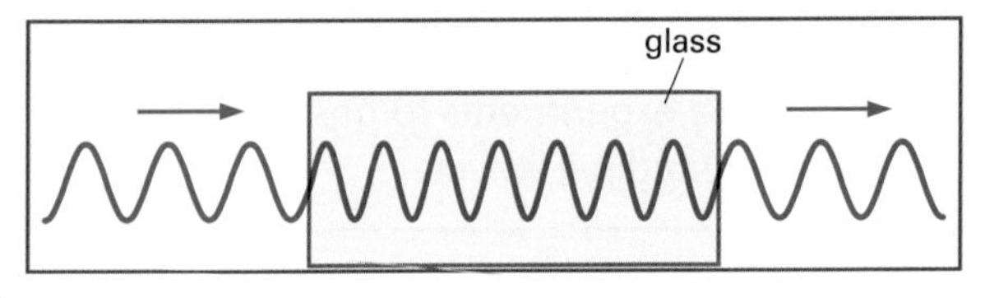

Figure 14.9 Waves change their wavelength when their speed changes. Their frequency remains constant. Here, light waves slow down when they enter glass and speed up when they return to the air.

QUESTIONS

8 Write down an equation relating speed, frequency and wavelength of a wave. Indicate the SI units of each quantity.

9 If 10 waves pass a point each second and their wavelength is 30 m, what is their speed?

10 All sound waves travel with the same speed in air. Which has the higher frequency, a sound wave of wavelength 2 m or one with wavelength 1 m?

11 Which have the longer wavelength, radio waves of frequency 90 MHz or 100 MHz?

12 Light slows down when it enters water from air.

a What happens to its speed?

b What happens to its wavelength?

c What happens to its frequency?

14.3 Reflection and refraction of waves

If we look at ripples on the surface of water in a ripple tank, we can begin to see why physicists say that light behaves as if it were a form of wave. The ripples are much more regular and uniform than waves on the sea, so they are a good **model** system to look at.

Reflection of ripples

Figure **14.10** shows what happens when a flat metal barrier is placed in the ripple tank. The photograph in Figure **14.10a** shows the pattern of the ripples observed, and Figure **14.10b** shows how the ripples are produced. Straight ripples ('plane waves') are reflected when they strike the flat surface of the barrier. The metal barrier acts like a mirror, and the ripples bounce off it. This shows an important thing about how waves behave. They pass through each other when they overlap.

In Figure **14.10c**, you can see the same pattern, this time as a drawing. This is an 'aerial view' of the ripples. The blue lines represent the tops of the ripples. These lines are known as **wavefronts**. The separation of the wavefronts is equal to the wavelength of the ripples. Figure **14.10c** also shows lines (the red arrows) to indicate how the direction of travel of the ripples changes. This diagram should remind you of the ray diagram for the law of reflection of light (Figure **13.5** on page **135**). The ripples are reflected by the metal barrier so that the angle of incidence equals the angle of reflection.

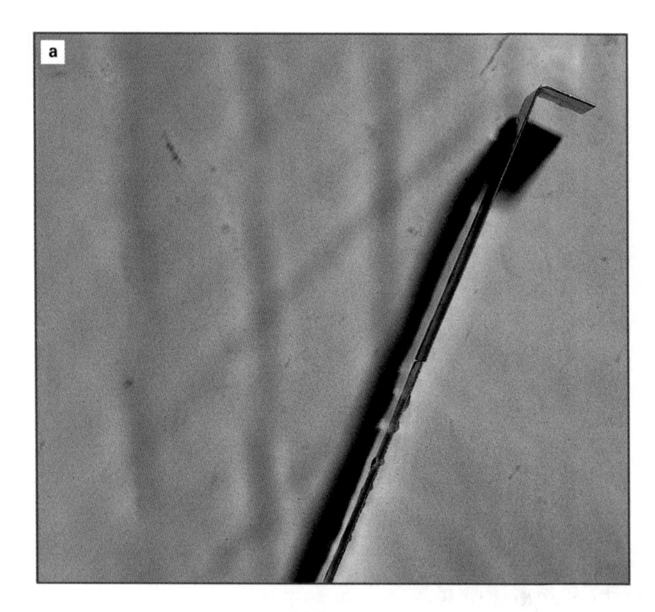

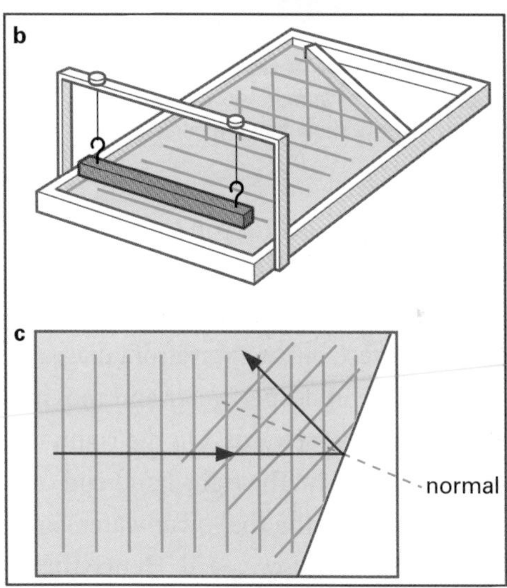

Figure 14.10 The reflection of plane waves by a flat metal barrier in a ripple tank. **a** This criss-cross pattern is observed as the reflected ripples pass through the incoming ripples. **b** How the ripples are produced. **c** The arrows show how the direction of the ripples changes when they are reflected. The angle of incidence is equal to the angle of reflection, just as in the law of reflection of light.

Refraction of ripples

Refraction occurs when the speed of light changes. We can see the same effect for ripples in a ripple tank (Figure **14.11**). A glass plate is immersed in the water, to make the water shallower in that part of the tank. There, the ripples move more slowly because they drag on the bottom of the tank (which is now actually the upper surface of the submerged glass plate).

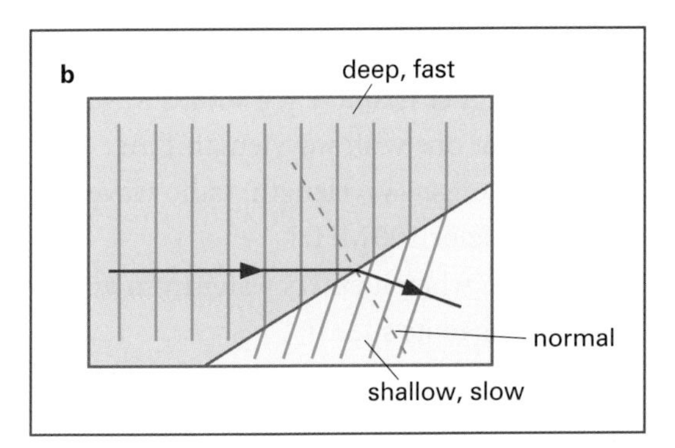

Figure 14.11 The refraction of plane waves by a flat glass plate in a ripple tank. **a** A submerged glass plate makes the water shallower on the right. In this region, the ripples move more slowly, so that they lag behind the ripples in the deeper water. **b** This wavefront diagram shows the same pattern of ripples. The rays show that the refracted ray is closer to the normal, just as when light slows down on entering glass.

In the photograph in Figure **14.11a**, you can see that these ripples lag behind the faster-moving ripples in the deeper water. Their direction of travel has changed. Figure **14.11b** shows the same effect, but as a wavefront diagram. On the left, the ripples are in deeper water and moving faster. They advance steadily forwards. On the right, the ripples are moving more slowly. The right-hand end of a ripple is the first part to enter the shallower water, so it has spent longest moving at a slow speed. Hence the right-hand end of each ripple lags furthest behind.

The rays (the red arrows) marked on Figure **14.11b** show the direction in which the ripples are moving. They are always at right angles to the ripples. They emphasise how the ripples turn so their direction is closer to the normal as they slow down, just as we saw with the refraction of light (Figure **13.10** on page **138**).

Activity 14.2 Ripple tank

Observe reflection and refraction of ripples in a ripple tank.

QUESTIONS

13 Draw a diagram to show what happens to plane waves when they strike a flat reflector placed at 45° to their direction of travel.

14 How can the speed of ripples in a ripple tank be changed?

E

Explaining reflection and refraction

The law of reflection of light **describes** how a ray of light behaves when it strikes a flat surface. However, it does not **explain** how light is reflected. We need the wave theory of light to provide an explanation.

We have seen that waves are reflected when they strike a surface. If we picture light as a form of wave, then we can predict that it will be reflected in the same way that ripples on the surface of water are reflected.

We have also seen that ripples are refracted when their speed changes. We know that the speed of light changes as light passes from one material to another. So the wave theory of light allows us to predict that light will be refracted.

E

If you study physics to a higher level, you will learn more about how light, sound and other phenomena can be described and explained using wave theory. Here, we will conclude this chapter by looking at another aspect of wave theory: the behaviour of waves when they pass through a gap or around an obstacle in their path.

14.4 Diffraction of waves

We can see an interesting phenomenon when we look at how ripples behave when they go through a gap in a barrier. Figure **14.12** shows what happens. As ripples pass through a gap in a barrier, they spread out into the space beyond the barrier.

This is an example of a phenomenon called **diffraction**. The effect is biggest when the gap is similar in size to the wavelength of the ripples (see Figure **14.13**).

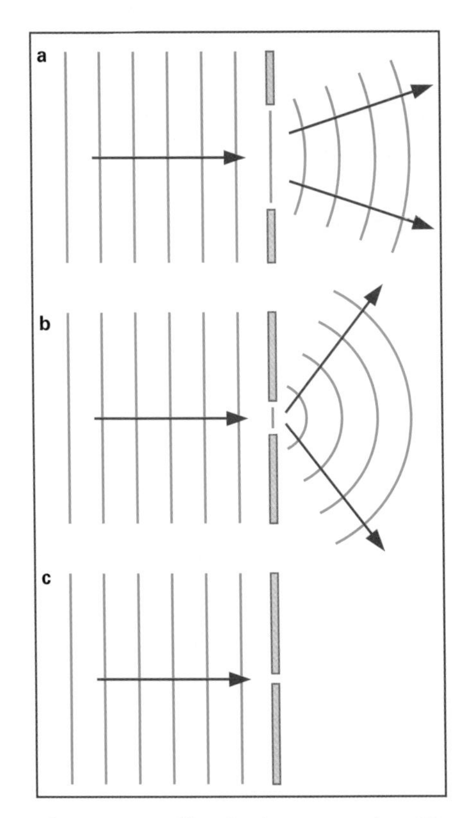

Figure 14.13 Diffraction is greatest when the width of the gap is similar to the wavelength of the waves being diffracted. When the gap is much smaller than the wavelength, the waves do not pass through at all.

Figure 14.12 Ripples are diffracted as they pass through a gap in a barrier – they spread into the space behind the barrier. The effect is greater in **a** than in **b** because the gap is narrower in **a**.

Diffraction at work

You might notice diffraction of water waves in a harbour. The waves enter the harbour mouth and spread around corners, so that no part of the harbour is entirely undisturbed. Boats bob up and down on the diffracted waves.

Sound waves have wavelengths between about 10 mm and 10 m. They are readily diffracted as they pass through doorways and open windows. We rely on the diffraction of sound every day to hear what is going on in the next room. This supports the idea that sound travels as a wave.

Light waves have a much shorter wavelength – less than a millionth of a metre. This means that very small gaps are needed to see light being diffracted. You might notice that, on a foggy night, street lamps and car headlights appear to be surrounded by a 'halo' of light. This is because their light is diffracted by the tiny droplets of water in the air. The same effect can also sometimes be seen around the Sun during the day (see Figure **14.14**).

Figure 14.14 Light from the Sun is diffracted as it passes through foggy air (which is full of tiny droplets of water), producing a halo of light.

QUESTIONS

15 What is observed when ripples pass through a gap in a barrier?

16 What can you say about the width of a gap if it is to produce the greatest diffraction effect?

Explaining diffraction

We can explain diffraction as follows. As the ripples arrive at the gap in the barrier, the water at the edge of the gap moves up and down. This sets off new circular ripples, which spread out behind the barrier.

If you look at the diffracted ripples in Figure **14.12b**, you will see that the central part of the ripple remains straight after it has passed through the gap. At the edges, the ripples have the shape of an arc of a circle.

QUESTION

17 Draw a diagram to show how a series of parallel, straight wavefronts are altered as they pass through a gap whose width is equal to the wavelength of the waves.

Summary

A wave is a regularly varying disturbance that travels from place to place.

Ripples on water can act as a model for the way in which waves travel.

In transverse waves, the disturbance varies from side to side, at right angles to the direction in which the wave is travelling.

In longitudinal waves, the disturbance is back and forth, along the direction of travel.

E Wave speed, frequency and wavelength are related by
speed = frequency × wavelength.

Waves can be reflected, when they reach a boundary between two different materials.

Waves can be refracted, when their speed changes.

Waves can be diffracted, when they pass through a gap.

E Reflection, refraction and diffraction can be explained using the wave model.

End-of-chapter questions

14.1 Look at the wave shown in Figure **14.15**.

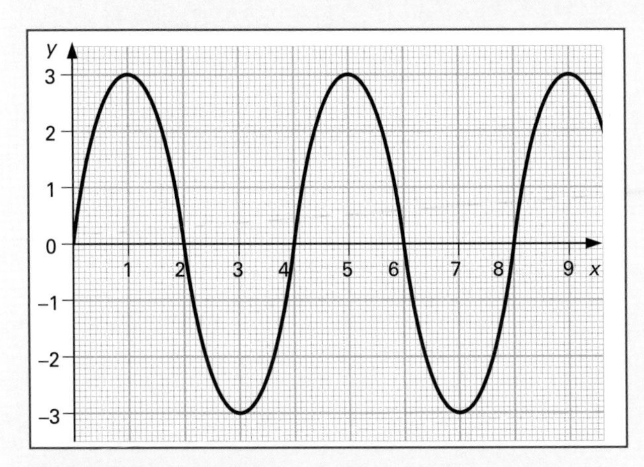

Figure 14.15 For Question **14.1**. The horizontal and vertical scales are in cm.

a What is its wavelength? [1]
b What is its amplitude? [1]
c If this wave is moving at a speed of 10 cm/s, what is its frequency? [3]
d On graph paper, with the same labelled and numbered axes as here, sketch a wave having **half** this amplitude and **twice** this wavelength. [2]

14.2 Copy and complete the diagrams in Figure **14.16** to show how the following effects appear in a ripple tank.

a Plane waves are reflected by a straight barrier. [2]
b Plane waves are diffracted as they pass through a narrow gap. [2]

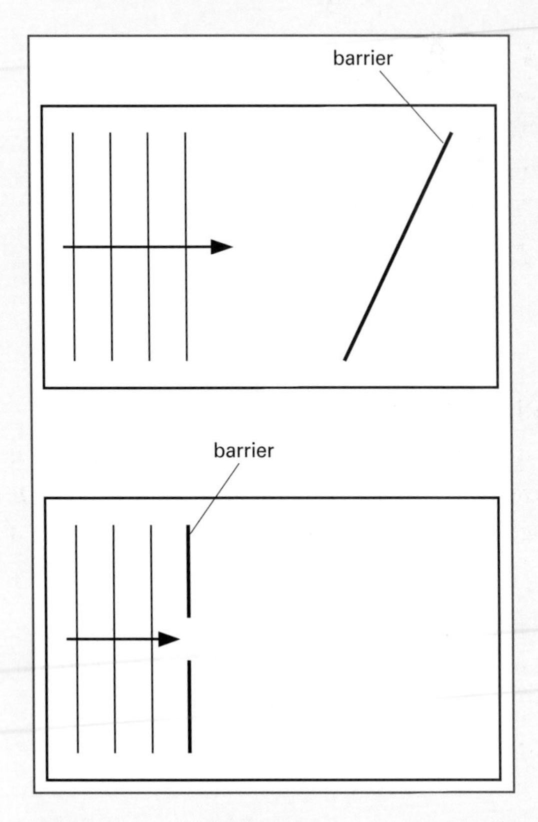

Figure 14.16 For Question **14.2**.

E

14.3 When light passes from air into glass, do the following quantities increase, decrease, or stay the same?

a speed [1]

b frequency [1]

c wavelength [1]

E

14.4 a Give an equation that relates the speed, frequency and wavelength of a wave. [1]

b Light waves of frequency 6×10^{14} Hz have a wavelength of 3.75×10^{-7} m in water. What is their speed in water? [2]

15 Spectra

Core	Describing the dispersion of light by a prism
Core	Describing the main features of the electromagnetic spectrum
Core	Stating that all electromagnetic waves travel at the same speed in vacuum
E Extension	Stating the value of the speed of electromagnetic waves

Light and colour

Diamonds are attractive because they sparkle. As you turn a cut diamond, light flashes from its different internal surfaces. As we have seen in Chapter **13**, this is a result of total internal reflection of light within the diamond. You may also notice that you can see all sorts of colours in the diamond, even though the diamond itself is likely to be colourless. (Some diamonds are slightly yellow, because they contain impurities.) Where do these varying colours come from?

Cut glass is a lot cheaper than diamonds, and has many more uses (Figure **15.1**). It is used for chandeliers, which move gently in the air. It is also used for glass ornaments. Your eye is caught by the changing colours as you walk past. Again, the glass itself is colourless, so where do these colours come from?

The underlying principle is shown in Figure **15.2**. A ray of white light is shone at a prism. It refracts as it enters and leaves the glass. At the same time, it is split into a **spectrum** of colours. You should notice that the colours merge into one another, and they are not all of equal widths in the spectrum.

Traditionally, we say that there are seven colours in the spectrum. The number seven was chosen because it had a mystical significance in the 17th century. It is very hard to distinguish between indigo and violet at the end of the spectrum, so you might say that there are really only six colours. Alternatively, you might suggest that there are many shades of red present, and of each of the other colours, so the spectrum shows many more than seven colours.

Figure 15.1 Cut glass is used for ornaments and chandeliers, because it shows all the colours of the rainbow as it moves in the light.

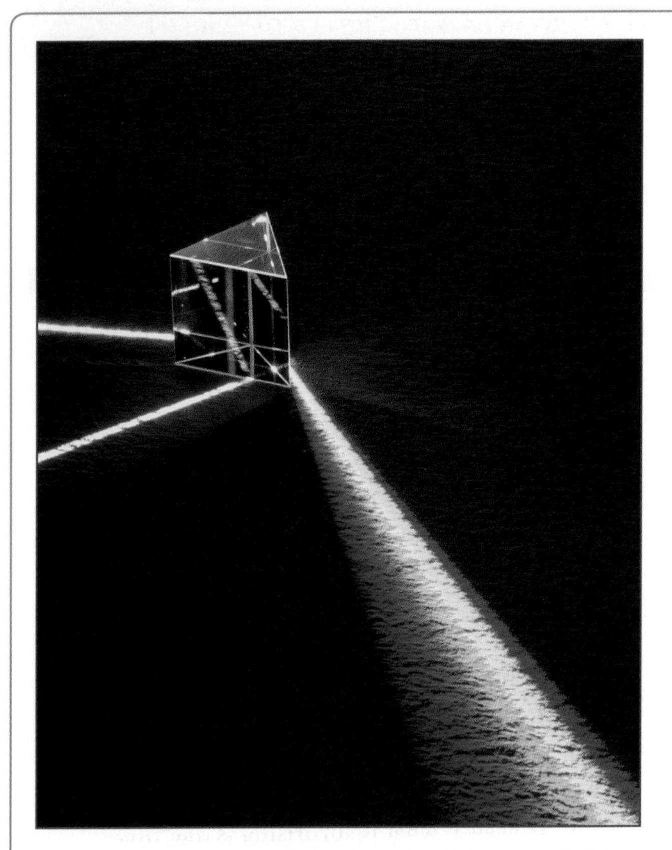

Figure 15.2 A spectrum can be produced by shining a ray of white light through a glass prism. The light is split up into a spectrum.

The standard list is as follows:

red orange yellow green blue indigo violet

There are different ways of remembering this list. One simple way is to remember the sequence of initial letters in the form of someone's name: Roy G Biv.

A rainbow is a naturally occurring spectrum. White light from the Sun is dispersed as it enters and leaves droplets of water in the air. It is also reflected back to the viewer by total internal reflection, which is why you must have the Sun behind you to observe a rainbow.

15.1 Dispersion of light

This splitting up of white light into a spectrum is known as **dispersion** ('spreading out'). Isaac Newton set out to explain how it happens. It had been suggested that light is coloured by passing it through a prism. Newton showed that this was the wrong idea by arranging for the spectrum to be passed back through another prism. The colours recombined to form white light again. He concluded that white light is a mixture of all the different colours of the spectrum.

So what happens in a prism to produce a spectrum? As the white light enters the prism, it slows down. We say that it is refracted and, as we have seen, its direction changes. Dispersion occurs because each colour is refracted by a different amount. Violet light slows down the most, and so it is refracted the most. Red light is least affected (Figure 15.3).

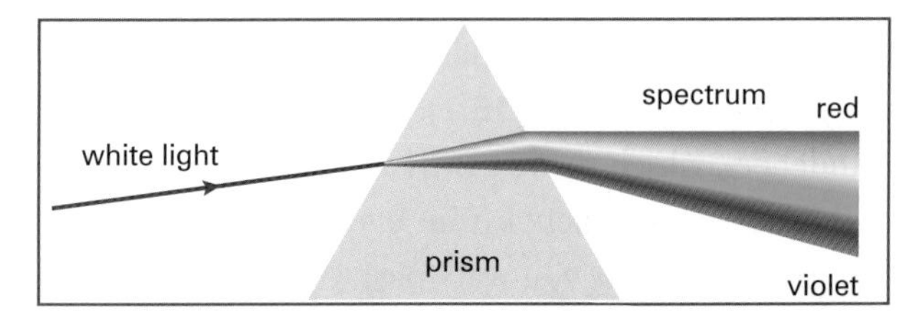

Figure 15.3 Violet light is dispersed more than red light as it passes through a prism.

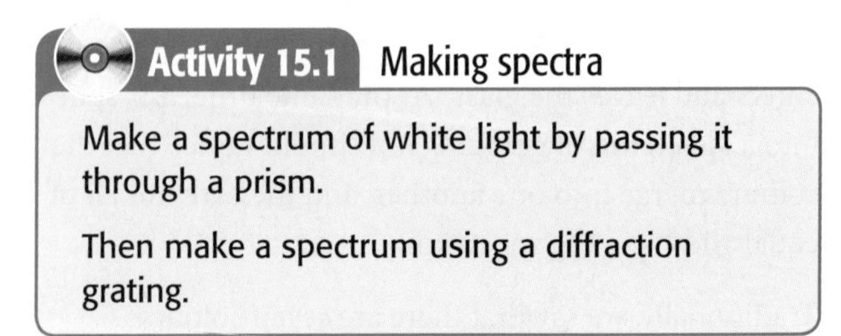

E

Laser light is not dispersed by a prism. It is refracted so that it changes direction, but it is not split up into a spectrum. This is because it is light of a single colour, and is described as **monochromatic** ('mono' = one, 'chromatic' = coloured).

QUESTIONS

1 What colours are next to green in the spectrum?
2 Draw a diagram to show how white light can be dispersed into a spectrum using a glass prism.
3 Why are some colours of light more strongly refracted than others when they enter glass?

15.2 The electromagnetic spectrum

In 1799, William Herschel was examining the spectrum of light from the Sun. He was an astronomer, German by birth but working at Slough, near London. He knew that the Sun was a star and wondered what he might find out about the Sun by looking at its spectrum. He shone the Sun's light through a prism to produce a spectrum, then placed a thermometer at different points in the spectrum, as shown in Figure 15.4. The reading on the thermometer rose, because objects get warm when they absorb light. Herschel noticed an interesting effect – the thermometer reading grew higher as he moved towards the red end of the spectrum. What would happen if he moved just beyond the end? To his surprise, he found that the reading was higher still. There was nothing to be seen beyond the red, but there was definitely something there. A little further, and the mercury in the thermometer rose higher still. Further still, and it started to fall.

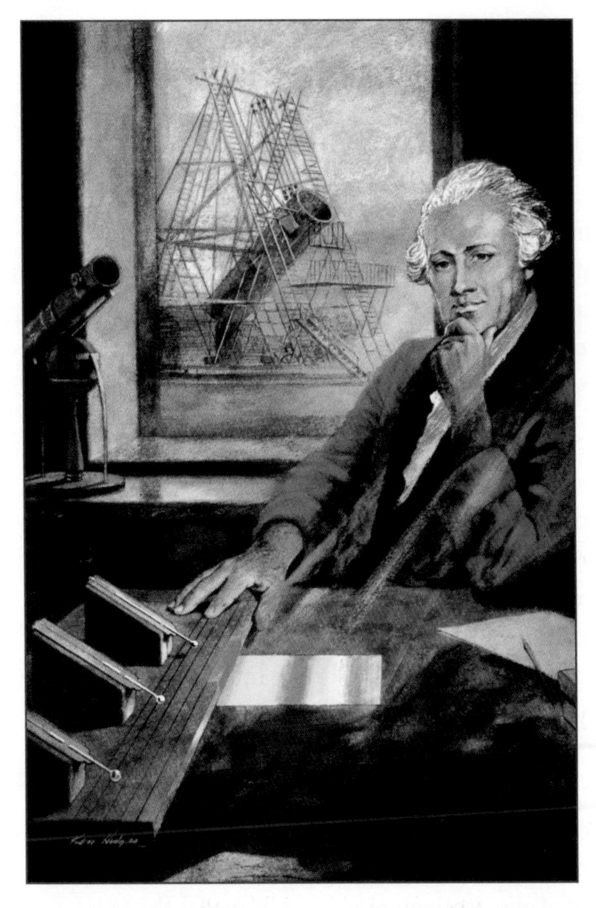

Figure 15.4 William Herschel, together with the apparatus he used to discover infrared radiation.

Herschel had discovered an invisible form of radiation, which he called **infrared radiation** ('infra' means 'below' or 'lower down'). You can experience infrared radiation for yourself, using a kettle that has recently boiled. With great care, hold the back of your hand near to the kettle. You feel the warmth of the kettle as it is absorbed by your skin. The kettle is emitting infrared radiation. (We sometimes call this 'heat radiation' – see Chapter 11 – but 'infrared radiation' is a better term.)

It is not surprising to learn that we receive heat from the Sun. However, what is surprising is that this radiation behaves in such a similar way to light. It is as if it is just an extension of the spectrum of visible light.

Beyond the violet

The discovery of radiation beyond the red end of the spectrum encouraged people to look beyond the violet end. In 1801, a German scientist called Johan Ritter used silver chloride to look for 'invisible rays'. Silver salts are blackened by exposure to sunlight (this is the basis of photography), so he directed a spectrum of sunlight onto paper soaked in silver chloride solution. The paper became blackened and, to his surprise, the effect was strongest beyond the violet end of the visible spectrum. He had discovered another extension of the spectrum, which came to be called **ultraviolet radiation** ('ultra' means 'beyond'). Although our eyes cannot detect ultraviolet radiation, sensitive photographic film can (see Figure 15.5).

Both infrared and ultraviolet radiations were discovered by looking at the spectrum of light from the Sun. However, they do not have to be produced by an object like the Sun. Imagine a lump of iron that you heat in a Bunsen flame. At first, it looks dull and black. Take it from the flame and you will find that it is emitting infrared radiation. Put it back in the flame and heat it more. It begins to glow, first a dull red colour, then more yellow, and eventually white hot. It is emitting visible light. When its temperature reaches about 1000 °C, it will also be emitting appreciable amounts of ultraviolet radiation.

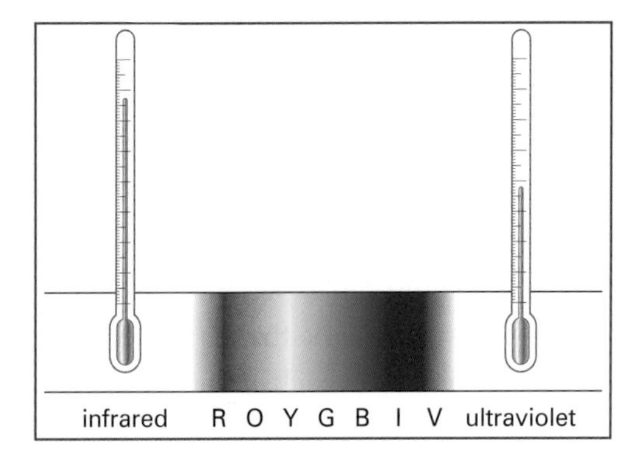

Figure 15.5 The spectrum of light from the Sun extends beyond the visible region, from infrared to ultraviolet.

This experiment should suggest to you that there is a connection between infrared, visible and ultraviolet radiations. A cool object emits only radiation at the cool end of the spectrum. The hotter the object, the more radiation it emits from the hotter end.

The Sun is a very hot object (Figure **15.6**). Its surface temperature is about 7000 °C, so it emits a lot of ultraviolet radiation. Most of this is absorbed in the atmosphere, particularly by the ozone layer. A small amount of ultraviolet radiation does get through to us. The thinning of the ozone layer by chemicals released by human activity means that this amount is increasing. This increased exposure is disturbing because it increases the risk of skin cancer.

Electromagnetic waves

In section **15.1**, we saw that a spectrum is formed when light passes through a prism because some colours are refracted more than others. The violet end of the spectrum is refracted most. Now we can deduce that ultraviolet radiation is refracted even more than violet light, and that infrared radiation is refracted less than red light.

To explain the spectrum, and other features of light, physicists developed the **wave model** of light. Just as sound can be thought of as vibrations or waves travelling through the air (or any other material), so we can think of light as being another form of wave. Sounds can have different pitches – the higher the frequency, the higher the pitch. We can think of a piano keyboard as being a 'spectrum' of sounds of different frequencies. Light can have different colours, according to its frequency. Red light has a lower frequency than violet light. Visible light occurs as a spectrum of colours, depending on its frequency.

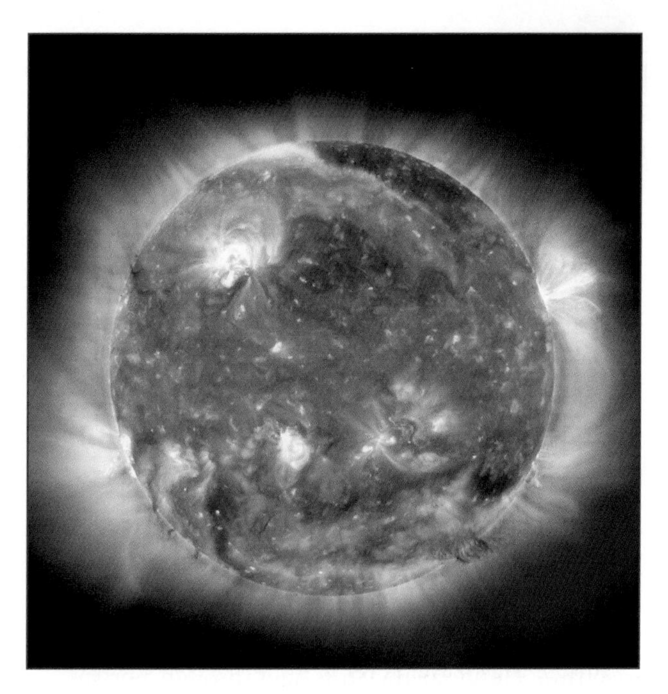

Figure 15.6 The Sun is examined by several satellite observatories. This image was produced by the SOHO satellite using a camera that detects the ultraviolet radiation given off by the Sun. You can see some detail of the Sun's surface, including giant prominences looping out into space. The different colours indicate variations in the temperature across the Sun's surface.

A Scottish physicist, James Clerk Maxwell, eventually showed in 1860 that light was in fact small oscillations in electric and magnetic fields, or **electromagnetic waves.** His theory allowed him to predict that they could have any value of frequency. In other words, beyond the infrared and ultraviolet regions of the spectrum, there must be even more types of electromagnetic wave. By the early years of the 20th century, physicists had discovered or artificially produced several other types of electromagnetic wave (see Table **15.1**), to complete the **electromagnetic spectrum.** Maxwell also predicted that all electromagnetic waves travel at the same speed through empty space, the speed of light (almost 300 000 000 m/s).

Type of electromagnetic wave	Discoverer	Date
infrared	William Herschel	1799
ultraviolet	Johan Wilhelm Ritter	1801
radio waves	Heinrich Hertz	1887
X-rays	Wilhelm Röntgen	1895
gamma (γ) rays	Henri Becquerel	1896

Table 15.1 Discoverers of electromagnetic waves.

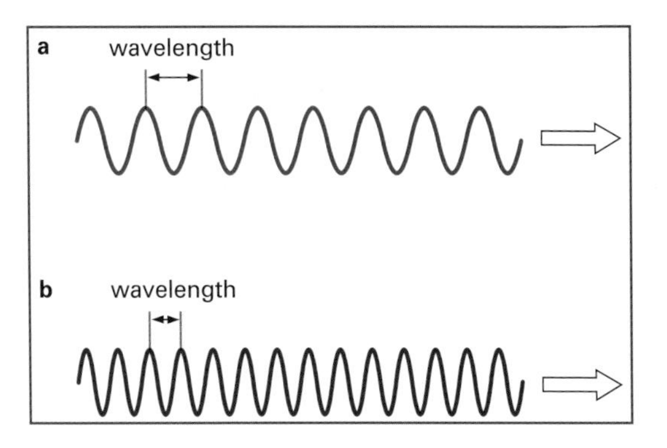

Figure 15.7 Comparing red and violet light waves. Both travel at the same speed, but red light has a longer wavelength because its frequency is less. The wavelength is the distance from one crest to the next (or from one trough to the next). Think of red light waves as long, lazy waves; violet light is made up of shorter, more rapidly vibrating waves.

E

The speed of electromagnetic waves

All types of electromagnetic wave have one thing in common: they travel at the same speed in a vacuum. They travel at the speed of light, whose value is close to 300 000 000 m/s (3×10^8 m/s). Like light, the speed of electromagnetic waves depends on the material through which they are travelling. They travel fastest through a vacuum.

Wavelength and frequency

We can represent light as a wave, just as we represented the small changes in air pressure as a sound wave (page **130**). Figure **15.7** compares red light with violet light. Red light has a greater wavelength than violet light – that is, there is a greater distance from one wave crest to the next. This is because both red light and violet light travel at the same speed (as predicted by Maxwell), but violet light has a greater frequency, so it goes up and down more often in the same length.

The waves that make up visible light have very high frequencies – over one hundred million million hertz, or 10^{14} Hz. Their wavelengths are very small, from 400 nm for violet light to 700 nm for red light. (One nanometre (1 nm) is one-billionth (one-thousand-millionth, 1/1 000 000 000th) of a metre, so 400 nm = 400×10^{-9} m.) So more than one million waves of visible light fit into a metre.

Figure **15.8** shows the complete electromagnetic spectrum, with the wavelengths and frequencies of each region. In fact, we cannot be very precise about where each region starts and stops. Even the ends of the visible light section are uncertain, because different

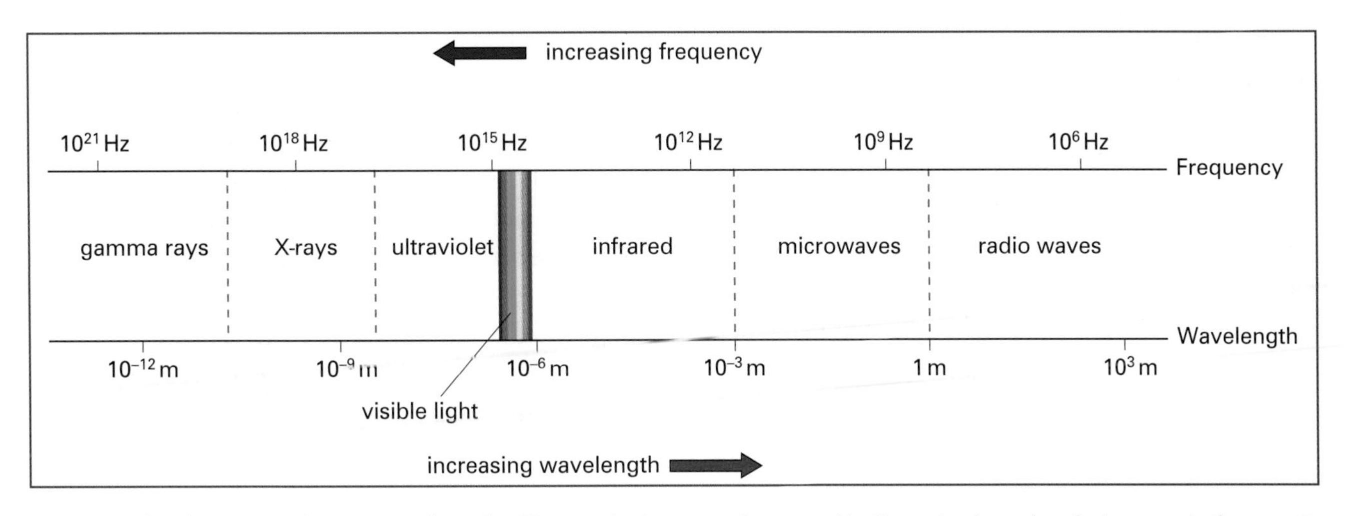

Figure 15.8 The electromagnetic spectrum. The scale of frequencies increases along one side. The scale of wavelengths increases in the opposite direction.

people can see slightly different ranges of wavelengths, just as they can hear different ranges of sound frequencies.

QUESTIONS

4 a Which has the longer wavelength, red light or violet light?
  b Which has the greater frequency?

5 a Which has the longer wavelength, red light or infrared radiation?
  b Which has the greater frequency?

6 Look at the spectrum shown in Figure **15.8**.
  a Which waves have the shortest wavelength?
  b Which have the lowest frequency?

7 a Which travels faster in empty space, violet light or red light?
  b Which travels faster in glass?

Uses of electromagnetic waves

Since the different regions of the electromagnetic spectrum were discovered, we have found many ways to make use of these waves. Here are some important examples.

Radio waves are used to broadcast radio and television signals. These are sent out from an aerial (a transmitter) a few kilometres away, to be captured by an aerial on the roof of a house.

Microwaves are used in satellite television broadcasting, because microwaves pass easily through the Earth's atmosphere as they travel up to a broadcasting satellite, thousands of kilometres away in space. Then they are sent back down to subscribers on Earth. Microwaves are also used to transmit mobile phone (cellphone) signals between masts, which may be up to 20 km apart.

Infrared radiation is used in remote controls for devices such as televisions and video recorders. A beam of radiation from the remote control carries a coded signal to the appliance, which then changes channel, starts to record, or whatever. You may be able to use a digital camera to observe this radiation, which would otherwise be invisible to our eyes. Grills and toasters also use infrared radiation, because it is thermal (heat) energy 'on the move'. Security alarms send out beams of infrared and detect changes in the reflected radiation – these may indicate the presence of an intruder.

X-rays can penetrate solid materials and so they are used in security scanners at airports (see Figure **15.9**). They are also used in hospitals and clinics to see inside patients without having to perform surgery.

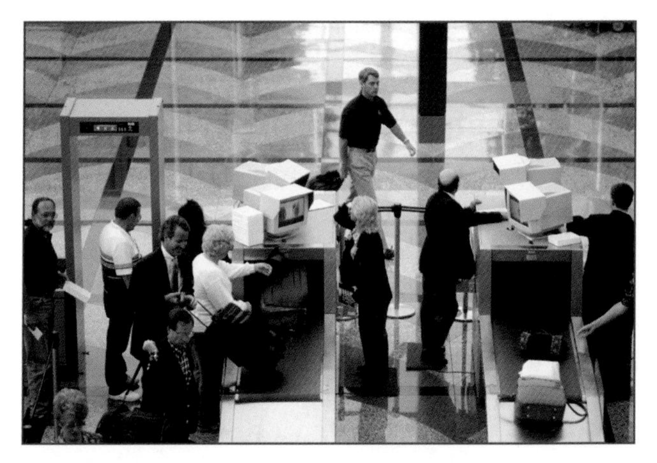

Figure 15.9 Two uses of electromagnetic radiation at the airport security check: X-rays are used to see inside the passengers' hand baggage, while radio waves detect metal objects as passengers walk through the arch.

Electromagnetic hazards

All types of radiation can be hazardous – even bright light shone into your eyes can blind you. So people who work with electromagnetic radiation must be careful and take appropriate precautions.

Microwaves are used to cook food in microwave ovens. This shows that they have a heating effect when absorbed. Telephone engineers, for example, must take care not to expose themselves to microwaves when they are working on the masts of a mobile phone (cellphone) network. Domestic microwave ovens must be checked to ensure that no radiation is leaking out.

People who work with X-rays must minimise their exposure. They can do this by standing well away when a patient is being examined, or by enclosing the equipment in a metal case, which will absorb X-rays.

QUESTIONS

8 Name **two** types of electromagnetic radiation that can be used for cooking food.

9 Explain how radio waves, microwaves and infrared radiation might all play a part when you watch a television show.

Activity 15.2 Using electromagnetic waves

Divide into groups and allocate a region of the electromagnetic spectrum to each group. Research uses of your region of the spectrum and stage a debate to decide whose region is the most useful.

Summary

White light can be dispersed by a prism to form a spectrum.

The electromagnetic spectrum extends to wavelengths and frequencies beyond those of the visible light spectrum.

Electromagnetic waves are varying electric and magnetic fields that travel through empty space at the speed of light.

Electromagnetic radiation has many uses. Care must be taken to ensure that the user does not come to any harm.

E

The speed of light in vacuum is approximately 300 000 000 m/s.

Laser light is light of a single wavelength, that is, it is monochromatic.

End-of-chapter questions

15.1 A glass prism can be used to show the dispersion of white light to form a spectrum.

a Draw a diagram to show how a ray of white light is dispersed as it passes through a prism. [2]

b Which colour of light is most strongly dispersed (deflected) as it passes through the prism? [1]

c Explain why some colours of light are more strongly dispersed than others. [2]

15.2 a Put the following regions of the electromagnetic spectrum in order, starting with the waves that have the greatest wavelength. [4]

visible light infrared radio waves gamma rays ultraviolet microwaves X-rays

b Which of these waves have the greatest frequency? [1]

c Which of these waves have the greatest speed in empty space (in vacuum)? [1]

E

15.3 At what speed do electromagnetic waves travel through a vacuum? [1]

15.4 Explain why white light is dispersed to form a spectrum when it passes through a glass prism but laser light is not. [3]

Block 4

Electricity and magnetism

At present, spacecraft are launched into space using rockets. These use burning fuel as their energy source to blast upwards against the force of gravity. However, things may be different in future. The picture shows a different way of launching a spacecraft, using a magnetic levitation (mag-lev) system. The idea is that a spacecraft will travel along a horizontal track, rather like an airport runway. The track will be made of electromagnets, which will do two things: they will support the spacecraft so that it will 'float' just above the track, and they will accelerate the spacecraft forwards.

Once it reaches a speed of 1000 km/h, it will leave the ground and a rocket will provide the final push needed to send it into orbit. A track like this could be used for several launches each day.

The red loops in the picture show the magnetic field produced by the electromagnet coils (gold). Of course, we cannot see a magnetic field like this, but we draw field lines to help us to show what is going on. In this block, you will study electricity and magnetism and learn some of the ways in which physicists picture 'invisible' processes such as magnetic and electric fields, electric current and so on.

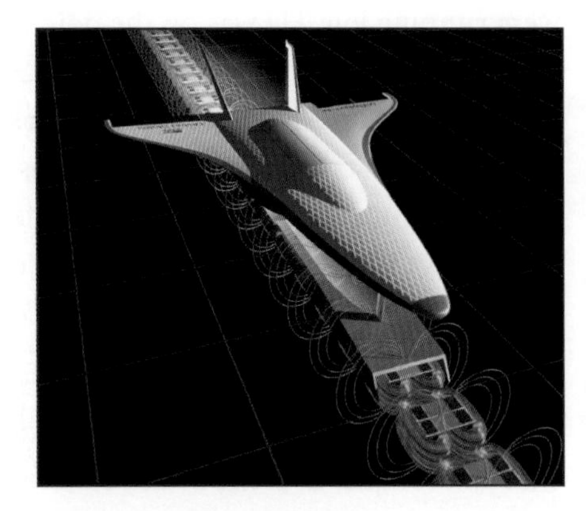

A computer-aided design for a mag-lev spacecraft launching system, being developed by engineers at NASA (USA) and the University of Sussex (UK).

16 Magnetism

Core Investigating permanent magnets and magnetic materials
Core Describing methods of magnetisation and demagnetisation
Core Explaining magnetic forces in terms of magnetic fields
Core Describing electromagnets and their uses

Setting a course

When Christopher Columbus set sail in 1492, he was hoping to find a new route to the East Indies by sailing west. To plot his course, he used a compass. He believed that the needle of a compass always points due north, and so it was easy to sail due west. In fact, he was wrong. A compass needle points towards the magnetic pole, and this is some distance away from the North Pole. This meant that Columbus's course across the Atlantic Ocean took him further south than he had intended to go. This had a happy consequence.

Columbus had been sailing for several weeks without sighting land. His crew were getting restless and stores were running low. He was on the point of turning back when the lookout sighted land. They had reached one of the Caribbean islands of the Bahamas (Figure **16.1**). It turns out that, if Columbus had realised that his compass pointed a few degrees away from north, he would have travelled further to the north. He would have turned back before he ever reached land.

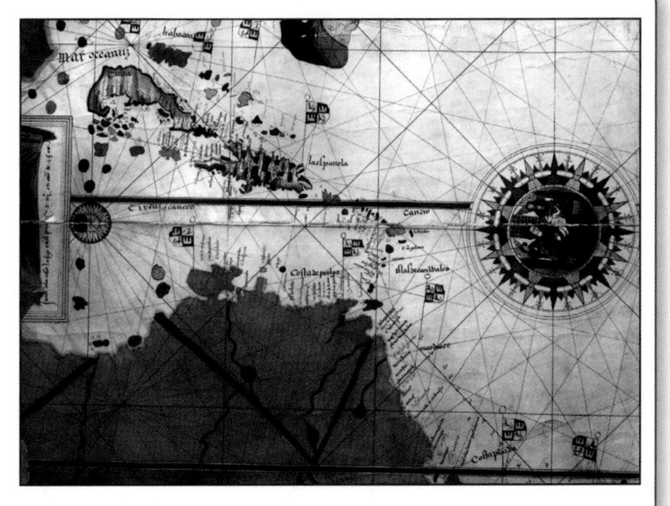

Figure 16.1 This map, published in 1506, shows the lands explored by Columbus. The Bahamas, where he first landed, are at the top, above the two large islands of Cuba and Hispaniola. On the right is a 'compass rose' used for navigation.

Compasses were vital instruments in the expansion of European nations as they looked for new places to trade with and conquer in the 15th and 16th centuries. With a compass and a reliable chart, you could set a steady course for your target port and have a good chance of reaching it. The first mention of a compass in European writing is in a book by Alexander Neckam, written in 1190:

> 'The sailors, as they sail the sea, when in cloudy weather they can no longer profit by the light of the Sun, or when the world is wrapped up in the darkness of the shades of night, and they are ignorant to what point of the compass their ship's course is directed, they touch the magnet with a needle. This then whirls round in a circuit until, when its motion ceases, its point looks direct to the north.'

Neckam is describing how an iron needle could be magnetised to make a compass, by touching or rubbing it on a 'magnet', a piece of a naturally magnetised material called lodestone. If you carried your lodestone and some needles with you, you could always make a new compass. Research in old documents has now shown that the Europeans learned about magnetism and compasses from the Chinese, who had been using compasses for over 1500 years before they reached Europe.

16.1 Permanent magnets

A compass needle is like a **bar magnet**. When it is free to rotate (Figure 16.2), it turns to point north–south. One end points north – this is the magnet's **north pole**, pointing roughly in the direction of the Earth's geographical North Pole. The other end is the magnet's **south pole**. (Sometimes, the north and south poles of a magnet are called the 'north-seeking' and 'south-seeking' poles, respectively.)

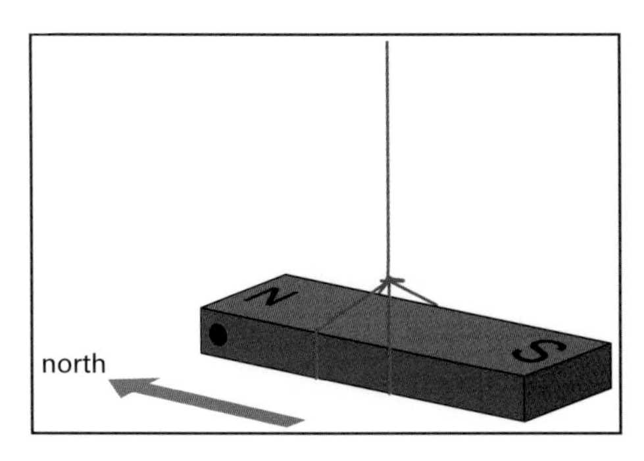

Figure 16.2 A freely suspended magnet turns so that it points north–south.

When two magnets are brought close together, there is a force between them. The north pole of one will attract the south pole of the other. Two north poles will repel each other, and two south poles will repel each other (Figure 16.3). This is summarised as follows:

- like poles repel
- unlike poles attract.

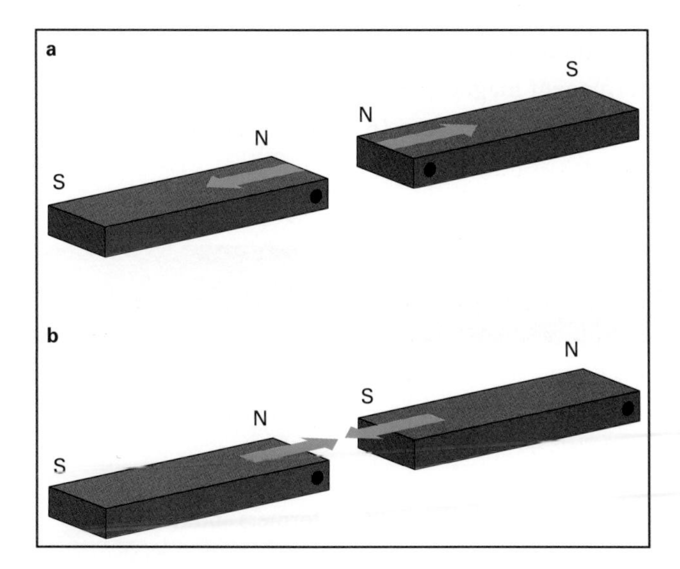

Figure 16.3 **a** Two like magnetic poles repel one another. **b** Two unlike magnetic poles attract each other.

('Like poles' means both north, or both south. 'Unlike poles' means opposite poles – one north and the other south. People often remember this rule more simply as 'opposites attract'.)

Since the north pole of the compass needle is attracted to the Earth's North Pole, it follows that there must be a magnetic south pole up there, under the Arctic ice! It is easy to get confused about this. In fact, for a long time, mediaeval scientists thought that compass needles were attracted to the Pole Star. Eventually, an English instrument-maker called Robert Norman noticed that, if he balanced a compass needle very carefully at its midpoint, it tilted downwards slightly, pointing into the Earth. Now we know that the Earth itself is magnetised, rather as if there was a giant bar magnet inside it.

Magnetic materials and magnetisation

A compass needle is a **permanent magnet**. Like many bar magnets, it is made of hard steel. You have probably come across another type of magnetic material, called ferrite. This is a ceramic material used for making fridge magnets and the magnets sometimes used to keep cupboard doors shut. There are also small 'rare-earth' magnets in the headphones used with MP3 players, based on elements such as neodymium.

Most magnetic materials (including steel and ferrite) contain iron, the commonest magnetic element. For this reason, they are known as **ferrous materials** (from the Latin word *ferrum* meaning 'iron'). Other magnetic elements include cobalt and nickel. (If a material contains iron, this is not a guarantee that it will be magnetic. Stainless steel contains a lot of iron but magnets will not stick to it.)

Usually, magnetic materials are in an unmagnetised state, and they must be magnetised. Four methods of doing this (called **magnetisation**) are listed below:

1 A piece of the material may be stroked with a permanent magnet. By stroking it consistently from one end to the other (never going in the reverse direction), it becomes magnetised (see Figure 16.4).

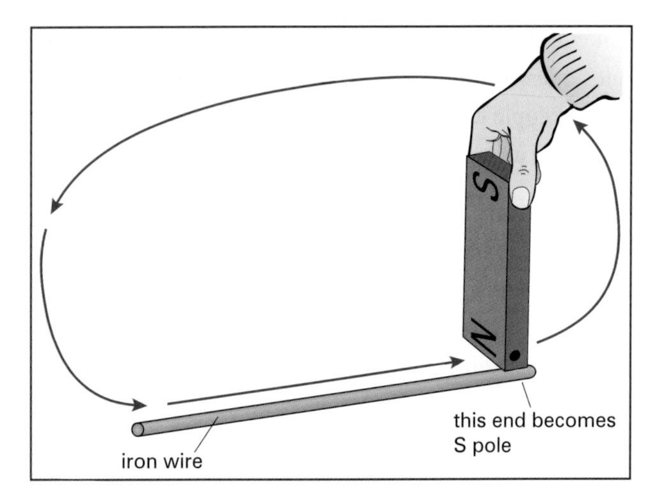

Figure 16.4 Magnetising an iron wire using a permanent magnet.

2 Place the material in a strong magnetic field, as produced by an electromagnet. It becomes magnetised.
3 Place a long thin piece of the material so that it lies north–south in the Earth's magnetic field. Heat it. When it cools, it will have been magnetised by the Earth's field.
4 Alternatively, hammer a piece of the material placed in a north–south direction and it will pick up the Earth's field. When steel ships are made, they are hammered as the metal plates are riveted together. The whole ship ends up magnetised, so that any compass on board will point the wrong way. The ship has to be demagnetised before it can set sail.

Magnetic materials may be classified as **hard** or **soft**. Table **16.1** summarises the difference. A soft magnetic material such as soft iron can be magnetised and demagnetised easily.

Demagnetisation

Just as there are several ways to magnetise a piece of magnetic material, there are several ways of demagnetising a magnet. Three methods of doing this (called **demagnetisation**) are listed below:

1 Hammer the magnet. When a magnet is placed in an east–west direction and hammered, it loses its magnetism. This explains why the magnets used in school labs gradually lose their magnetism if they are repeatedly dropped and bashed about.
2 Heat the magnet. If its temperature goes above a certain temperature, it will lose its magnetism.
3 Place the magnet in the field of an electromagnet (see below) that is connected to an alternating current supply. The magnetic field will vary back and forth. Gradually reduce the current to zero. The magnet will be demagnetised.

Induced magnetism

A bar magnet is an example of a permanent magnet. It can remain magnetised. Its magnetism does not get 'used up'. Permanent magnets are made of hard magnetic materials.

A permanent magnet can attract or repel another permanent magnet. It can also attract other **unmagnetised** magnetic materials. For example, a bar magnet can attract steel pins or paper clips, and a fridge magnet can stick to the steel door of the fridge.

What is going on here? Steel pins are made of a magnetic material. When the north pole of a permanent magnet is brought close to a pin, the pin is attracted (see Figure **16.5**). The attraction tells us

Type of magnetic material	Description	Examples	Uses
hard	retains magnetism well, but difficult to magnetise in the first place	hard steel	permanent magnets, compass needles, loudspeaker magnets
soft	easy to magnetise, but readily loses its magnetism	soft iron	cores for electromagnets, transformers and radio aerials

Table 16.1 Hard and soft magnetic materials. 'Hard steel' is both hard to the touch and difficult to magnetise and demagnetise. 'Soft iron' is both easy to bend and easy to magnetise and demagnetise.

that the end of the pin nearest the magnetic pole must be a magnetic south pole, as shown in Figure **16.5**. This is known as **induced magnetism**. When the permanent magnet is removed, the pin will return to its unmagnetised state (or it may retain a small amount of magnetism).

Figure 16.5 A steel pin is temporarily magnetised when a permanent magnet is brought close to it.

Activity 16.1 Making magnets

Make and test a magnet. Try to demagnetise it.

Use the idea of induced magnetism to explain how a piece of iron or steel can be magnetised.

QUESTIONS

1 Two bar magnets are placed side by side as shown in Figure **16.6**.

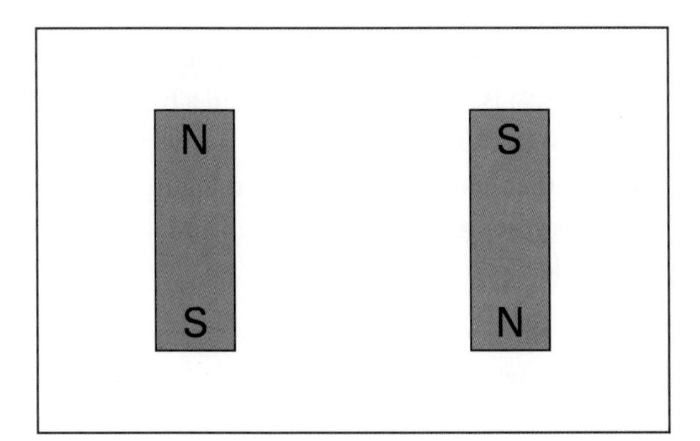

Figure 16.6 For Question **1**.

- **a** Copy the diagram and show the forces the two magnets exert on each other. State whether they will attract or repel each other.
- **b** One of the magnets is reversed so that its north pole is where its south pole was. Draw this situation and show the forces the two magnets exert on each other.

2 Iron is often described as a 'soft' magnetic material. Many types of steel are described as 'hard' magnetic materials.

- **a** Explain the difference between these two types of material.
- **b** Explain why a permanent magnet should be made of steel rather than iron.

16.2 Magnetic fields

A magnet affects any piece of magnetic material that is nearby. We say that there is a **magnetic field** around the magnet. You have probably done experiments with iron filings or small compasses to show up the magnetic field of a magnet. Figure **16.7** shows the field of a bar magnet as revealed by iron filings.

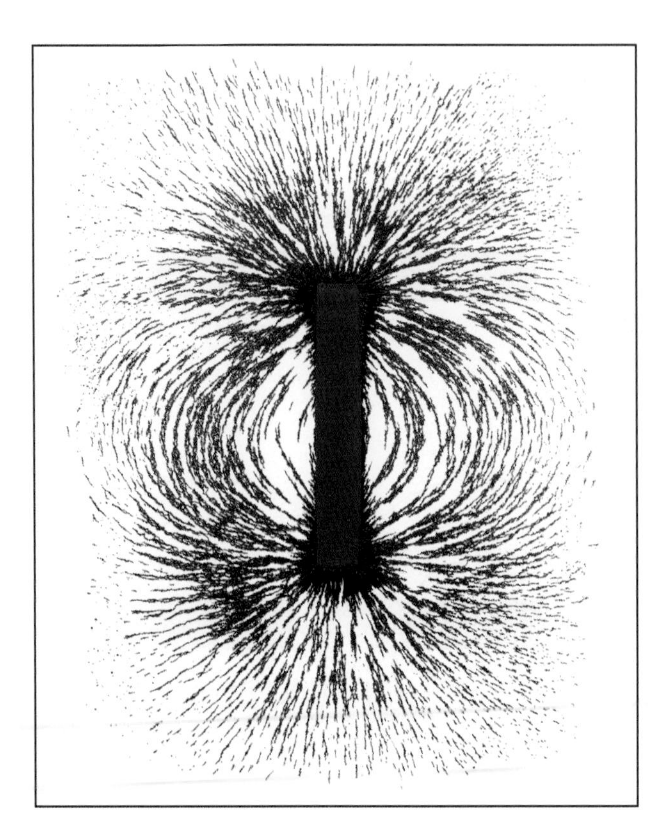

Figure 16.7 The magnetic field pattern of a bar magnet is shown up by iron filings. The iron filings cluster most strongly around the two poles of the magnet. This is where the field is strongest.

Figure **16.8a** shows how we represent the magnetic field of a single bar magnet, using **magnetic field lines**. Of course, the field fills all the space around the magnet, but we can only draw a selection of typical lines to represent it. The pattern tells us two things about the field:

- **Direction**. If you were to place a tiny compass at a point in the field, it would align itself along the field line at that point. We use a convention that says that field lines come out of north poles and go into south poles.
- **Strength**. Lines close together indicate a strong field.

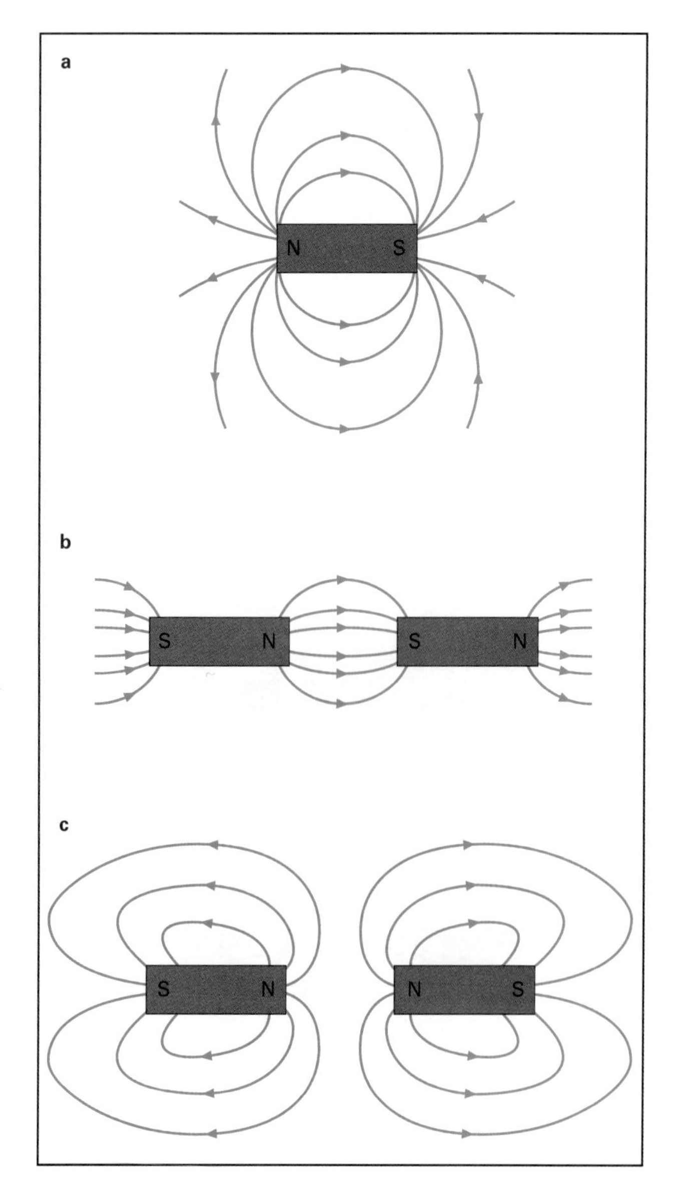

Figure 16.8 **a** Field lines are used to represent the magnetic field around a bar magnet. **b** The attraction between two opposite magnetic poles shows up in their field pattern. **c** The field pattern for two like poles repelling each other.

We can also show the field patterns for two magnets attracting (Figure **16.8b**) and repelling (Figure **16.8c**) each other. Notice that there is a point between the two repelling magnets where there is no magnetic field.

Plotting field lines

Iron filings can show up the pattern of the magnetic field around a magnet. Place the magnet under a stiff sheet of plain paper or (preferably) clear plastic. Sprinkle filings over the paper or plastic. Tap the paper or plastic to allow the filings to move slightly so that they line up in the field. You should obtain a pattern similar to that shown in Figures **16.7** and **16.8a**. (An alternative method of doing this uses small compasses called plotting compasses.)

Electromagnets

Using magnetic materials is only one way of making a magnet. An alternative method is to use an **electromagnet**. A typical electromagnet is made from a coil of copper wire. A coil like this is sometimes called a **solenoid**. When a current flows through the wire, there is a magnetic field around the coil (Figure **16.9**). Copper wire is often used, because of its low resistance, though other metals will do. The coil does not have to be made from a magnetic material. The point is that it is the electric current that produces the magnetic field.

You can see that the magnetic field around a solenoid (Figure **16.9**) is similar to that around a bar magnet (Figure **16.8a**). One end of the coil is a north pole, and the other end is a south pole. In Figure **16.9**, the field lines emerge from the left-hand end, so this is the north pole.

There are three ways to increase the strength of an electromagnet:

- increase the current flowing through it – the greater the current, the greater the strength of the field
- increase the number of turns of wire on the coil – this does not mean making the coil longer, but packing more turns into the same space to concentrate the field
- add a soft iron core.

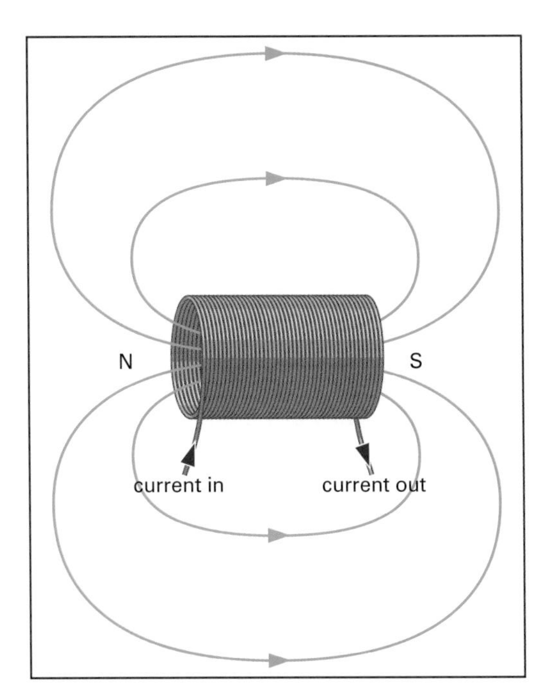

Figure 16.9 A solenoid. When a current flows through the wire, a magnetic field is produced. The field is similar in shape to that of a bar magnet. Note that the field lines go all the way through the centre of the coil.

An iron core becomes strongly magnetised by the field, and this makes the whole magnetic field much stronger.

Electromagnets have the great advantage that they can be switched on and off. Simply switch off the current and the field around the coil disappears. This is the basis of a number of applications – for example, the electromagnetic cranes that move large pieces of metal and piles of scrap around in a scrapyard (Figure 16.10). The current is switched on to energise the magnet and pick up the scrap metal. When it has been moved to the correct position, the electromagnet is switched off and the metal is released.

Electromagnets are also used in electric doorbells, loudspeakers, electric motors, relays and transformers.

Activity 16.2 Plotting field lines

Plot the magnetic field pattern around a bar magnet and around an electromagnet.

Figure 16.10 Using an electromagnet in a scrapyard. A steel object or pile of scrap can be lifted and moved. Then the current is switched off to release it.

QUESTIONS

3 Draw a diagram to show the field pattern between two magnets of equal strength whose south poles are placed close together.

4 Describe how an electromagnet could be used to separate copper from iron in a scrapyard.

The field around a solenoid

When an electric current flows through a solenoid, a magnetic field is produced inside and outside the coil (see Figure 16.9). This field is similar to that around a bar magnet:

- One end of the solenoid is the north pole and the other end is the south pole. Field lines emerge from the north pole and go into the south pole.

E

- The field lines are closest together at the poles, showing that this is where the magnetic field is strongest.
- The lines spread out from the poles, showing that the field is weaker in these regions.

The strength of the field can be increased by increasing the current. The field can be reversed by reversing the direction of the current.

QUESTION

5 **a** Sketch a diagram of the magnetic field pattern of a solenoid.

b How would the pattern change if the current through the solenoid was reversed?

Summary

Like poles repel, opposite poles attract.

Soft magnetic materials are easily magnetised and demagnetised. Hard magnetic materials retain their magnetism.

Magnetic fields are represented by field lines.

The pattern of field lines around a magnet can be shown up using iron filings or plotting compasses.

Electromagnets have the advantage over permanent magnets that they can be switched on and off.

End-of-chapter questions

16.1 Figure **16.11** shows four permanent magnets arranged to form a square.

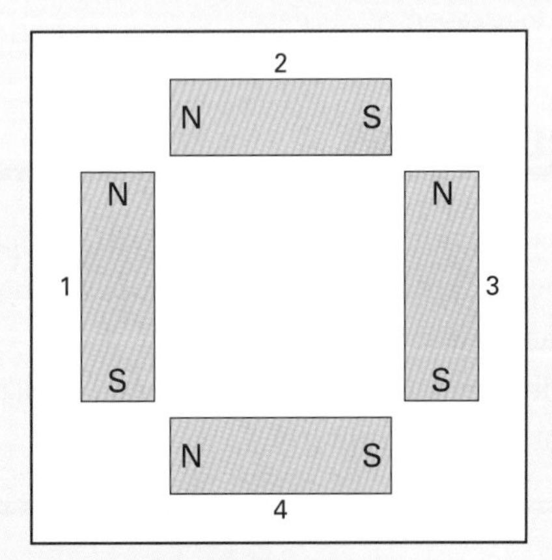

Figure 16.11 For Question **16.1**.

a Copy the diagram and indicate which pairs of magnets will attract one another and which will repel. [4]

b Draw a second diagram in which the four magnets are arranged in a square so that each magnet attracts the two other magnets to which it is closest. [2]

16.2 An electromagnet is a coil of wire through which a current can be passed.

a State **three** ways in which the strength of the electromagnet can be increased. [3]

b An electromagnet can be switched on and off. Suggest **one** situation where this would be an advantage over the constant field of a permanent magnet. [1]

16.3 **a** What is the difference between a **hard** magnetic material and a **soft** magnetic material? [3]

b Explain which you would choose for a permanent magnet. [2]

c Explain which you would choose for the core of an electromagnet. [2]

E

16.4 A solenoid has a magnetic field similar to that of a bar magnet.

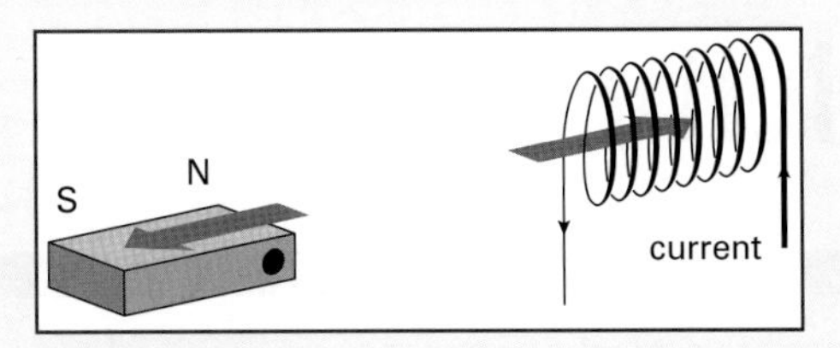

Figure 16.12 For Question **16.4**.

E

a Copy the diagram in Figure **16.12**, which shows a bar magnet and a solenoid arranged so that they repel one another. Label the poles of the solenoid. [2]

b Add field lines on your diagram to represent the magnetic fields of the bar magnet and the solenoid. [4]

17 Static electricity

Core	Investigating the forces between positive and negative electric charges
Core	Explaining electric forces in terms of electric fields
E Extension	Explaining charging by induction
Extension	Explaining static electricity in terms of electrons

A bright spark

Benjamin Franklin was an American, born in Boston in 1706. He was a scientist, as well as many other things – politician, printer, economist, musician and publisher, among various other occupations. His most famous experiment (Figure **17.1**), carried out in 1752, involved him in a most dangerous activity, flying a kite in a thunderstorm. He was investigating lightning as part of his studies of static electricity.

Figure 17.1 Benjamin Franklin, flying a kite in an attempt to capture a bolt of lightning. Franklin showed that lightning is similar to the sparks produced in experiments on static electricity. Shortly after, a Swedish scientist called Richtmann was killed when he tried to repeat Franklin's experiment. His body was dissected to discover the effect of electricity on his organs.

Franklin believed that lightning was a form of static electricity. He pointed out that a lightning flash was similar in shape and colour to the sparks that could be produced in the laboratory. In the demonstration shown in Figure **17.1**, Franklin attached a sharp-pointed metal wire to the top of a kite. He expected to draw down a spark from a lightning bolt. To avoid being electrocuted, he included a metal key at the bottom of the kite string, and attached a length of ribbon to the key. Holding the ribbon, he was relatively safe from electrocution (although other people were killed when they repeated his experiment). As a bolt of lightning struck the kite, Franklin saw the fibres of the kite string stand on end and a spark jumped from the key to the ground.

Franklin noticed that electrical sparks tend to jump from sharp points. He made use of this when he devised the lightning conductor. Today, most tall buildings have a sharply pointed metal rod projecting from their roofs, with a continuous metal rod running down the side of the building and into the ground. When lightning strikes, it is most likely to hit a lightning conductor and be safely channelled to the ground. Franklin's invention was enormously popular with insurance companies (who required the buildings they insured to install them), and the number of fires caused by lightning decreased dramatically.

Franklin made great progress in developing theories of electricity. Many of the terms we use today were first used by him – positive and negative charge, battery and conductor, among others.

17.1 Charging and discharging

As well as lightning flashes, we experience **static electricity** in a number of ways in everyday life. You may have noticed tiny sparks when taking off clothes made of synthetic fibres. You may have felt a small shock when getting out of a car. An **electrostatic charge** builds up on the car and then discharges through you when you touch the metal door. You have probably rubbed a balloon on your clothes or hair and seen how it will stick to a wall or ceiling.

If you rub a plastic ruler with a cloth, both are likely to become electrically charged. You can tell that this is so by holding the ruler and then the cloth close to your hair – they attract the hair. (If your hair is not attracted, try some tiny scraps of paper instead.) You have observed that static electricity is generated by rubbing. You have also observed that a charged object may attract uncharged objects.

Now we have to think systematically about how to investigate this phenomenon. First, how do two charged objects affect one another? Figure **17.2** shows one way of investigating this. A plastic rod is rubbed with a cloth so that both become charged. The rod is hung in a cradle so that it is free to move. When the cloth is brought close to it, the rod moves towards the cloth (Figure **17.2a**). If a second rod is rubbed in the same way and brought close to the first one, the hanging rod moves away (Figure **17.2b**). Now we have

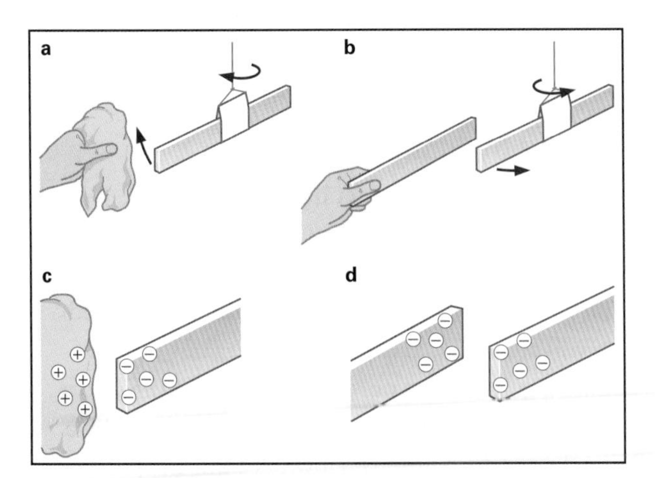

Figure 17.2 Two experiments to show the existence of two, opposite, types of static electricity. **a** The charged rod and cloth attract one another. **b** The two charged rods repel one another. **c** The rod and the cloth have opposite electric charges. **d** The two rods have electric charges of the same sign.

seen both **attraction** and **repulsion**, and this suggests that there are two types of static electricity. Both rods have been treated in the same way, so we expect them to have the same type of electricity. The cloth and the rod must have different types.

The two types of static electricity are referred to as **positive charge** and **negative charge**. We can explain the experiments shown in Figure **17.2** by saying that the process of rubbing gives the rods one type of electric charge (say, negative), while the cloth is given the opposite type (say, positive). Figures **17.2c** and **17.2d** show the two experiments with the charges marked.

From these experiments, we can also say something about the forces that electric charges exert on each other:

- like charges repel
- unlike charges attract.

('Like charges' means both positive, or both negative. 'Unlike charges' means opposite charges – one positive and the other negative. People often remember this rule as 'opposites attract'.)

You can see that this rule is similar to the rule we saw for magnetic poles in Chapter **16**. But do not confuse magnetism with static electricity! Magnetism arises from magnetic poles – static electricity arises from electric charges. When you rub a plastic rod, you are not making it magnetic.

Electric fields

A charged object can affect other objects, both charged and uncharged, without actually touching them. For example, a charged plastic rod can exert a force on another charged rod placed close by.

We say that there is an **electric field** around a charged object. Any charged object placed in the field will experience a force on it.

Try out some basic experiments to find out about static electricity.

QUESTIONS

1 Two positively charged polystyrene spheres are held close to one another. Will they attract or repel one another?

2 A polythene rod is rubbed using a woollen cloth. The rod gains a negative charge.
  a What can you say about the charge gained by the cloth?
  b Will the rod and the cloth attract or repel each other?

3 Here are some things you may have noticed:
  - If you rub a comb through your hair, your hair is attracted to the comb.
  - After combing, your hair is light and fluffy – the individual hairs repel each other.

  What do these observations tell you about the electric charges on your hair and on the comb?

17.2 Explaining static electricity

Before Benjamin Franklin and other scientists started carrying out their systematic experiments on static electricity, little was known about it. It had been known for centuries that, when rubbed, amber could attract small pieces of cloth or paper. Amber is a form of resin from trees, which has become fossilised. It looks like clear, orange plastic. The Greek name for amber is *elektron*, and this is where we get the name of the tiny charged particles (electrons) that account for electricity.

Franklin and those who worked on the problem at the same time as him had no idea about electrons – these particles were not discovered until a hundred years later. However, that did not stop them from developing a good understanding of static electricity. In the discussion that follows, we will talk about electrons. After all, they were discovered over a century ago, and they make it much easier to understand what is going on in all aspects of electricity.

Friction and charging

It is the force of friction that causes charging. When a plastic rod is rubbed on a cloth, friction transfers tiny particles called **electrons** from one material to the other. If the rod is made of polythene, it is usually the case that electrons are rubbed off the cloth and onto the rod.

Electrons are a part of every atom. They are negatively charged, and they are found on the outside of the atom. Since they are relatively weakly held in the atom, they can be readily pulled away by the force of friction. An atom has no electric charge – we say that it is **neutral**. When an atom has lost an electron, it becomes positively charged.

Since a polythene rod becomes negatively charged when it is rubbed with a silk cloth, we can imagine electrons being rubbed from the cloth onto the rod (see Figure **17.3**). It is difficult to explain why one material pulls electrons from another. The atoms that make up polythene contain positive charges, and these must attract electrons more strongly than those of the silk cloth.

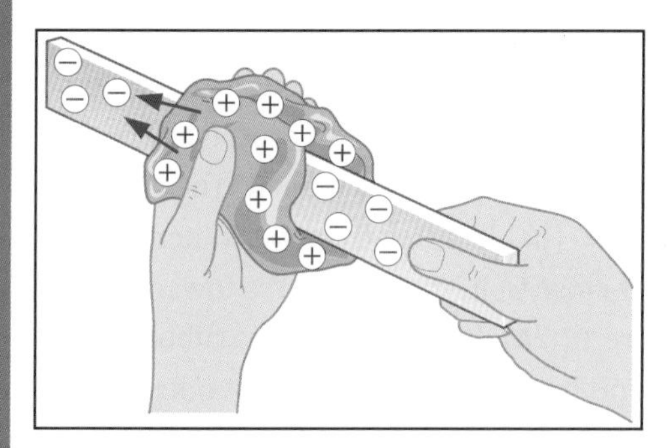

Figure 17.3 When a polythene rod is rubbed with a silk cloth, electrons are transferred from the silk to the polythene. The silk is left with a positive charge.

Charging by induction

A charged object can attract uncharged objects. For example, scatter some tiny pieces of paper on the bench. Rub a polythene rod on a woollen cloth (Figure **17.4**). Both the charged rod and the charged cloth will attract the paper. This is the same effect as rubbing a balloon on your clothes and sticking it to a wall. An uncharged object (the wall) is attracted by a charged one (the balloon). How does this happen?

Suppose the balloon has a positive charge. It must be attracted to a negative charge in the wall. The wall itself is neutral (uncharged), but its atoms are made up

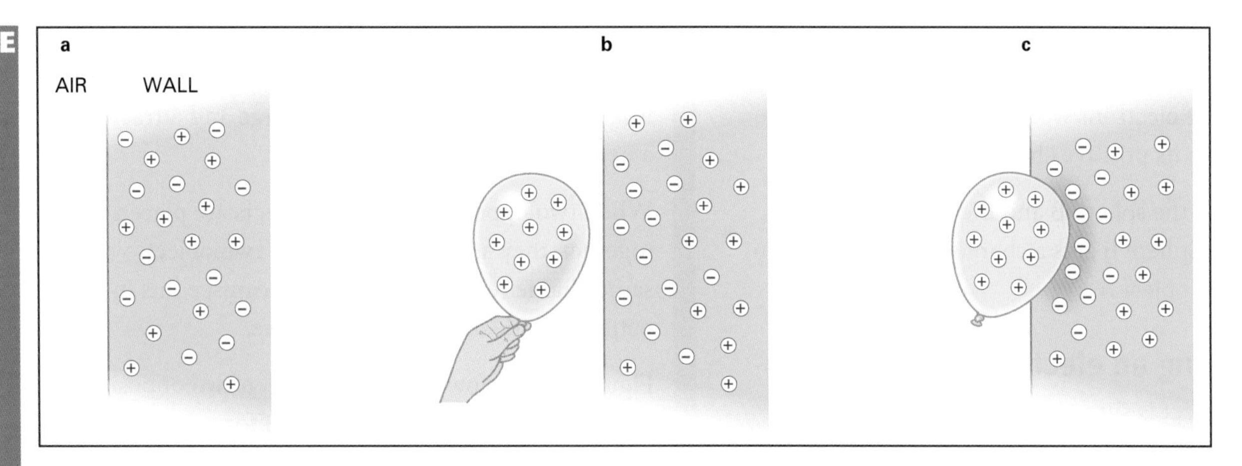

Figure 17.4 **a** The wall is neutral, because it has equal amounts of positive and negative charge. **b** The charged balloon attracts the negative charges in the wall, so that they move towards it. **c** The positive balloon and the negative surface of the wall stick together.

of positively and negatively charged particles (Figure **17.4a**). When the balloon is brought close to the wall, its negative charges (electrons) move towards the balloon, because they are attracted by it (Figure **17.4b**). They may not move very far, but the effect is enough to give the surface of the wall a negative charge, which attracts the balloon (Figure **17.4c**).

We say that a negative charge has been **induced** on the wall. This process is known as charging by **induction**. The same process occurs when the charged rod and cloth attract scraps of paper. The negative rod induces a positive charge on the paper, by repelling electrons away. The positive cloth attracts the electrons.

We can use charging by induction to charge a metal object, as shown in Figure **17.5**. We start with two objects: an object A with a large negative charge, and an uncharged metal sphere B on an insulating stand. The method is as follows.

a Object A has a large negative charge. When the metal sphere B is placed near it, electrons in the sphere are repelled away. The front of the sphere (near A) has an induced positive charge.

b Now the sphere is touched, either by a hand or by a wire connected to earth. This allows electrons to escape from the sphere.

c The connection is removed. Now the sphere has a positive charge.

d Finally, the sphere B is taken away from object A. Sphere B has a uniformly distributed positive charge all over it.

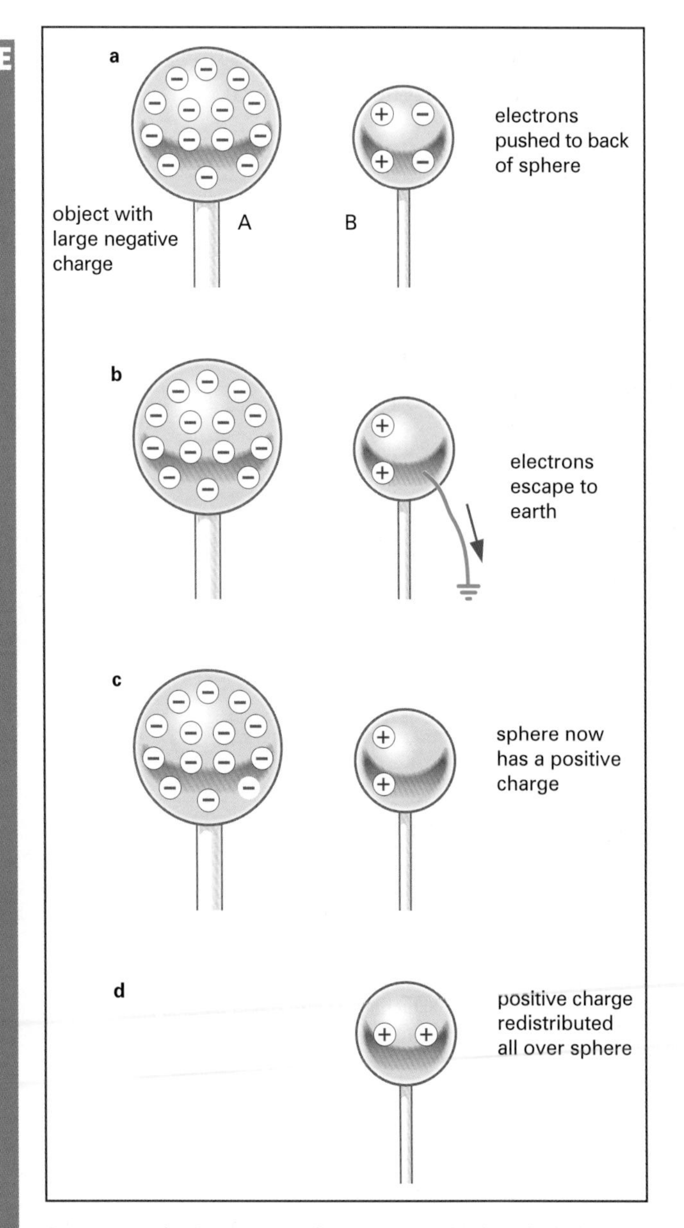

Figure 17.5 The four steps in charging a metal sphere by induction.

Note that the connection to earth must be disconnected **before** B is moved away from A. Otherwise, the electrons would simply run back up to B to neutralise its positive charge.

Note also that the sphere B and the charged object A **never touch**. Sphere B gets a charge that is opposite in sign to that of object A.

Representing an electric field

A charged object is surrounded by an electric field. If a charged object moves into the electric field of a charged object, it will experience a force – it will be attracted or repelled. Figure **17.6** shows how we represent an electric field by **lines of force** (or **electric field lines**), in a similar way to the representation of a magnetic field by magnetic field lines.

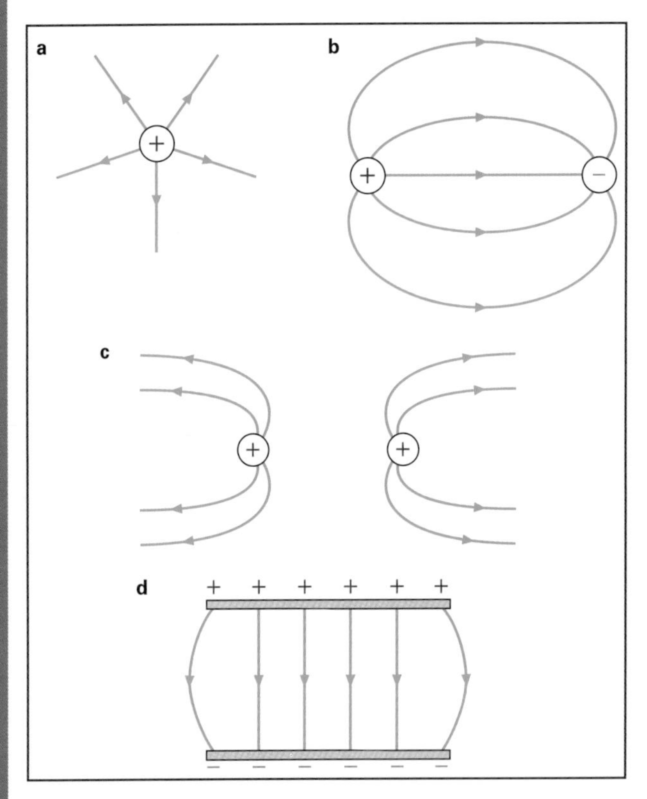

Figure 17.6 The electric field around a charged object is represented by lines of force: **a** an isolated positive charge; **b** two oppositely charged objects; **c** two like charges; and **d** two parallel plates with opposite charges.

The lines of force are shown coming out of a positive charge and going into a negative charge. This is because the lines indicate the direction of the force on a **positive** charge placed in the field. A positive charge is repelled by another positive charge and attracted by a negative charge.

When two oppositely charged objects are placed close together, they attract one another. Two objects with the same charge repel each other. You can see this in the patterns made by their lines of force.

Figure **17.6d** shows the field between two oppositely charged parallel plates. The lines of force between the plates are straight and parallel to one another (except at the edges).

Activity 17.2 Charging by induction

Use the process of induction to charge an object. Then test its charge.

What is electric charge?

In physics, we find it relatively easy to answer questions like 'What is a rainbow?' or 'How does an aircraft fly?' It is much harder to answer an apparently simple question like 'What is electric charge?' We have to answer it by saying how objects with electric charge behave. Objects with the same sign of charge repel one another. Objects with opposite charge attract. This is not a very satisfying answer, because magnetic poles behave in the same way: north poles repel north poles and attract south poles. Because electric charge is a fundamental property of matter, we have to get a feel for it, rather than having a clear definition.

The electric force between two charged objects is one of the fundamental forces of nature. (The force of gravity between two masses is another fundamental force.) The electric force holds the particles that make up an atom together. It holds atoms together to make molecules, and it holds molecules together to make solid objects. Just think: whenever you stand on the floor, it is the electric force between molecules that prevents you from falling through the floor. It is a very important force.

Charged particles

We have already seen that **electrons** are the charged particles that are transferred from one object to another when they are rubbed together. Electric charge is a property of the particles that make up atoms.

Charge is measured in **coulombs** (C), named after Charles-Augustin de Coulomb, a French physicist who worked on static electricity at about the same time as Benjamin Franklin. He discovered that the force between two charged objects depends on how big their charges are and on how far apart they are.

An electron is a negatively charged particle. It is much smaller than an atom, and only weakly attached to the outside of the atom. It is held there by the attraction of the positively charged nucleus of the atom. The nucleus is positively charged because it contains positively charged particles called **protons**.

An electron has a very tiny amount of electric charge. The **electron charge** is so small that it takes over 6 million million million electrons to make 1 C of charge:

$$\text{electron charge} = -0.000\,000\,000\,000\,000\,000\,16\,\text{C}$$
$$= -1.6 \times 10^{-19}\,\text{C}$$

A proton has exactly the same charge as electrons, but positive, so the **proton charge** is:

$$\text{proton charge} = +0.000\,000\,000\,000\,000\,000\,16\,\text{C}$$
$$= +1.6 \times 10^{-19}\,\text{C}$$

No-one knows why these values are **exactly** the same (or even if they **are** exactly the same), but it is fortunate that they are because it means that an atom that contains, say, six protons and six electrons is electrically neutral. If all the objects around us were made of charged atoms, we would live in a shocking world!

QUESTIONS

4 Draw a diagram to show how a negatively charged polythene rod can attract an uncharged scrap of paper.

5 a What charge does an electron have, positive or negative?
  b Would two electrons attract or repel one another?

6 Two identical metal spheres are placed close to one another. One is given a large negative charge. The two are then connected by a wire. Use the idea of electric force to explain what happens next.

7 a Draw a diagram to represent the electric field around a positively charged sphere.
  b Draw a diagram to represent the electric field between two horizontal parallel plates. The upper plate has a positive charge and the lower plate has a negative charge.

Summary

When one object is rubbed against another, they may gain opposite electrostatic charges.

Charged objects exert forces on each other: like charges repel; unlike charges attract.

An electric field exists anywhere where a charged object experiences a force.

Objects gain an electrostatic charge when they gain or lose electrons.

Electrons have a negative charge, so an object that gains electrons becomes negatively charged. The object that loses electrons becomes positively charged.

A charged object may attract an uncharged object. Electrons in the uncharged object move slightly, so that the object becomes charged by induction.

Electric charge is measured in coulombs, C.

End-of-chapter questions

17.1 When a Perspex rod is rubbed on a woollen cloth, the rod acquires a negative electric charge.

a What type of electric charge does the cloth acquire? [1]

b What can you say about the amounts of charge on the two charged items? [1]

c If you had two Perspex rods charged up in this way, how could you show that they both have electric charges of the same sign? [2]

E

17.2 a If you rub a balloon on woollen clothing, the balloon gains a negative electrostatic charge. Use the idea of the **movement of electrons** to explain this. [3]

b If the balloon is held close to an uncharged scrap of paper, the paper and the balloon will attract each other. Use the idea of **charging by induction** to explain this. [3]

18 Electrical quantities

Core Identifying conductors and insulators
Core Understanding electric current
E Extension Relating electric current to electron flow
Core Measuring and calculating electrical resistance
E Extension Calculating energy and power in electric circuits

Model circuits

It is very likely that you will have made electric circuits in the lab, and looked at some real-life circuits. The circuits that you have experimented with are models for circuits that have real purposes in the world. They are **models** because they present a simplified view of how circuits work. Practical circuits are usually more complex, but it makes sense to start with simple circuits to build up a picture of how electric current flows.

The photographs show two rather different types of electric circuit. Figure **18.1** shows part of the electric circuit that carries power from a large generating station to the industrial complex where it is used. Electric current is flowing through thick metal cables, held above the ground by tall pylons.

Figure 18.1 These cables carry large electric currents. Energy is being transferred from the power station to an industrial plant, where it will be used to turn machinery, split chemicals and so on.

Figure **18.2** shows the electric circuits inside a computer. There are several 'chips' (integrated

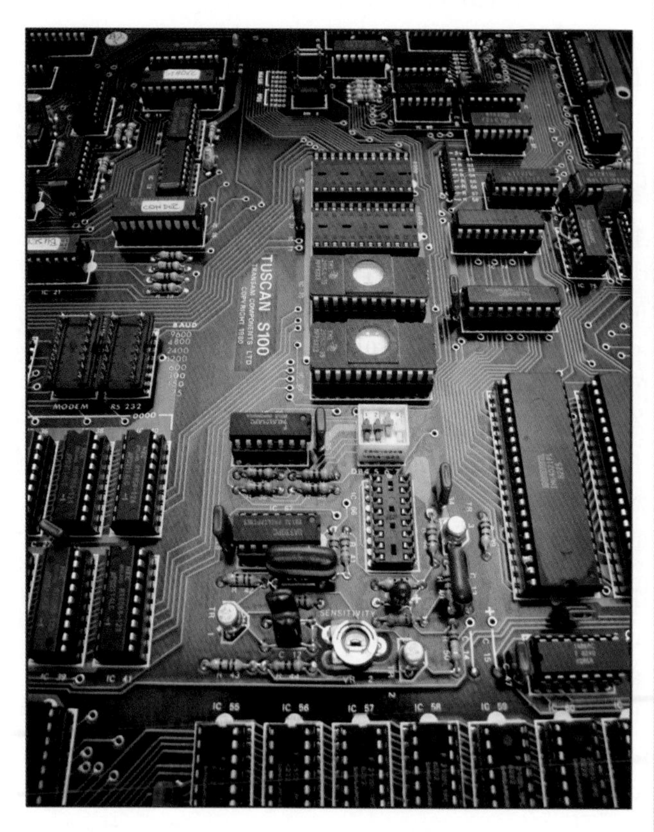

Figure 18.2 The electric circuits of a computer are highly engineered. Each of the chips (the rectangular objects) contains many millions of electric circuits, which work to process information at high speed.

circuits) in the computer. In these, electric current flows through silicon, a material that is not such a good conductor as a metal. Engineers design chips to be as small as possible. This is because, although electric current flows quickly, it is not instantaneous, and so current takes less time to flow around a small component than a larger one.

These two pictures illustrate the two general uses we have for electric circuits:

- Electricity can be used to transport **energy** from place to place. For such a use, the circuit contains devices for transforming energy. Think of a simple circuit like a torch. Energy is transferred electrically from the battery to the bulb, where it is transformed into light and heat.
- Electricity can be used to transport **information** from one place to another. Digital information comes into the computer, and its circuits then manipulate the information to produce pictures, sounds and new data. We even have electric circuits in our bodies for handling information – our brain and nerves work electrically, and it is possible to trace the flow of electricity around our bodies.

In this chapter and the next, we will look at electric circuits in detail. We shall extend your understanding of the different components that are used in circuits to control the current that flows and the energy that is transferred or transformed.

18.1 Current in electric circuits

If an electric **current** is to flow, two things are needed: a complete circuit for it to flow around, and something to push it around the circuit. The 'push' might be provided by a **cell**, battery or power supply. A **battery** is simply two or more cells connected end-to-end. In most familiar circuits, metals such as copper or steel provide the circuit for the current to flow around. Figure **18.3a** shows how a simple circuit can be set up in the lab. Once the switch is closed, there is a continuous metal path for the current to flow along. Current flows from the positive terminal of the battery (or cell). It flows through the switch and the filament lamp, back to the negative terminal of the battery. Such a current that flows in the same direction all the time is called **direct current** (**d.c.**).

Figure **18.3b** shows the same circuit as represented by a circuit diagram. Each component has its own standard symbol. If you imagine the switch being pushed so that it closes, it is clear from the diagram that there is a continuous path for the current to flow around the circuit.

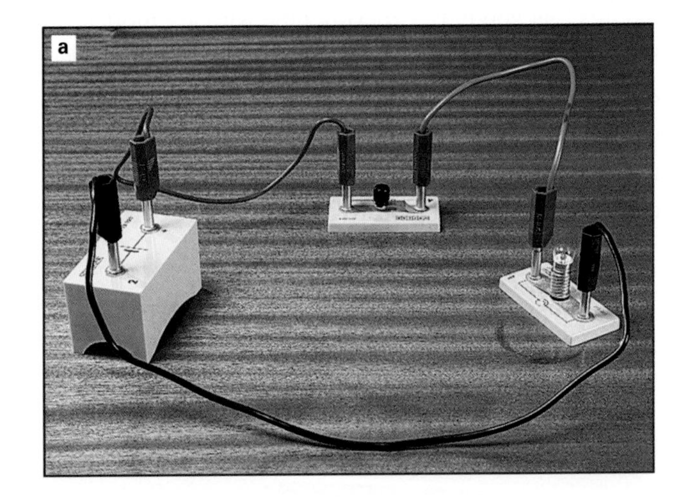

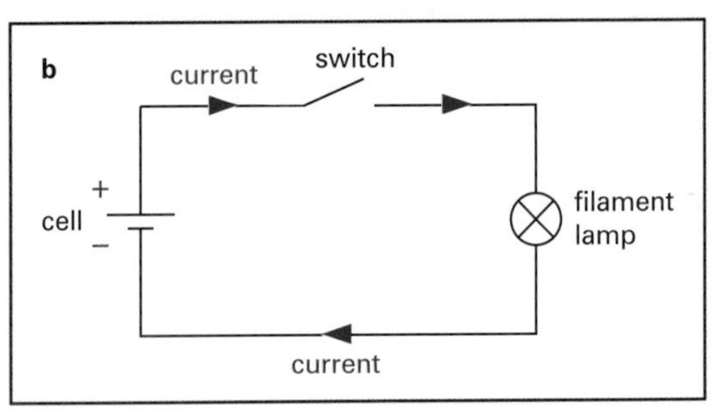

Figure 18.3 **a** A simple electric circuit, set up in a lab. **b** The same circuit represented as a circuit diagram.

It is obvious how the switch in Figure **18.3a** works. You push the springy metal downwards until it touches the other metal contact. Then the current can flow through it. Most switches work by bringing two pieces of metal into contact with one another, though you cannot usually see this happening. It is worth having a look inside some switches to see how they work.

Similarly, take a look at some filament light bulbs, like the one in Figure **18.3a**. Every bulb has two metal contacts, for the current to flow in and out. Inside, one fine wire carries the current up to the filament (which is another wire), and a second wire carries the current back down again. Notice also how the circuit symbols for these and many other components have two connections for joining them into a circuit.

Good conductors, bad conductors

The wires we use to connect up circuits are made of metal because metals are good **conductors** of electric current. The metal is usually surrounded by plastic, so that, if two wires touch, the electric current cannot pass directly from one to another (a short circuit). Plastics (polymers) are good electrical **insulators**.

- Good conductors: most metals, including copper, silver, gold, steel.
- Good insulators: polymers (such as Perspex or polythene), minerals, glass.

In between, there are many materials that do conduct electricity, but not very well. For example, liquids may conduct, but they are generally poor conductors.

People can conduct electricity – that is what happens when you get an electric shock. A current passes through your body and, if it is big enough, it makes your muscles contract violently. Your heart may stop, and burns may also result. Our bodies conduct because the water in our tissues is quite a good electrical conductor.

What is electric current?

When a circuit is complete, an electric current flows.

> Current flows from the positive terminal of the supply, around the circuit, and back to the negative terminal.

What is actually travelling around the circuit? The answer is electric **charge**. The battery or power supply in a circuit provides the push needed to make the current flow. This 'push' is the same force that causes electric charges to attract or repel one another.

> A current is a flow of electric charge.

Measuring electric current

To measure electric current, we use an **ammeter**. There are two types, as shown in Figure **18.4**.

- An **analogue meter** has a needle, which moves across a scale. With this type of meter, it is easy to see when the current flowing is increasing or decreasing. You have to make a judgement of the position of the needle against the scale.
- A **digital meter** gives a direct read-out in figures. There is no judgement involved in taking a reading. Data-loggers and personal computers use digital data.

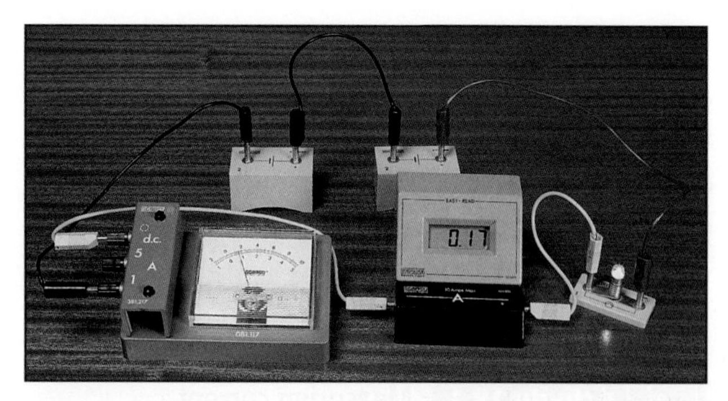

Figure 18.4 Ammeters measure electric current, in amps (A). There are two types: analogue (on the left) and digital (on the right).

An ammeter is connected into a circuit in series – that is to say, the current flows in through one terminal (red, positive) and out through the other (black, negative). If the meter is connected the wrong way round, it will give negative readings. To add an ammeter to a circuit, the circuit must be broken (see Figure **18.5**).

In a simple series circuit like the one shown in Figure **18.5**, it does not matter where the ammeter is added, since the current is the same all the way round the circuit. It does not get used up as it flows through the lamp or other components in the circuit.

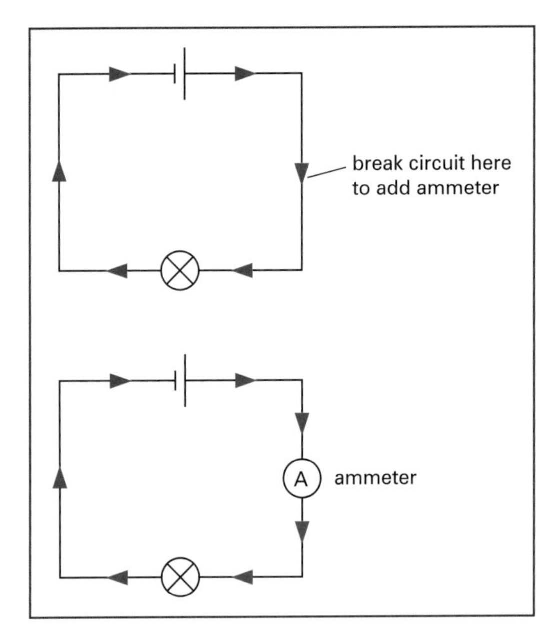

Figure 18.5 Adding an ammeter to a circuit. The ammeter is connected in series, so that the current can flow through it.

The reading on an ammeter is in **amps (A)**. The ampere (shortened to amp) is the SI unit of current. Smaller currents may be measured in milliamps (mA) or microamps (μA):

1 milliamp = 1 mA = 0.001 A = 10^{-3} A
1 microamp = 1 μA = 0.000 001 A = 10^{-6} A

Activity 18.1 Measuring current

Investigate some materials to discover which are good conductors.

Measure the current flowing in a simple circuit.

QUESTIONS

1 a What instrument is used to measure electric current?
  b How should it be connected in a circuit?
  c Draw its circuit symbol.

2 A circuit is set up in which a cell makes an electric current flow through a lamp. Two ammeters are included, one to measure the current flowing into the lamp, the other to measure the current flowing out of the lamp.
  a Draw a circuit diagram to represent this circuit.
  b Add an arrow to show the direction of the current around the circuit.
  c What can you say about the readings on the two ammeters?

3 a Name **two** materials that are good electrical conductors.
  b Name **two** materials that are good electrical insulators.

E

Two pictures: current and electrons

Metals are good electrical conductors because they contain electrons that can move about freely. (These have already been mentioned in Chapters **11** and **17**.) The idea is that, in a bad conductor such as most polymers, all of the electrons in the material are tightly bound within the atoms or molecules, so that they cannot move. Metals are different. While most of the electrons in a metal are tightly bound within their atoms, some are free to move about within the material. These are called **conduction electrons** (see Figure **18.6**). A voltage, such as that provided by a battery or power supply, can start these conduction electrons moving in one direction through the metal, and an electric current flows. Since electrons have a negative electric charge, they are attracted to the positive terminal of the battery.

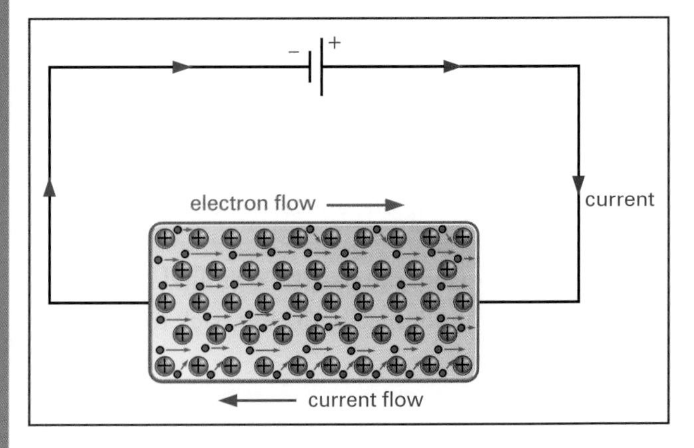

Figure 18.6 In a metal, some electrons are free to move about. These are known as conduction electrons. In copper, there is one conduction electron for each atom of the metal. The atoms, having lost an electron, are positively charged ions. A battery pushes the conduction electrons through the metal. The force is the attraction between unlike charges that was discussed in Chapter **17**.

Electric current flows from positive to negative. Figure **18.7** shows the direction of flow of charge around a simple circuit. We picture positive charge flowing out of the positive terminal, around the circuit and back into the cell at the negative terminal. Now, we know that in a metal it is the negatively charged electrons that move. They leave the negative terminal of the cell, and flow around to the positive terminal, in the opposite direction to the current. Hence we have two different pictures of what is going on in a circuit.

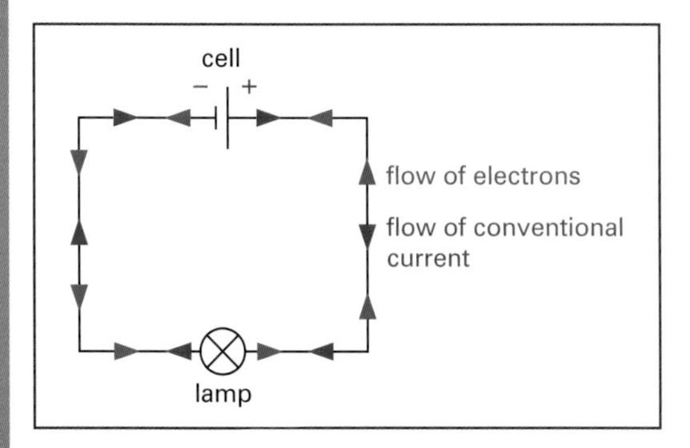

Figure 18.7 Two ways of picturing what happens in an electric circuit: conventional current flows from positive to negative; electrons flow from negative to positive.

We can think of **conventional current**, a flow of positive charge, moving from positive to negative. Conventional current is rather like a fluid moving through the wires, just like water moving through pipes. This picture does not tell us anything about what is going on inside the wires or components of a circuit. However, it is perfectly good for working out many things to do with a circuit: what the voltage will be across a particular component, for example, or how much electrical energy will be transferred to a particular **resistor**.

Alternatively, we can think of **electron flow**, a movement of conduction electrons, from negative to positive. As we will see shortly, this picture can allow us to think about what is going on inside the components of a circuit: why a resistor gets warm when a current flows through it, for example, or why a diode allows current to flow in one direction only.

The electron flow picture is a **microscopic model**, since it tells us what is going on at the level of very tiny particles (electrons and ions). The conventional current picture is a **macroscopic model** (a large-scale model).

The electrons in a circuit flow in the opposite direction to the electric current. It is a nuisance to have to remember this. It stems from the early days of experiments on static electricity. Benjamin Franklin realised that there were two types of electric charge, which he called positive and negative. He had to choose which type he would call positive, and his choice was to say that, when amber was rubbed with a silk cloth, the amber acquired a negative charge. Franklin was setting up a **convention**, which other scientists then followed – hence the term **conventional current**. He had no way of knowing that electrons were being rubbed from the silk to the amber, but his choice means that we now say that electrons have a negative charge.

Current and charge

An ammeter measures the rate at which electric charge flows past a point in a circuit – in other words, the amount of charge that passes per second. We can write this relationship between current and charge as an equation, using the symbols shown in Table **18.1**:

$$\text{current (A)} = \frac{\text{charge (C)}}{\text{time (s)}} \qquad I = \frac{Q}{t}$$

So a current of 10 A passing a point means that 10 C of charge flows past the point every second. You may find it easier to recall this relationship in the following form:

$$\text{charge (C)} = \text{current (A)} \times \text{time (s)} \qquad Q = It$$

So if a current of 10 A flows around a circuit for 5 s, 50 C of charge flows around the circuit.

Quantity	Symbol for quantity	Unit	Symbol for unit
current	I	amps	A
charge	Q	coulombs	C
time	t	seconds	s

Table 18.1 Symbols and units for some electrical quantities.

Worked example 1 shows how to calculate the charge that flows in a circuit.

Worked example 1

A current of 150 mA flows around a circuit for 1 minute. How much electric charge flows around the circuit in this time?

Step 1: Write down what you know, and what you want to know. Put all quantities in the units shown in Table **18.1**.

$I = 150\,\text{mA} = 0.15\,\text{A}$ (or $150 \times 10^{-3}\,\text{A}$)

$t = 1\,\text{minute} = 60\,\text{s}$

$Q = ?$

Step 2: Write down an appropriate form of the equation relating Q, I and t. Substitute values and calculate the answer.

$Q = It$

$Q = 0.15\,\text{A} \times 60\,\text{s} = 12\,\text{C}$

So 12 coulombs of charge flow around the circuit.

QUESTIONS

4 **a** In which direction does conventional current flow around a circuit?
b In which direction do electrons flow?

5 **a** What is the unit of electric current?
b What is the unit of electric charge?

6 **a** How many milliamps are there in 1 amp?
b How many microamps are there in 1 amp?

7 Which of the following equations shows the correct relationship between electrical units?

$1\,\text{A} = 1\,\text{C/s}$

$1\,\text{C} = 1\,\text{A/s}$

8 If 20 C of charge pass a point in a circuit in 1 s, what current is flowing?

9 A current of 4 A flows around a circuit for 10 s. How much charge flows around the circuit in this time?

18.2 Electrical resistance

If you use a short length of wire to connect the positive and negative terminals of a cell (a battery) together, you can do a lot of damage. The wire and the cell may both get hot, as a large current will flow through them. There is very little **electrical resistance** (usually simply called **resistance**) in the circuit, so the current is large. Power supplies are protected by trip switches, which cause them to cut out if too large a current flows.

The current flowing in a circuit can be controlled by adding components with electrical resistance to the circuit. The greater the resistance, the smaller the current that will flow. Figure **18.8** shows a circuit in which a cell pushes a current through a resistor. The cell provides the **voltage** needed to push the current through the resistor. Here, 'voltage' is a rather loose term, and we should say that there is a **potential difference (p.d.)** across the resistor. Potential difference is another term for voltage, and is measured in volts. It indicates that there is a difference in electrical potential across the resistor. This is rather like the difference in height that makes a ball roll downhill.

There is a special name for the p.d. across a cell. It is called the **e.m.f.** of the cell, and is also measured in volts. (The letters e.m.f. stand for **electro-motive force**, but this can be misleading since e.m.f. is a voltage, not a force.) Any component that pushes a current around a circuit is said to have an e.m.f. – cells, batteries, power supplies, dynamos and so on.

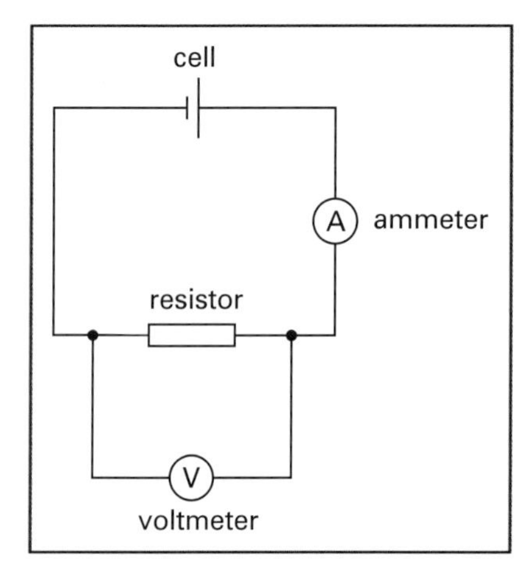

Figure 18.8 The cell provides the p.d. needed to push the current around the circuit. The amount of current depends on the p.d. and the resistance of the resistor. The ammeter measures the current flowing through the resistor. The voltmeter measures the p.d. across it. This circuit can thus be used to find the resistance of the resistor.

QUESTIONS

10 **a** What do the letters p.d. stand for?
b What meter is used to measure p.d.?
c Draw the symbol for this meter.

11 **a** What name is given to the p.d. across a cell or battery?
b What unit is this measured in?

Defining resistance

How much current can a cell push through a resistor? This depends on the resistance of the resistor. The greater its resistance, the smaller the current that will flow through it. The resistance of a component is measured in **ohms** (**Ω**) and is defined by this equation:

$$\text{resistance } (\Omega) = \frac{\text{potential difference (V)}}{\text{current (A)}} \qquad R = \frac{V}{I}$$

The circuit shown in Figure **18.8** illustrates how we can measure the resistance of a resistor (or of any other component). We need to know the current flowing through the resistor, measured by the ammeter. We also need to know the p.d. across it, and this is measured by the **voltmeter** connected across it.

The reading on a voltmeter is in **volts** (**V**). Smaller 'voltages' may be measured in millivolts (mV) or microvolts (μV). Take care not to confuse (italic) *V* as the symbol for an unknown potential difference or voltage with (upright) V for the unit, volts. In books, the first of these is shown in *italic* type (as here), but you cannot tell the difference when they are written.

A voltmeter is always connected across the relevant component, because it is measuring the potential difference between the two ends of the component.

- Ammeters are connected in series, so that the current can flow through them.
- Voltmeters are connected across a component, to measure the p.d. across the component.

Worked example **2** shows how to calculate the resistance of a resistor from measurements of current and p.d.

Worked example 2

A resistor allows a current of 0.02 A to flow through it when there is a p.d. of 10.0 V between its ends. What is its resistance?

Step 1: Write down what you know, and what you want to know. (You may prefer to write these quantities on a sketch of the situation – see Figure **18.9**.)

current $I = 0.02$ A
p.d. $V = 10.0$ V
resistance $R = ?$

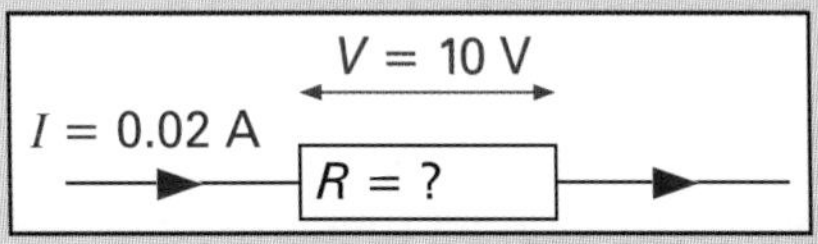

Figure 18.9 The quantities involved in Worked example **2**. Notice that we can show the current as an arrow entering (or leaving) the resistor. The p.d. is shown by a double-headed arrow, to indicate that it is measured across the resistor. The resistance is simply shown as a label on or next to the resistor – it does not have a direction.

Step 2: Write down the equation for *R*. Substitute values and calculate the answer.

$$R = \frac{V}{I} \qquad R = \frac{10.0\text{ V}}{0.02\text{ A}} = 500\,\Omega$$

So the resistance of the resistor is 500 Ω.

What is an ohm?

Let us think about the equation

$$R = \frac{V}{I}$$

that defines what we mean by resistance. We can see that it takes a p.d. of 10 V to make a current of 1 A flow through a 10 Ω resistor. It takes 20 V to make 1 A flow through a 20 Ω resistor, and so on. Hence resistance (in Ω) tells us how many volts are needed to make 1 A flow through that resistor. To put it another way:

one ohm is one volt per amp $\qquad 1\,\Omega = 1\text{ V/A}$

In the case of Worked example **2**, it would take 500 V to make 1 A flow through the 500 Ω resistor.

Changing current

You can think of an electric circuit as an obstacle race. The current (or flow of charge) comes out of the positive terminal of the cell and must travel around the circuit to the negative terminal. Along the way, it must pass through the different components. The greater their resistance, the harder it will be for the charge to flow, and so the current will be smaller.

> The greater the resistance in the circuit, the smaller the current that flows.

However, we can make a bigger current flow by increasing the p.d. that pushes it. A bigger p.d. produces a bigger current.

> The greater the p.d. in a circuit (or across a component), the greater the current that flows.

Through thick and thin

The idea of an obstacle race can help us to think about the resistance of wires of different shapes. A long, thin wire has more resistance than a short, fat one. Imagine an obstacle course that includes pipes of different sizes through which the runners have to pass. It is easy to get through a short pipe with a large diameter. It is much harder when the pipe is long and narrow.

> - The longer a wire, the greater is its resistance.
> - The greater the diameter of a wire, the less is its resistance.

QUESTIONS

12 **a** What is the resistance of a lamp if a current of 2.0 A flows through it when it is connected to a 12 V supply?

b If the p.d. across the lamp is increased, will the current flowing increase or decrease?

13 A student cuts two pieces of wire, one long and one short, from a reel.

a Which piece of wire will have the greater resistance?

b Draw a circuit diagram to show how you would check your answer by measuring the resistances of the two pieces of wire.

Measuring resistance

The circuit shown above in Figure **18.8** can be used to find the resistance of a resistor. However, the circuit would only provide a single value each for the p.d. V and the current I. A better technique is shown in Figure **18.10**. In place of the cell is a power supply, which can be adjusted to give several different values of p.d. For each value, the current is measured, and results like those shown in Table **18.2** are found. The last column in the table shows values for R calculated using $R = V/I$. These can be averaged to find the value of R.

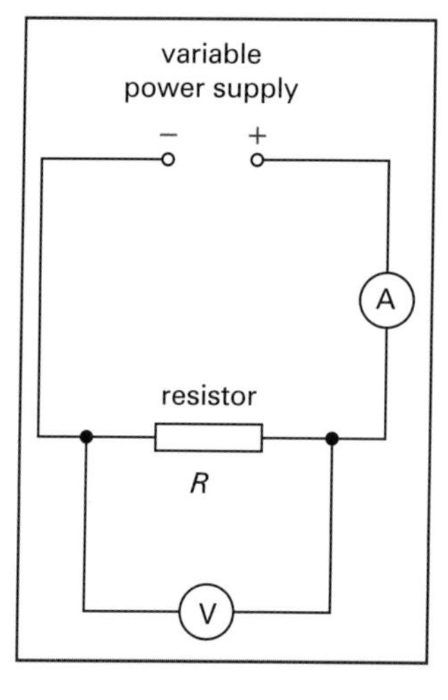

Figure 18.10 A circuit for investigating the current against voltage characteristic of a resistor. The power supply can be adjusted to give a range of values of p.d. (typically from 0 V to 12 V). For each value of p.d., the current is recorded.

P.d. V / V	Current I / A	Resistance R / Ω
2.0	0.08	25.0
4.0	0.17	23.5
6.0	0.24	25.0
8.0	0.31	25.8
10.0	0.40	25.0
12.0	0.49	24.5

Table 18.2 Typical results for an experimental measurement of resistance. The values of resistance are calculated using $R = V/I$.

Measuring resistance

Carry out some experiments to measure the resistance of some different electrical components.

Resistance calculations

The equation $R = V/I$ is used to calculate the resistance of a component in a circuit. We can rearrange the equation in two ways so that we can calculate current or p.d.:

$$I = \frac{V}{R}$$

$$V = IR$$

So, for example, we can calculate the current that flows through a 20 Ω resistor when there is a p.d. of 6.0 V across it. The current I is

$$I = \frac{6.0}{20} = 0.3\text{ A}$$

QUESTIONS

14 What p.d. is needed to make a current of 1 A flow through a 20 Ω resistor?

15 **a** What is the resistance of a resistor if a p.d. of 20 V across it causes a current of 2 A to flow through it?

b What p.d. would cause a current of 3 A to flow through the resistor?

16 What current flows when a p.d. of 14.5 V is connected across a 1000 Ω resistor?

Length and area

We have seen that the resistance of a wire depends on its length and diameter. In fact, it is the cross-sectional area of the wire that matters.

- The resistance of a wire is proportional to its length.
- The resistance of a wire is inversely proportional to its cross-sectional area.

Suppose that we have a 4.0 m length of wire. Its resistance is 100 Ω. What will be the resistance of a 2.0 m length of wire with twice the cross-sectional area? (Notice that making the wire shorter will reduce its resistance, and increasing its area will also reduce its resistance.)

Half the length gives half the resistance = 50 Ω
Doubling the area halves the resistance again = 25 Ω

QUESTION

17 A 1.0 m length of wire is found to have a resistance of 40 Ω.

a What would be the resistance of a piece of the same wire of length 2.0 m?

b What would be the resistance of a 2.0 m wire with half the cross-sectional area, made of the same material?

18.3 Electricity and energy

We use electricity because it is a good way of transferring energy from place to place. In most places, if you switch on an electric heater, you are getting the benefit of the energy released as fuel is burned in a power station, which may be over 100 km away. Mains electricity is **alternating current** (a.c.).

When you plug in an appliance to the mains supply, you are connecting up to quite a high voltage – something like 110 V or 230 V, depending on where you live. This high voltage is the e.m.f. of the supply. Recall that e.m.f. is the name given to the p.d. across a component such as a cell or power supply that pushes current around a circuit.

Why do we use high voltages for our mains supply? The reason is that a supply with a high e.m.f. gives a lot of energy to the charge that it pushes around the circuit. A 230 V mains supply gives 230 J of energy to each coulomb of charge that travels round the circuit.

The greater the e.m.f. of a supply, the greater the energy it supplies to the charges it drives round the circuit.

Electrical power

Most electrical appliances have a label that shows their power rating. An example is shown in Figure **18.11**. Power ratings are indicated in watts (W) or kilowatts (kW). The power rating of an appliance shows the rate at which it transforms energy.

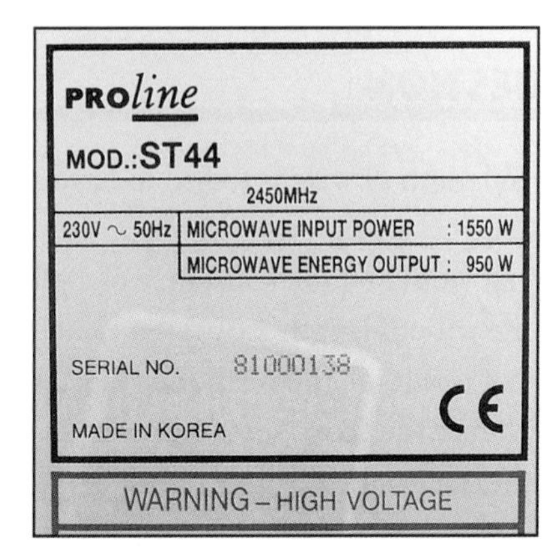

Figure 18.11 This label is fixed to the back of a microwave oven. The power rating indicates the maximum power it draws from the mains supply when the oven is operating at full power.

Power is the rate at which energy is transferred (from place to place) or transformed (from one form to another):

$$\text{power (W)} = \frac{\text{energy transformed (J)}}{\text{time taken (s)}} \qquad P = \frac{\Delta E}{t}$$

If you have studied Chapter **8**, you will recognise this definition of power. It applies to all energy transfers and transformations, not just electrical ones. We use the symbol ΔE to represent energy transferred or transformed.

This equation also reminds us of the definition of the unit of power, the watt:

one watt is one joule per second $\qquad 1\,\text{W} = 1\,\text{J/s}$

Voltage and energy

We have seen that the e.m.f. (voltage) of a supply tells us about how much energy it transfers to charges flowing around the circuit. The greater the current flowing around the circuit, the faster that energy is transferred. Hence the rate at which energy is transferred in the circuit (the power P) depends on both the e.m.f. V of the supply and the current I that it pushes round the circuit. The following equation shows how to calculate the power:

$$\text{power (W)} = \text{current (A)} \times \text{p.d. (V)} \qquad P = IV$$

You may prefer to remember this as an equation relating units:

watts = amps × volts

Calculating energy

Since energy transformed = power × time, we can modify the equation $P = IV$ to give an equation for energy transformed ΔE:

$$\text{energy transformed (J)} = \text{current (A)} \times \text{p.d. (V)} \times \text{time (s)}$$

$$\Delta E = IVt$$

Worked example 3

An electric fan runs from the 230 V mains. The current flowing through it is 0.4 A. At what rate is electrical energy transformed by the fan? How much energy is transformed in 1 minute?

Step 1: First, we have to calculate the rate at which electrical energy is transformed. This is the power, P. Write down what you know and what you want to know.

$V = 230\,\text{V}$

$I = 0.4\,\text{A}$

$P = ?$

Step 2: Write down the equation for power, which involves V and I, substitute values and solve.

$P = IV$

$P = 0.4\,\text{A} \times 230\,\text{V} = 92\,\text{W}$

Step 3: To calculate the energy transformed in 1 minute, use $\Delta E = Pt$ (or $\Delta E = IVt$). Recall that time t must be in seconds.

$\Delta E = 92 \times 60 = 5520\,\text{J}$

So the fan's power is 92 W, and it transforms 5520 J of energy each minute.

E

Activity 18.3 Using electrical power

Determine the power of some electrical components.

QUESTIONS

18 Write down an equation linking watts, volts and amps.

19 A 10 V power supply pushes a current of 5 A through a resistor. At what rate is energy transferred to the resistor?

20 A tropical fish tank is fitted with an electric heater, which has a power rating of 30 W. The heater is connected to a 12 V supply. What current flows through the heater when it is switched on?

21 How much energy is transformed by an electric lamp in 100 s if a current of 0.22 A flows through it when it is connected to a 120 V supply?

Summary

Current is a flow of charge from positive to negative.

Current (in amps) is measured using an ammeter connected in series in a circuit.

E

Conventional current is the flow of positive charge. In metals, electrons flow from negative to positive.

Charge, current and time are related by

charge = current × time $Q = It$

The potential difference (in volts, V) across a component is measured using a voltmeter, connected across the component.

Resistance (in ohms, Ω) is defined by

$$\text{resistance} = \frac{\text{p.d.}}{\text{current}} \qquad R = \frac{V}{I}$$

E

Power (in watts, W) is defined by

power (W) = current (A) × p.d. (V) $P = IV$

Energy transformed (in joules, J) is defined by

energy transformed (J) = power (W) × time (s)

$\Delta E = Pt$

energy transformed (J) = current (A) × p.d. (V) × time (s)

$\Delta E = IVt$

End-of-chapter questions

18.1 **a** Draw a circuit to show how a cell can be connected to a switch and a lamp, so that the lamp lights up when the switch is closed. Label the components in your circuit. [4]

b Add arrows to your circuit to show the direction in which electric current flows in the circuit. [2]

c Name the device you would use to measure the e.m.f. of the cell. [1]

d What unit is e.m.f. measured in? [1]

18.2 To determine the resistance *R* of a resistor, an ammeter and a voltmeter can be used.

a Draw a circuit diagram to show how you would use these instruments, together with a variable power supply, to determine *R*. [5]

b What quantity does the ammeter measure? [1]

c What quantity does the voltmeter measure? [1]

d If the voltmeter gave a reading of 6.5 V and the ammeter gave a reading of 1.25 A, what would be the value of R? [3]

E

18.3 Electrical appliances are used to transform electrical energy into other, more useful, forms of energy.

a Into what useful form of energy does a filament lamp transform electrical energy? [1]

b Into what other, less useful, form is electrical energy transformed by the lamp? [1]

c A lamp is labelled '12 V, 36 W'. This indicates that it should be used with a 12 V supply. What other information does the label provide? [1]

d How much electrical energy does the lamp transform in 1 minute? [3]

e The lamp is connected to a 12 V supply. Use the relationship $P = I V$ to calculate the current that flows through it. [3]

18.4 An electric heater is connected to a 10 V supply.

a In 20 s, 30 C of electric charge flows through the heater. Calculate the current flowing. [3]

b Calculate the energy transferred by the heater in 20 s. [4]

19 Electric circuits

Core Constructing and interpreting circuit diagrams
Core Predicting currents and voltages in series and parallel circuits
E **Extension** Describing the behaviour of electronic circuits
Extension Predicting the effects of logic gates
Core Describing and explaining electrical safety measures

An international language

The technicians in the photograph (Figure **19.1**) are checking the quality of some circuit boards. These boards carry many electrical components connected together in complex circuits.

Circuits like these are used in many different applications – in cars, radios, computers, washing machines and so on. They may be designed by electronic engineers in one country, constructed in another country, and put to use in a third country. Everyone involved in the process must understand what is required. That is why we have an internationally agreed set of circuit symbols to represent the different components used in circuits.

You are already familiar with some of these symbols – for cell, lamp, resistor, switch, ammeter and voltmeter. Figure **19.2** shows these, together with some of the others that you will learn about in this and later chapters.

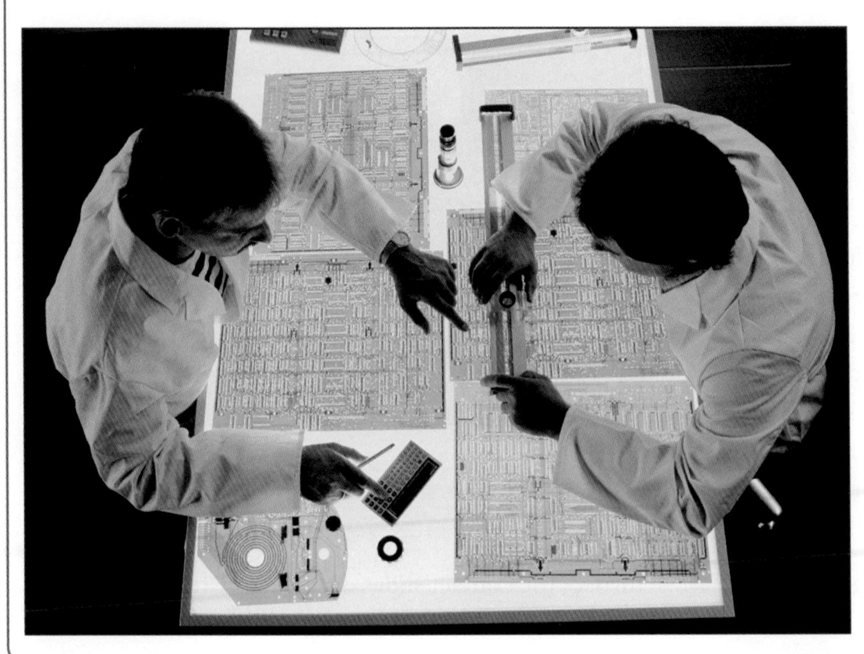

Figure 19.1 Checking circuit boards. The boards are placed on a light box or light table and technicians use magnifiers to see the fine detail.

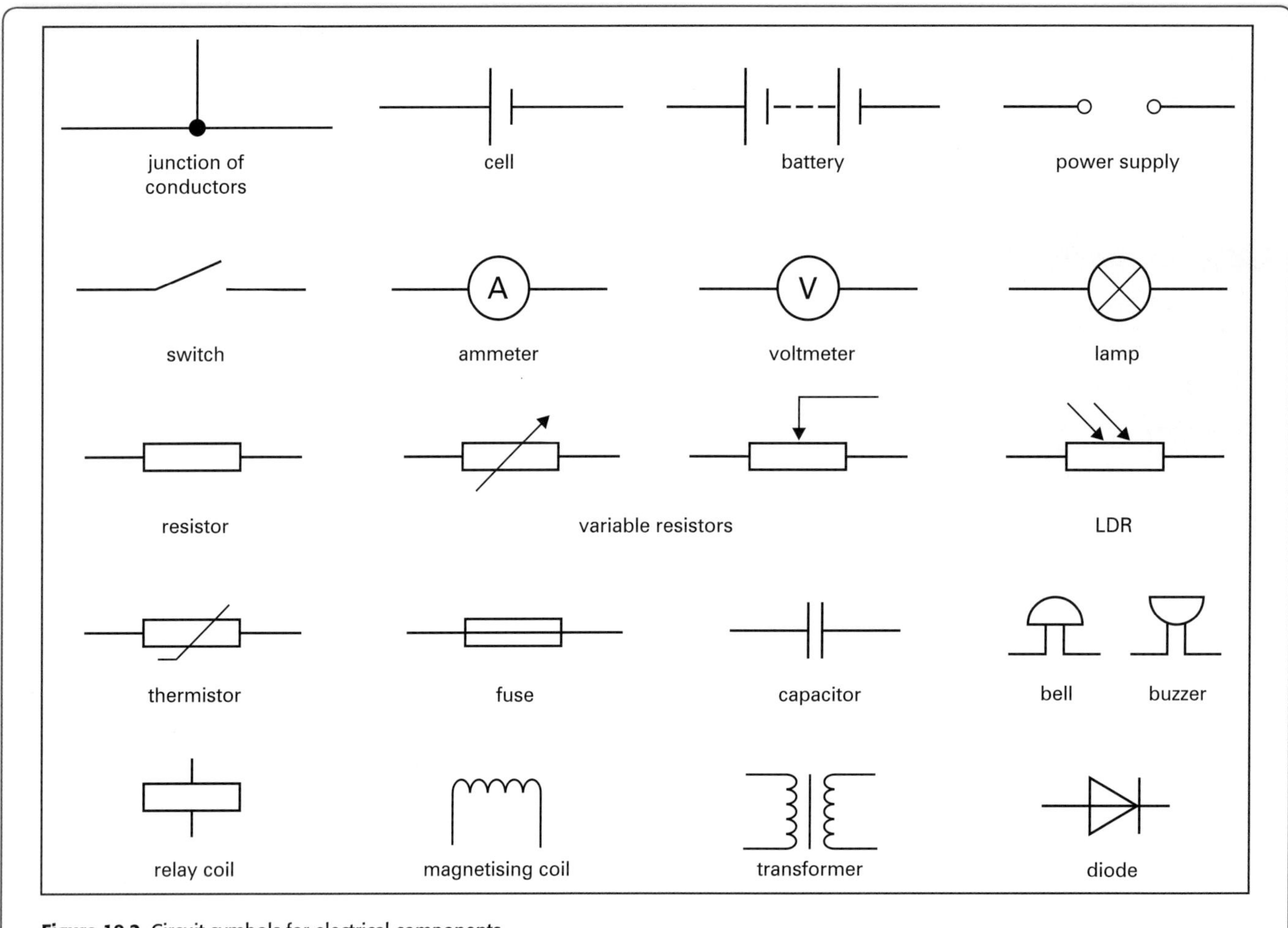

Figure 19.2 Circuit symbols for electrical components.

19.1 Circuit components

Resistors

A **resistor** (Figure 19.3) can be used to control the amount of current flowing around a circuit. A resistor has two terminals, so that the current can flow in one end and out the other. They may be made from metal wire (usually an alloy – a mixture of two or more metals with a high resistance) or from carbon. Carbon (like the graphite 'lead' in a pencil) conducts electricity, but not as well as most metals. Hence high-resistance resistors tend to be made from graphite, particularly as it has a very high melting point.

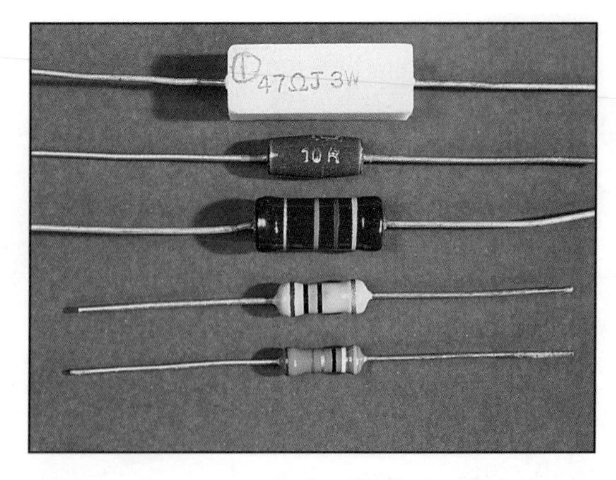

Figure 19.3 A selection of resistors. Some have colour-coded stripes to indicate their value, and others use a number code.

A **variable resistor** (sometimes called a potentiometer) can be used to alter the current flowing in a circuit. Figure 19.4a shows the inside of a variable resistor – notice that it has three terminals. As the control is turned, the contact slides over the resistive track. The current enters at one end and flows through the track until it reaches the contact, where it leaves the resistor. The amount of track that it flows through depends on

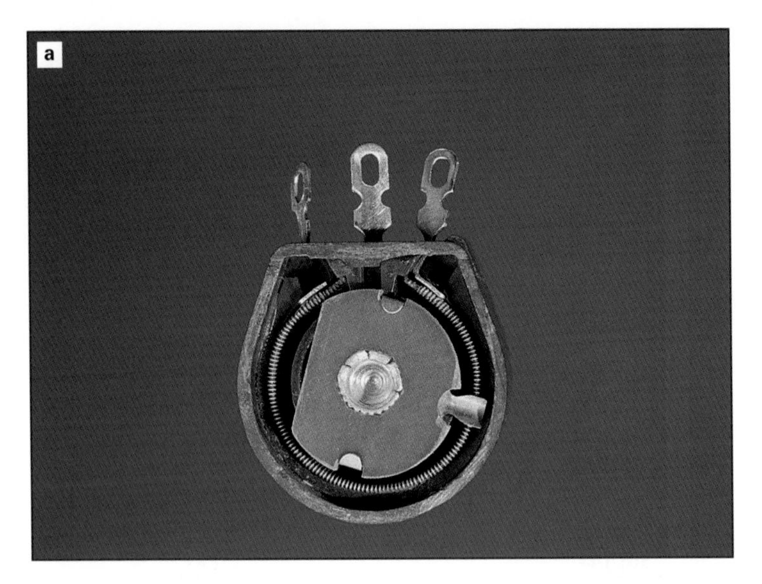

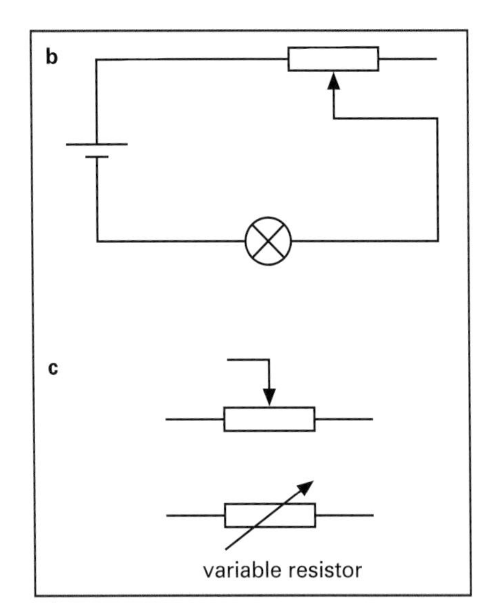

Figure 19.4 **a** A variable resistor. The resistance is provided by a 'track' of resistive wire or carbon. The resistance in the circuit depends on the position of the sliding contact. **b** The current flowing around this circuit depends on the position of the slider on the variable resistor. Imagine sliding the arrow to the right. The current will then have to flow through more resistance, and so it will decrease. **c** Symbols for a variable resistor.

the position of the contact. Variable resistors like this are often used for the volume control of a radio or stereo system. (You may have come across a **rheostat**, which is a lab version of a variable resistor.)

Figure **19.4b** shows an example of a circuit that contains a variable resistor, and Figure **19.4c** shows two different circuit symbols for a variable resistor. Note that the upper symbol has three terminals (like the resistor itself), but this circuit only makes use of two of them.

Transducers

We use many different electric circuits to make things happen automatically. For example, if the temperature falls, we may want a heater to come on automatically. If someone is moving around inside a bank at night, the burglar alarm must sound.

Electronic systems like these depend on devices described as **transducers**. A system might be represented like this:

input transducer → circuit → output transducer

An **input transducer** responds to a change in the environment (for example, a change in light or temperature) and produces a voltage. The electrical circuit to which the input transducer is connected then provides the voltage needed to operate the **output transducer**.

For a burglar alarm system, the input transducer might be a light sensor, and the output transducer could be a bell or flashing light:

light sensor → circuit → bell

We will now look at some devices that can act as input transducers.

Light-dependent resistors

A **light-dependent resistor (LDR)** is a type of 'variable resistor' whose resistance depends on the amount of light falling on it (Figure **19.5**). An LDR is made of a material that does not normally conduct well. In the dark, an LDR has a high resistance, often over 1 MΩ.

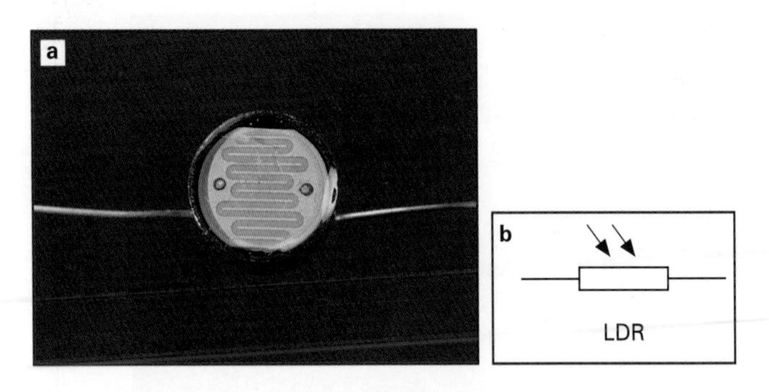

Figure 19.5 **a** A light-dependent resistor. The interlocking silver 'fingers' are the two terminals through which the current enters and leaves the resistor. In between (yellow-coloured) is the resistive material. **b** In the circuit symbol, the arrows represent light shining on the LDR.

However, light can provide the energy needed to allow a current to flow. Shine light on an LDR and its resistance decreases. In bright light, its resistance may fall to 400 Ω.

LDRs are used in circuits to detect the level of light, for example in security lights that switch on automatically at night. Some digital clocks have one fitted. When the room is brightly lit, the display is automatically brightened so that it can be seen against its bright surroundings. In a darkened room, the display need only be dim.

Thermistors

A **thermistor** (Figure **19.6**) is another type of resistor whose resistance depends on its environment. In this case, its resistance depends on its temperature. The resistance changes by a large amount over a narrow range of temperatures.

For some thermistors, the resistance **decreases** as they are heated – perhaps from 2 kΩ at room temperature to 20 Ω at 100 °C. These thermistors are thus useful for temperature probes – see the discussion of thermometers on pages **102–3**.

For other thermistors, the resistance **increases** over a similar temperature range. These are included in circuits where you want to prevent over-heating. If the current flowing is large, components may burn out. With a thermistor in the circuit, the resistance increases as the temperature rises, and the high current is reduced.

Activity 19.1 Investigating resistive components

Find out more about thermistors and light-dependent resistors.

Capacitors

A **capacitor** (Figure **19.7a**) is a component used in circuits to store energy. You can think of it as being similar to a rechargeable battery. Connect it to a power supply and it will gradually 'charge up', storing energy until the p.d. across it is equal to the e.m.f. of the supply.

Many computers contain capacitors as a form of emergency power supply. The capacitor charges up when the computer is in use. Then, if the computer's battery or mains supply fails, the energy stored in the capacitor will keep the computer operating for long enough for it to shut down safely (or to keep running until the normal supply recovers). In this way, the user does not lose any work if the battery goes flat or if the mains supply fails.

The graph in Figure **19.7b** shows how the p.d. across the capacitor increases as it charges up. It does not charge up instantaneously (unless there is no resistance in the circuit). It charges rapidly at first, and then more and more slowly. This means that it can be used in a time delay circuit.

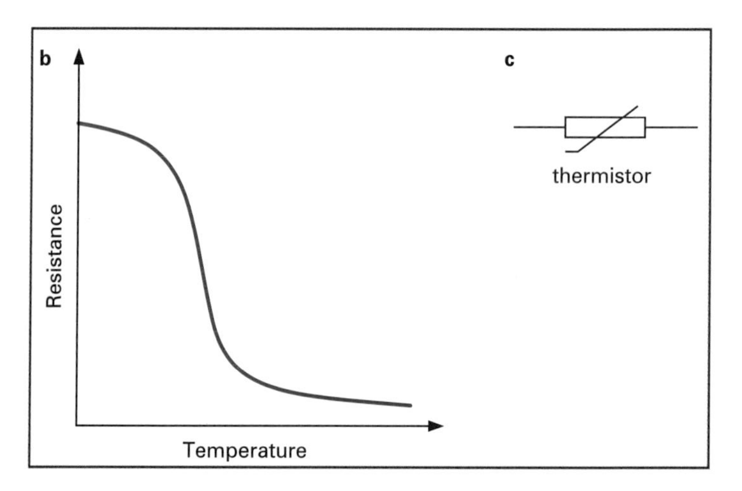

Figure 19.6 **a** A thermistor. **b** The resistance of a thermistor depends on the temperature. In this case, in the middle of the curve, its resistance drops a lot as the temperature increases by a small amount. **c** In the circuit symbol, the line through the resistor indicates that its resistance is not fixed but depends on an external factor (in this case, the temperature).

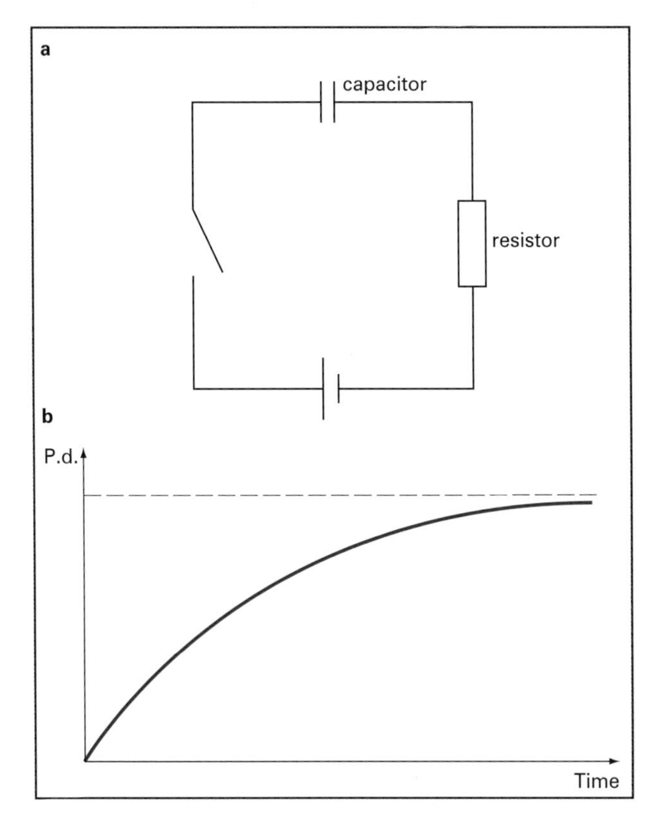

Figure 19.7 **a** A capacitor in a simple circuit. **b** The voltage across the capacitor increases as it charges up.

Suppose you want the interior light in a car to switch off 20 s after the driver has closed the door. A circuit with a capacitor will start to charge up as soon as the door is closed. After 20 s the voltage across the capacitor reaches a certain level, which operates the switch to turn off the light.

Relays

A **relay** is a type of switch that works using an electromagnet. The diagram (Figure **19.8**) shows that, when a relay is used, there are **two** circuits:

- the electromagnet coil of the relay (represented by a rectangle) is in one circuit
- the switch is in the other circuit.

When a current flows through the relay coil in the first circuit, it becomes magnetised. It pulls on the switch in the second circuit, causing it to close, and allowing a current to flow in the second circuit.

The second circuit often involves a large voltage, which would be dangerous for an operator to switch, or which could not be switched by a normal electronic circuit (because these work at low voltage).

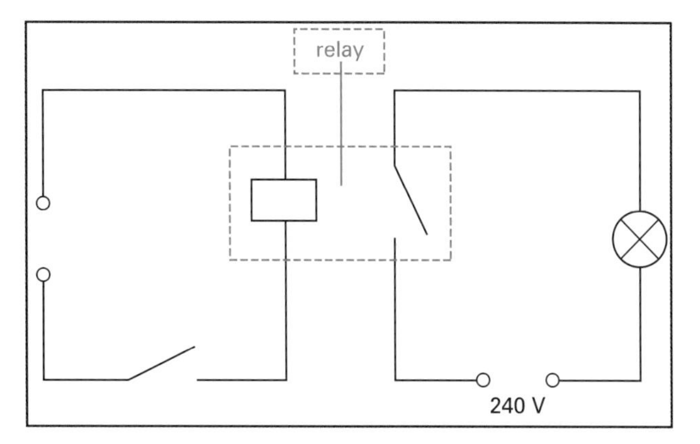

Figure 19.8 A relay is used to link two circuits together. The relay is composed of a coil and a switch (shown in the blue dashed box).

Diodes

A **diode** is a component that allows electric current to flow in one direction only. Its circuit symbol (Figure **19.9a**) represents this by showing an arrow to indicate the direction in which current can flow. The bar shows that current is stopped if it tries to flow in the opposite direction. It can help to think of a diode as being a 'waterfall' in the circuit (Figure **19.9b**). Charge can flow over the waterfall, but it cannot flow in the opposite direction, which would be uphill. Some diodes give out light when a current flows through them (Figure **19.9c**). A diode that does this is called a **light-emitting diode** (**LED**). Again, it can help to think of the waterfall. As the charge flows over the waterfall, some of the energy it loses is given out as light.

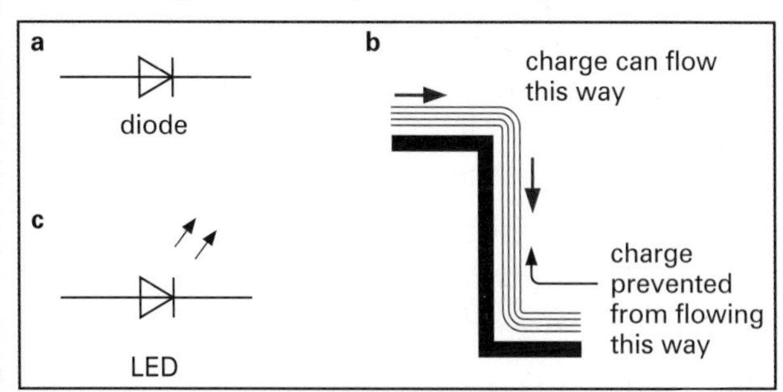

Figure 19.9 **a** Circuit symbol for a diode. A diode allows current to flow in one direction only – in the direction of the arrow. **b** A diode is rather like a waterfall. Charge can flow downhill, but is prevented from flowing back uphill. **c** Circuit symbol for a light-emitting diode. The arrows represent the light that is emitted when a current flows through it.

Diodes are useful for converting alternating current (which varies back and forth) into direct current (which flows in one direction only). This process is known as rectification and the diode acts as a **rectifier**. Rectification is necessary, for example, in a radio that operates from the mains supply. Mains electricity is alternating current (a.c.) but the radio works using direct current (d.c.).

Light-emitting diodes are familiar in many pieces of electronic equipment. For example, they are used as the small indicator lights that show whether a stereo system or television is on. Modern traffic lights often use arrays of bright, energy-efficient LEDs in place of filament bulbs. These LED arrays use very little power, so they are much cheaper to run than traditional traffic lights. Also, they require little maintenance, because, if one LED fails, the remainder still emit light.

QUESTIONS

1 a Draw the circuit symbol for a resistor.
  b Draw the circuit symbol for a variable resistor.
2 a What does LDR stand for?
  b Draw its circuit symbol.
  c What happens to the resistance of an LDR when light is shone on it?
3 a Draw the circuit symbol for a thermistor.
  b Give **one** use for a thermistor.
  c Explain why a thermistor is suitable for this use.

19.2 Combinations of resistors

If you have two resistors, there are two ways they can be connected together in a circuit: **in series** and **in parallel**. This is illustrated for two 10 Ω resistors in Figure **19.10**. It is useful to be able to work out the total resistance of two resistors like this. What is their **combined resistance** or **effective resistance**?

a For the two 10 Ω resistors in series. The current has to flow through two resistors instead of one. The resistance in the circuit is doubled, so the combined resistance is 20 Ω.

b For the two 10 Ω resistors in parallel. There are two possible paths for the current to flow along, instead of just one. The resistance in the circuit is halved, so the effective resistance is 5 Ω.

(We have not really proved these values for the combined or effective resistance, but you should see that they are reasonable values.)

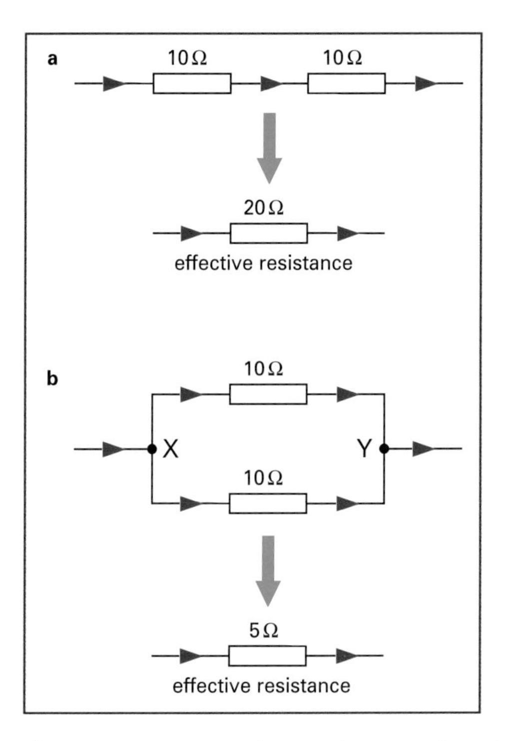

Figure 19.10 Two ways of connecting two resistors in a circuit: **a** in series and **b** in parallel.

To recognise when two resistors are connected in series, trace the path of the current around the circuit. If all the current flows through one resistor and then through the other (Figure **19.10a**), the resistors are connected in series. They are connected end-to-end. For resistors in parallel, the current flows differently. It flows around the circuit until it reaches a point where the circuit divides (point X in Figure **19.10b**). Then some of the current flows through one resistor, and some flows through the other. Then the two currents recombine (point Y in Figure **19.10b**) and return to the cell. Resistors in parallel are connected side-by-side.

Resistors in series

If several resistors are connected in series, then the current must flow through them all, one after another. The combined resistance R in the circuit is simply the sum of all the separate resistances. For three resistors in series (Figure **19.11a**), the formula for their combined resistance is:

$$R = R_1 + R_2 + R_3$$

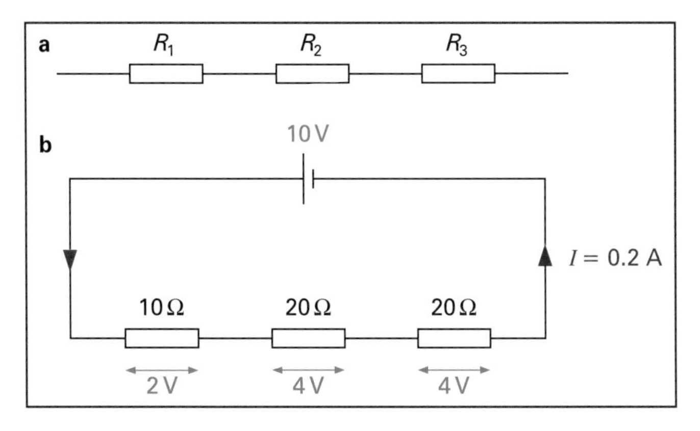

Figure 19.11 **a** Three resistors connected in series. **b** Values of current and p.d. in a series circuit. The same current I flows through each of the three resistors.

Figure **19.11b** shows the same current I flowing through three resistors – remember, current cannot be used up. We can calculate the combined resistance for this circuit:

$$\text{combined resistance} = 10\,\Omega + 20\,\Omega + 20\,\Omega = 50\,\Omega$$

So the three resistors could be replaced by a single 50 Ω resistor and the current in the circuit would be the same.

So, for **resistors in series:**

- the combined resistance is equal to the sum of the resistances
- the current is the same at all points around the circuit.

Resistors in parallel

The lights in a conventional house are connected in parallel with one another. The reason for this is that each one requires the full voltage of the mains supply to work properly. If they were connected in series, the p.d. would be shared between them and they would be dim. In parallel, each one can be provided with its own switch, so that it can be operated separately. If one bulb fails, the others remain lit.

The effective resistance of several resistors connected in parallel is less than that of any of the individual resistors. This is because it is easier for the current to flow. You can see this for three resistors in parallel in

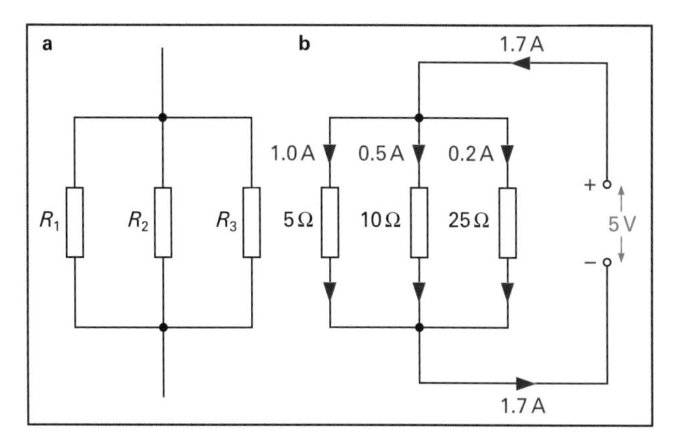

Figure 19.12 **a** Three resistors connected in parallel. **b** Values of current and p.d. in a parallel circuit. The current flowing from the supply is shared between the resistors.

Figure **19.12a**. The current flowing from the source divides up as it passes through the resistors. Figure **19.12b** shows the current from the power supply splitting up and passing through three resistors in parallel.

So, for two **resistors in parallel:**

- the effective resistance is less than the resistance of either resistor
- the current from the source is greater than the current through either resistor.

QUESTIONS

4 Name the following electrical components.

   a Its resistance decreases when light shines on it.
   b It stores energy in an electric circuit.
   c Its resistance changes rapidly over a narrow range of temperatures.
   d A current in one circuit operates an electromagnetic switch to control another circuit.

5 What component could be used as an input transducer for a fire alarm circuit? Explain your choice.

6 What is the combined resistance of two 20 Ω resistors connected in series?

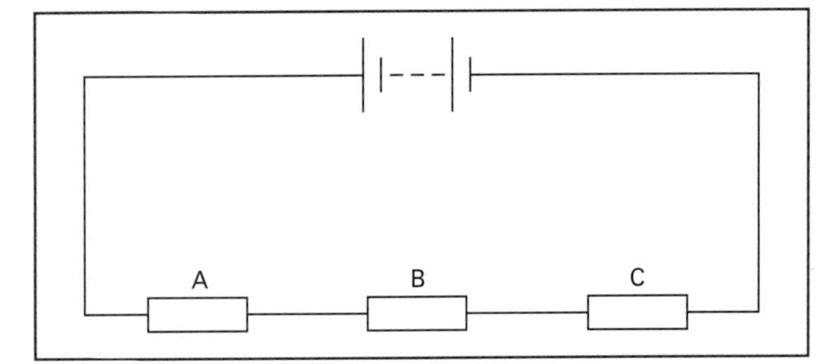

Figure 19.13 For Question **7**.

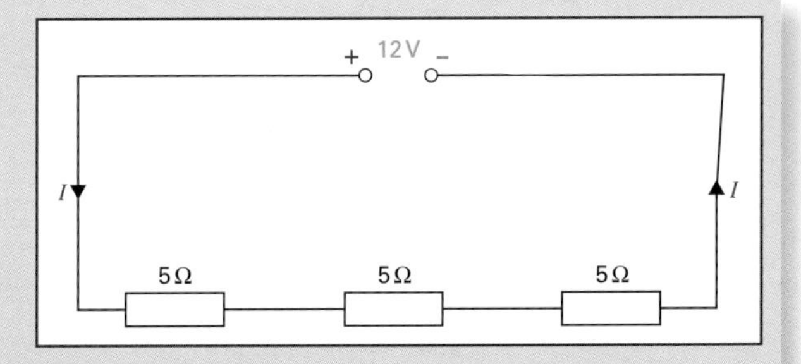

Figure 19.14 For Worked example **1**.

7 Three resistors are connected in series with a battery, as shown in Figure **19.13**. Resistor A has the greatest resistance of the three. The current through A is 1.4 A. What can you say about the currents through B and C?

8 What is the combined resistance of three 30 Ω resistors connected in series?

Voltage in series circuits

There is a p.d. across each resistor. From the numerical example shown in Figure **19.11b**, you can see that adding up the p.d.s across the three separate resistors gives the p.d. of the power supply. In other words, the p.d. of the supply is shared between the resistors. We can write this as an equation:

$$V = V_1 + V_2 + V_3$$

Festive lights, such as those used on Christmas trees, are often wired together in series. This is because each bulb works on a small voltage. If a single bulb was connected to the mains supply, the p.d. across it would be too great. By connecting them in series, the mains voltage is shared out between them. The disadvantage of this is that, if one bulb fails (its filament breaks), they all go out because there is no longer a complete circuit for the current to flow around.

Worked example 1

Three 5 Ω resistors are connected in series with a 12 V power supply. Calculate their combined resistance, the current that flows in the circuit, and the p.d. across each resistor.

Step 1: Draw a circuit diagram and mark on it all the quantities you know (see Figure **19.14**). Add arrows to show how the current flows.

Step 2: Calculate the combined resistance.

$$R = R_1 + R_2 + R_3$$
$$R = 5\,\Omega + 5\,\Omega + 5\,\Omega$$
$$R = 15\,\Omega$$

Step 3: Calculate the current flowing. A p.d. of 12 V is pushing current through a resistor of 15 Ω resistance. So:

$$\text{current } I = \frac{V}{R} = \frac{12\,\text{V}}{15\,\Omega} = 0.8\,\text{A}$$

Step 4: Calculate the p.d. across an individual 5 Ω resistor when a current of 0.8 A flows through it.

$$\text{p.d. } V = IR = 0.8\,\text{A} \times 5\,\Omega = 4\,\text{V}$$

Hence each resistor has a p.d. of 4 V across it. Note that the 12 V of the supply is shared out equally between the resistors, since each has the same resistance. We could have worked this out without knowing the current.

Current and resistance in parallel circuits

From Figure **19.12b**, you can see that the current divides up to pass through the branches of a parallel circuit. Adding up the currents through the three separate resistors gives the current flowing out of the power supply.

In other words, the current from the supply is the sum of the currents flowing through the resistors:

$$I = I_1 + I_2 + I_3$$

Because the resistors are connected side by side, each feels the full push of the supply.

To calculate the effective resistance R for three resistors in parallel, we use this formula:

$$\frac{1}{R} = \frac{1}{R_1} + \frac{1}{R_2} + \frac{1}{R_3}$$

There are two ways to calculate this type of sum: either use a calculator, or add up the fractions by finding their lowest common denominator. Worked example 2 shows how to use this formula, and how to work out the sum by finding the lowest common denominator.

Worked example 2

Two 40 Ω resistors and a 20 Ω resistor are all connected in parallel with a 12 V power supply. Calculate their effective resistance, and the current through each. What current flows from the supply?

Step 1: Draw a circuit diagram and mark on it all the quantities you know (see Figure **19.15**). Add arrows to show how the current flows.

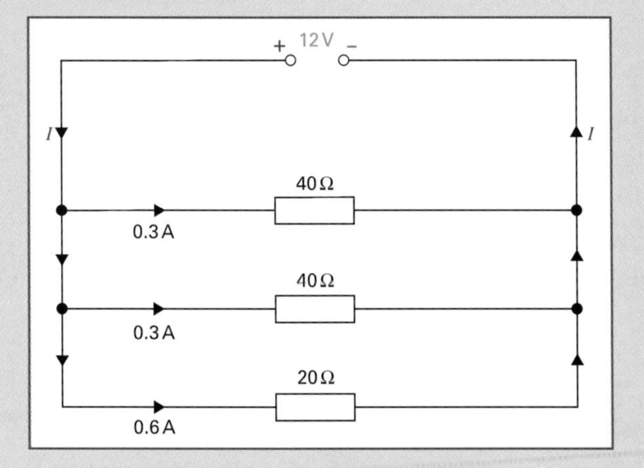

Figure 19.15 For Worked example **2**.

Step 2: Calculate the effective resistance.

$$\frac{1}{R} = \frac{1}{R_1} + \frac{1}{R_2} + \frac{1}{R_3}$$

$$\frac{1}{R} = \frac{1}{40\,\Omega} + \frac{1}{40\,\Omega} + \frac{1}{20\,\Omega}$$

$$\frac{1}{R} = \frac{1}{40\,\Omega} + \frac{1}{40\,\Omega} + \frac{2}{40\,\Omega}$$

$$\frac{1}{R} = \frac{4}{40\,\Omega}$$

$$\frac{1}{R} = \frac{1}{10\,\Omega}$$

$$R = 10\,\Omega$$

So the three resistors together have an effective resistance of 10 Ω.

Step 3: Each resistor has a p.d. of 12 V across it. Now we can calculate the currents using the equation

$$I = \frac{V}{R}$$

We get the following results for the currents.

current through 20 Ω resistor $= \frac{12\,\text{V}}{20\,\Omega} = 0.6\,\text{A}$

current through 40 Ω resistor $= \frac{12\,\text{V}}{40\,\Omega} = 0.3\,\text{A}$

These values have been marked on the diagram. Notice that, as you might expect, the smaller (20 Ω) resistor has a bigger current flowing through it than the larger (40 Ω) resistors.

Step 4: The current I flowing from the supply is the sum of the currents flowing through the individual resistors.

$$I = 0.6\,\text{A} + 0.3\,\text{A} + 0.3\,\text{A} = 1.2\,\text{A}$$

We could have reached the same result using the effective resistance (10 Ω) of the circuit that we found in Step 2.

$$I = \frac{12\,\text{V}}{10\,\Omega} = 1.2\,\text{A}$$

This is a useful way to check that you have calculated the effective resistance correctly.

E

Activity 19.2 Resistor combinations

Connect up some combinations of resistors in series and in parallel.

Measure their combined or effective resistances and compare them with calculated values.

QUESTIONS

9 Use the idea of resistors in series to explain why a long wire has more resistance than a short wire (of the same thickness and material).

10 Use the idea of resistors in parallel to explain why a thick wire has less resistance than a thin wire (of the same length and material).

11 A 10 Ω resistor is connected in series with a 20 Ω resistor and a 15 V power supply.

a Calculate the current flowing around the circuit.

b Which resistor will have the larger share of the p.d. across it?

12 What will be the effective resistance of three 60 Ω resistors connected together in parallel?

13 Two resistors of values 30 Ω and 60 Ω are connected in parallel. Calculate their effective resistance.

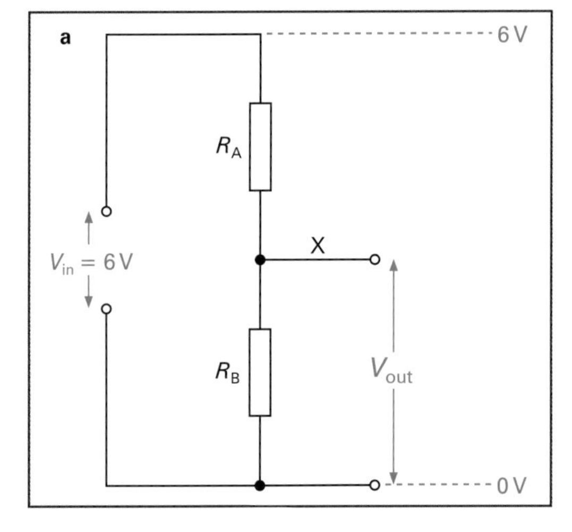

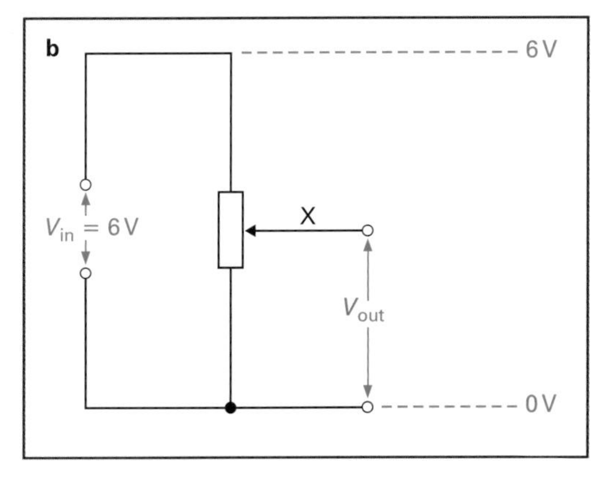

Figure 19.16 **a** A simple potential-divider circuit. The output voltage is a fraction of the input voltage. The input voltage is divided according to the relative values of the two resistors. **b** A variable resistor is used to create a potential-divider circuit, which gives an output voltage that can be varied.

Potential-divider circuits

Often, a power supply or a battery provides a fixed potential difference. To obtain a smaller p.d., or a variable p.d., this fixed p.d. must be split up using a circuit called a **potential divider**. Figure **19.16** shows two forms of potential divider.

In the circuit shown in Figure **19.16a**, two resistors R_A and R_B are connected in series across the 6 V power supply. The p.d. across the pair is thus 6 V. (It helps to think of the bottom line as representing 0 V and the top line as 6 V.) The p.d. at point X, between the two resistors, will be part-way between 0 V and 6 V, depending on the values of the resistors. If the resistors are equal, the p.d. at X will be 3 V. The p.d. of the supply will have been divided in half – hence the name **potential divider**.

To produce a variable output, we replace the two resistors with a variable resistor, as shown in Figure **19.16b**. By altering the resistance of the variable resistor, the voltage at X can have any value between 0 V and 6 V.

QUESTION

14 **a** Two resistors are to be connected to form a potential-divider circuit. Should they be connected in series or in parallel with each other?

b State briefly the function of a potential-divider circuit.

19.3 Electronic circuits

Electronic circuits (such as those found in phones, radios, mp3 players and television sets) make use of a number of other components to control the way that current flows in a circuit. In this section, we will look at transistors and logic gates.

Transistors

A **transistor** is an electronic component with three terminals or connections. Figure **19.17** shows some examples, together with the circuit symbol. The three terminals are called the collector (c), base (b) and emitter (e). In the circuits in which we are interested, a transistor acts as a device for controlling the flow of current. In the following paragraph we describe how it works.

The transistor is connected so that there is a p.d. of 6 V between the collector and the emitter. A current wants to flow from the collector to the emitter. The arrow indicates the direction of current flow. The transistor has a very high resistance, so no current flows. However, if a small current flows into the base, this greatly reduces the transistor's resistance, and so a large current flows from the collector to the emitter. Hence a small base current I_b permits a large current I_c to flow into the collector. The two currents join together and flow out through the emitter. We can show this as a formula:

base current + collector current = emitter current

$$I_b + I_c = I_e$$

Hence a small current (I_b) controls a large current (I_c). When no current flows through the transistor, we say that it is OFF. When a current flows, it is ON. So the transistor acts as a switch, controlled by the base current.

A practical switching circuit

Now we will look at an example of a circuit in which sufficient current can flow through a transistor to operate a buzzer (Figure **19.18**). This circuit sounds an alarm when the temperature rises – it could be part of a fire alarm system.

The input to the transistor is a potential-divider circuit (on the left), consisting of a fixed resistor and a thermistor, connected in series. The resistance of the thermistor decreases if it gets hot.

To switch the transistor ON, we need a current to flow into the base. For this to happen, the voltage at point X

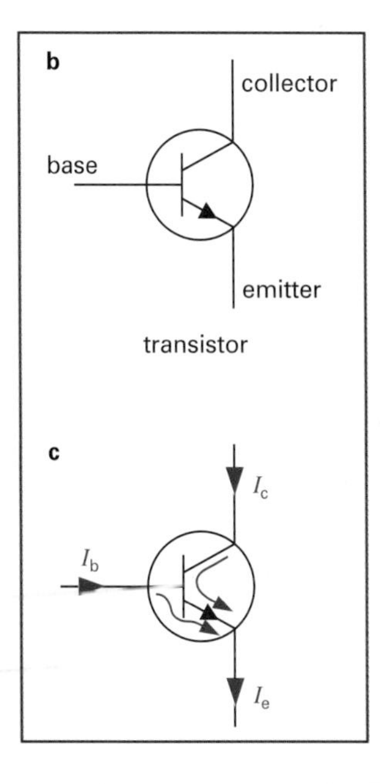

Figure 19.17 **a** Some transistors. **b** The circuit symbol for a transistor, along with the names of the three terminals. **c** With no current flowing into the base, no current can flow through the transistor. When a small current flows into the base, a large current flows into the collector, through the transistor and out through the emitter. (The arrows show the direction of current flow. The emitter emits electrons in the opposite direction to the flow of conventional current, and the collector collects them.)

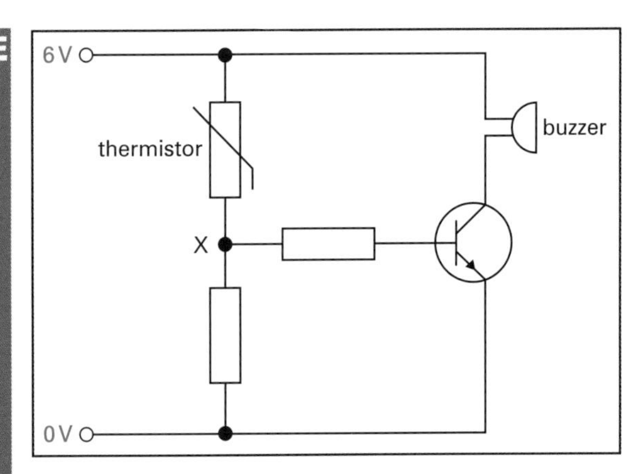

Figure 19.18 A transistor acting as a switch – it switches the buzzer on when the temperature rises.

must be high. This will happen when the resistance of the thermistor is low, that is, when it is hot. So here is how the circuit works.

- When the thermistor is cold, its resistance is high. The voltage at point X is low, and so the base current I_b flowing into the transistor is too small to cause a current to flow through the transistor. The transistor is OFF.
- When the thermistor is hot, its resistance is low. The voltage at point X will be close to 6 V, and this voltage makes a small current flow into the base of the transistor. (The resistor is there to make sure the current does not get too big.)
- This allows a large current I_c to flow through the transistor, which is ON. The current also flows through the buzzer, which buzzes.

Figure **19.19** shows how we could achieve the same thing using a relay. In this case, the rising voltage at X

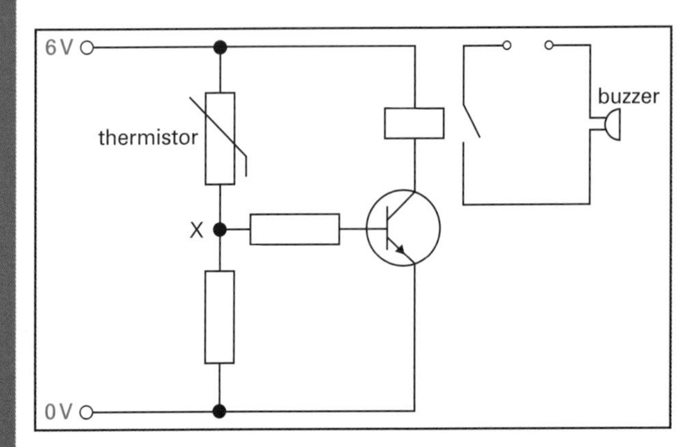

Figure 19.19 Using a relay in place of a transistor in an automatic fire alarm circuit. Compare this with the circuit in Figure **19.18**.

makes a current flow through the coil of the relay. This causes the switch to close in the buzzer circuit.

Light sensor circuits

We could set up similar circuits using transistors and relays to respond to changes in light levels. In this case, the thermistor would be replaced by a light-dependent resistor (LDR). As the brightness of the light increased, the resistance of the LDR would decrease. Eventually, this would switch the transistor ON, and the current through the relay would close the switch. This could be used, for example, to warn a glasshouse operator when the sunlight was becoming too bright and might damage the growing plants.

If you set up the potential-divider circuit with the resistor and the LDR as shown in Figure **19.20**, it would operate in a slightly different way. This could be used to switch a light on automatically at night. In the dark, the LDR would have a high resistance. The voltage at X would be high, the transistor would be switched ON and the light would switch on. During the day, the resistance of the LDR would be low and so would the voltage at X, switching the transistor OFF.

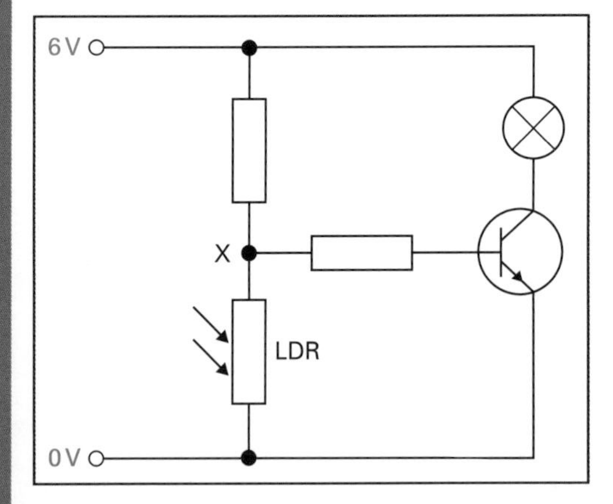

Figure 19.20 A transistor circuit in which a lamp is switched on when the LDR is in the dark.

QUESTIONS

15 Draw the circuit symbol for a transistor. Label each of the three terminals with its correct name.

16 Draw a circuit to do the following: when bright light falls on a light-dependent resistor, a transistor switches on a buzzer.

Analogue and digital

Circuits in which a transistor acts as a switch are described as **digital**. The transistor is either ON or OFF – there is no in-between state. Digital electronic systems are very useful for storing and transferring information. Computers, mobile phones and mp3 players all work digitally.

In digital systems, each piece of information (such as a number or a letter) is represented in binary form as a sequence of 1s (ones) and 0s (zeros). In a circuit, a 1 corresponds to a high voltage, or to a transistor being switched ON. A 0 corresponds to a low voltage (close to 0 V) or to a transistor being switched OFF.

The opposite of a digital system is an **analogue** system. In an analogue system, voltages can have any value, positive or negative.

Logic gates

Rather than working with individual transistors, electronic engineers prefer to use logic gates. Each logic gate has a specific function, and many can be combined together to produce complex effects. However, inside each logic gate there are a number of transistors working as switches, together with other components.

A **logic gate** is a device that receives one or more electrical input signals, and produces an output signal that depends on those input signals. These signals are voltages:

- a high voltage is referred to as ON, and is represented by the symbol 1
- a low voltage is referred to as OFF, and is represented by the symbol 0.

It is easiest to understand this by looking at three specific examples: the **AND**, **OR** and **NOT** gates, whose circuit symbols are shown in Figure **19.21**. The first two symbols have two inputs on the left and a single output on the right. The third one has one input on the left and one output on the right.

a An AND gate functions like this: its output is ON if both input 1 **and** input 2 are ON.

b An OR gate functions like this: its output is ON if either input 1 **or** input 2 **or** both is ON.

c A NOT gate functions like this: its output is ON if its input is **not** ON.

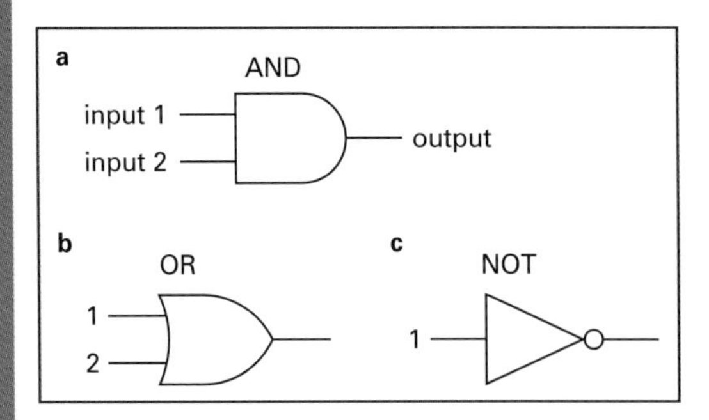

Figure 19.21 Circuit symbols for three logic gates: **a** AND, **b** OR and **c** NOT.

Let us look at a practical example. An OR gate might be useful in a heating system for the rooms in a house. There might be temperature sensors in two rooms. If a room was cold, the sensor would send an ON signal to the OR gate. If either room was cold, the output of the gate would be ON, and this would switch on the heaters.

The way in which these three gates operates is clear from their names. Another way to remember how they operate is by learning their truth tables, shown in Figure **19.22** (next page). In a truth table, we use 0 and 1 to stand for OFF and ON.

A **truth table** shows all the possible combinations of inputs, and the output that results from each combination. The NOT gate (Figure **19.22c**) has only one input, which can be ON or OFF, so this is the simplest table. The AND gate (Figure **19.22a**) and OR gate (Figure **19.22b**) both have two inputs. So there are four possible combinations of inputs, and there is a corresponding output for each. For example, you can see from the last line in the truth table for the AND gate that two input 1s give an output 1. For all other combinations of inputs, the output is 0. You should check that you understand how these truth tables represent the same information as in the sentences above that describe these gates.

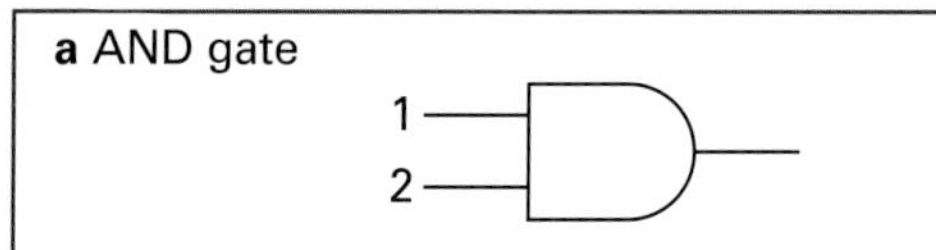

input 1	input 2	output
0	0	0
1	0	0
0	1	0
1	1	1

b OR gate

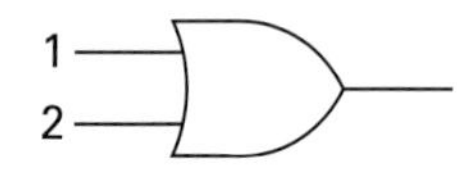

input 1	input 2	output
0	0	0
1	0	1
0	1	1
1	1	1

c NOT gate

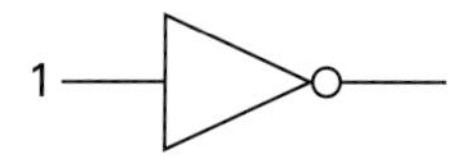

input 1	output
0	1
1	0

Figure 19.22 Truth tables for three logic gates: **a** AND, **b** OR and **c** NOT. In a truth table, 0 stands for OFF or a low voltage; and 1 stands for ON or a high voltage.

Combining logic gates

Computer chips (microprocessors) are made up of many millions of logic gates. They combine together to produce outputs that depend on many different inputs.

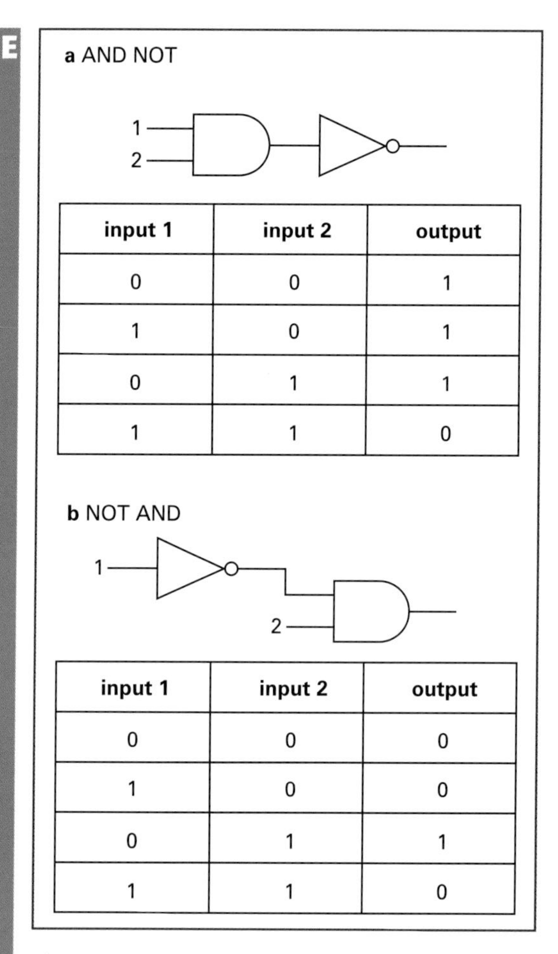

a AND NOT

input 1	input 2	output
0	0	1
1	0	1
0	1	1
1	1	0

b NOT AND

input 1	input 2	output
0	0	0
1	0	0
0	1	1
1	1	0

Figure 19.23 Two ways of combining a NOT gate with an AND gate, together with the resulting truth tables: **a** AND NOT and **b** NOT AND.

We will restrict ourselves to some simple examples involving just a few gates, to illustrate the principles involved.

Figure **19.23a** shows an AND gate with a NOT gate connected to its output. We can work out the truth table for this combination by realising that the **output** of the AND gate is the **input** of the NOT gate. When the AND gate output is 1, the NOT gate turns this into a 0.

Figure **19.23b** shows the same gates but connected together differently, along with the resulting truth table. This shows that the order in which gates are connected together is important. By combining the same gates in different orders, we can achieve different effects.

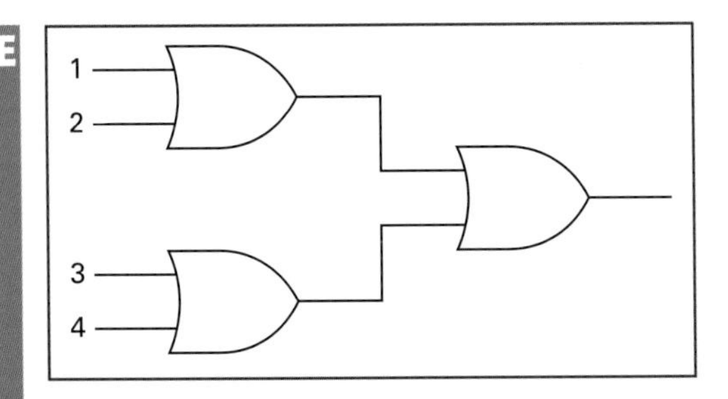

Figure 19.24 Three OR gates connected together. The combination has four inputs, so there are 16 different possible combinations of input signals. If one or more inputs is ON, the output is ON.

Figure 19.24 shows a combination of three OR gates. Let us look at a practical example of how this might function. A building has smoke detectors in four different places. Their outputs are connected via this combination of OR gates to a single alarm siren. If any detector gives an ON signal, the siren will be switched on. This saves the expense of a separate siren for each detector.

a NAND gate

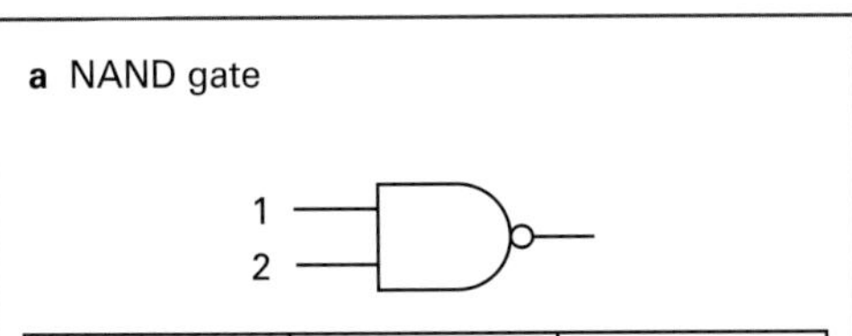

input 1	input 2	output
0	0	1
1	0	1
0	1	1
1	1	0

b NOR gate

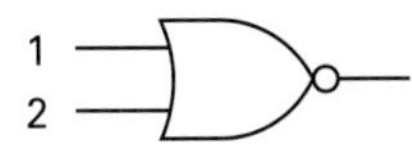

input 1	input 2	output
0	0	1
1	0	0
0	1	0
1	1	0

Figure 19.25 Symbols for **a** NAND gate and **b** NOR gate, together with their truth tables. The little circle on each symbol is like the circle on the NOT gate symbol.

Two more logic gates

Figure **19.25** shows the symbols for two more logic gates, the **NAND** and **NOR** gates, each of which has two inputs and a single output. Their truth tables are also shown. From the truth tables, you should see that these gates can be described as follows.

a A NAND gate functions like this: its output is ON if both input 1 **and** input 2 are **not** ON.

b A NOR gate functions like this: its output is ON if neither input 1 **nor** input 2 is ON.

You could construct a NAND gate by connecting a NOT gate to the output of an AND gate, so AND + NOT = NAND. Similarly, you could construct a NOR gate by connecting a NOT gate to the output of an OR gate, so OR + NOT = NOR.

Activity 19.3 Logical thinking

Solve some problems involving logic gates.

QUESTIONS

17 The output of a NOT gate is connected to the input of another NOT gate.
  a Draw up a truth table for this arrangement.
  b Write a sentence to describe its effect.

18 **a** Draw the symbol for a NOR gate.
  b Draw a truth table to represent the operation of this gate.
  c Write a sentence summarising its operation.

19 Look at the combination of gates shown in Figure **19.23b**, together with its truth table.
  a Draw the same combination of gates, but with the NOT gate connected to the other input of the AND gate.
  b Draw up the corresponding truth table for this new arrangement.

20 **a** Using the correct symbols, show the following. The outputs of two AND gates are connected to the inputs of a third AND gate.
  b When is the output of this third gate ON?
  c Suggest a use for this combination of gates.

19.4 Electrical safety

Mains electricity is hazardous, because of the large voltages involved. If you come into contact with a bare wire at 240 V, you could get a fatal electric shock. Here, we will look at some aspects of the design of electrical systems and see how they can be used safely.

Electrical cables

The cables that carry electric current around a house are carefully chosen. Figure **19.26** shows some examples. For each, there is a maximum current that it is designed to carry. A 5 A cable (Figure **19.26a**) is relatively thin. This might be used for a lighting circuit, since lights do not require much power, so the current flowing is relatively small. The wires in a 30 A cable (Figure **19.26c**) are much thicker. This might be used for an electric cooker, which requires much bigger currents than a lighting circuit.

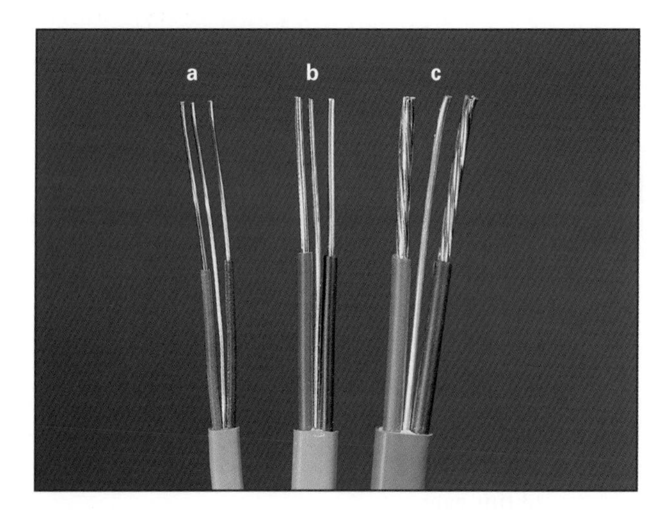

Figure 19.26 Cables of different thicknesses are chosen according to the maximum current that they are likely to have flowing through them: **a** 5 A, **b** 15 A and **c** 30 A. Each cable has live, neutral and earth wires, which are colour coded. In this case, the earth wire does not have its own insulation.

The wires in each cable are insulated from one another, and the whole cable has more protective insulation around the outside. If this insulation is damaged, there is a chance that the user will touch the bare wire and get an electric shock. There is also a chance that current will flow between two bare wires, or from one bare wire and any piece of metal it comes into contact with. Often, the metal case of an electrical appliance is **earthed** by connecting it to the earth wire to reduce the chances of a fatal electric shock.

Another hazard can arise if an excessive current flows in the wires. They will heat up and the insulation may melt, causing it to emit poisonous fumes or even catch fire. Thus it is vital to avoid using appliances that draw too much current from the supply. Fuses help to prevent this from happening – see below.

When using electricity, it is important to avoid damp or wet conditions. Recall that water is an electrical conductor (see page **187**). So, for example, if your hands are wet when you touch an electrical appliance, the water may provide a conductive path for current to flow from a live wire through you to earth. That could prove fatal.

Fuses

Fuses are included in circuits to stop excessive currents from flowing. If the current gets too high, cables can burn out and fires can start. A **fuse** contains a thin section of wire, designed to melt and break if the current gets above a certain value. Usually, fuses are contained in cartridges, which make it easy to replace them, but some fuses use fuse wire, as shown in Figure **19.27**. The thicker the wire, the higher the current that is needed to make it 'blow'. A fuse represents a weak link in the electricity supply chain. Replacing a fuse is preferable to having to rewire a whole house.

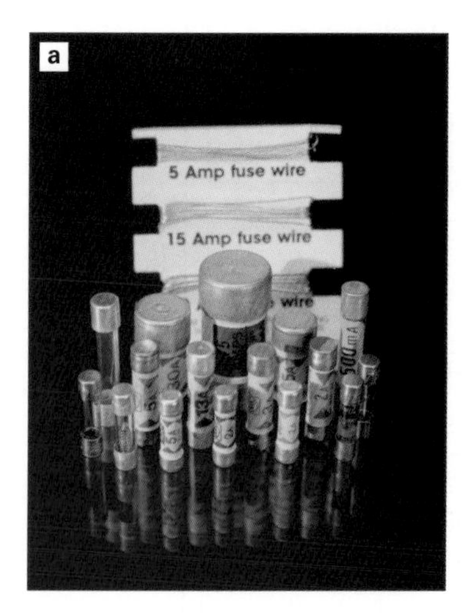

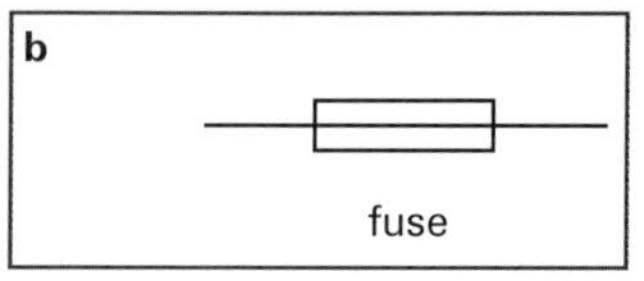

Figure 19.27 **a** Cartridge fuses and fuse wire. The thicker the wire, the higher the current that causes it to blow. **b** The circuit symbol for a fuse.

It is important to choose a fuse of the correct value in order to protect an appliance. The current rating of the fuse should be just above the value of the current that flows when the appliance is operating normally (see Worked example 3).

Worked example 3

A 2kW heater works on a 230V mains supply. The current flowing through it in normal use is 8.7 A. What current rating would a suitable fuse have? Choose from 3 A, 13 A and 30 A.

Step 1: The 3 A fuse has a current rating that is too low, and it would blow as soon as the heater was switched on.

Step 2: The 30 A fuse would not blow, but it is unsuitable because it would allow an excessive current (say, 20 A) to flow, which could cause the heater to overheat.

Step 3: The 13 A fuse is the correct choice, because it has the lowest rating above the normal operating current.

Circuit breakers

There are two types of **circuit breaker** used in electrical safety – try not to confuse them. Both work using electromagnets, but we will not consider the detail of their construction here.

A **trip switch** can replace a fuse. When the current flowing through the trip switch exceeds a certain value, the switch 'trips', breaking the circuit. Some modern house wiring systems use trip switches instead of fuses in the fuse box (Figure **19.28**). You have probably come across trip switches on lab power supplies. If too much current starts to flow, the supply itself might overheat and be damaged. The trip switch jumps out, and you may have to wait a short while before you can reset it.

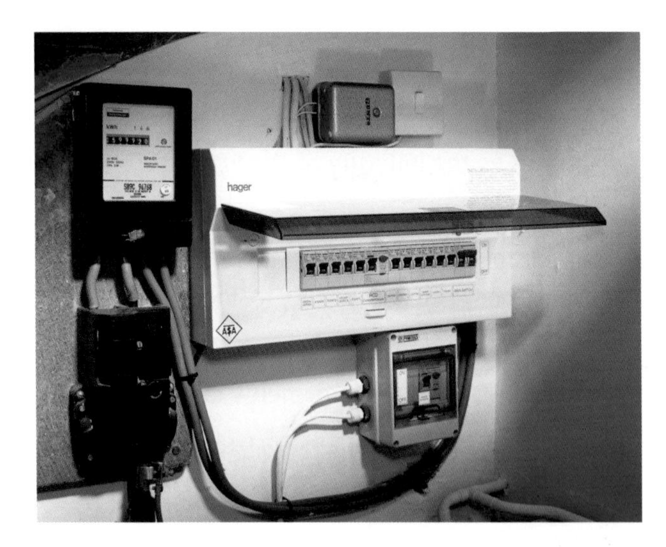

Figure 19.28 This is where the mains electricity supply enters a house. On the left is the meter. The white box contains a trip switch for each circuit in the house, together with an RCD, which protects the users of any circuit.

A **residual-current device** (**RCD**) protects the user rather than an appliance or cable. In normal circumstances, the currents flowing in the live and neutral wires are the same, because they form part of a series circuit. However, suppose that there is a fault. Someone cutting the lawn with an electric lawnmower has accidentally damaged the live wire, by running over the flex with the mower. Some current then flows through the user, rather than along the neutral wire. Now more current is flowing in the live wire than in the neutral. The RCD detects this and switches off the supply. Houses often have RCDs fitted next to the fuse box. School labs usually have one too, to protect students and teachers.

Activity 19.4 Electrical safety

Find out more about electrical hazards.

QUESTIONS

21 In normal use, a current of 3.5 A flows through a hairdryer. Choose a suitable fuse from the following: 3 A, 5 A, 13 A, 30 A. Explain your choice.

22 **a** Why are fuses fitted in the fuse box of a domestic electricity supply?

b What device could be used in place of the fuses?

23 What hazards can arise when the current flowing in an electrical wire is too high?

Summary

A light-dependent resistor has a resistance that decreases as the intensity of the light falling on it increases.

A thermistor has a resistance that changes (increases or decreases) as its temperature increases.

A capacitor is used to store energy in a circuit.

A relay is an electromagnetic switch.

E A diode allows current to flow through it in one direction only.

Resistors in series: the same current flows through all resistors in series.

Resistors in series: their combined resistance is the sum of their individual resistances:
$R = R_1 + R_2 + R_3$

E Resistors in series: the p.d. of the supply is equal to the sum of the p.d.s across the resistors:
$V = V_1 + V_2 + V_3$

Resistors in parallel: all resistors in parallel have the same p.d. across them.

Resistors in parallel: the current from the supply is shared between resistors in parallel.

Resistors in parallel: the effective resistance of two resistors in parallel is less than the resistance of either resistor.

E Resistors in parallel: the current from the supply is equal to the sum of the currents in each of the separate branches:
$I = I_1 + I_2 + I_3$

Resistors in parallel: their effective resistance is calculated from the formula:

$$\frac{1}{R} = \frac{1}{R_1} + \frac{1}{R_2} + \frac{1}{R_3}$$

A potential-divider circuit consists of two resistors connected in series across the input p.d., and can be used to produce a lower variable p.d.

A potential-divider circuit can include a light-dependent resistor (LDR), so that the output p.d. depends on the light level.

A potential-divider circuit can include a thermistor, so that the output p.d. depends on the temperature.

E A transistor can be used as an electronic switch.

Logic gates are combined to give digital control circuits.

Fuses and circuit-breakers protect the mains wiring in a house and the people who use them.

End-of-chapter questions

19.1 Draw the circuit symbol for each of the following electrical components.
a resistor **b** lamp **c** bell **d** fuse [4]

19.2 **a** Draw a circuit in which two resistors are connected in series with each other, and with a switch and a 6 V supply. [4]

b If the two resistors in part **a** have values of 10 Ω and 40 Ω, what will be their combined resistance? [2]

c If the current flowing from the supply is 0.12 A, what current will flow through the 10 Ω resistor? [1]

d What current will flow back to the supply? [1]

19.3 Name the following electrical components.

a It stores energy in a circuit. [1]

b Its resistance decreases when light shines on it. [1]

c It acts as an electromagnetic switch. [1]

19.4 An electric circuit is designed to carry a current of 10 A.

a What problem may arise if the current rises above this value? [1]

b Name **two** devices that could be fitted into the circuit to protect the circuit if the current becomes dangerously high. [2]

c If the circuit is required to carry a higher current, how should the wiring be changed? Explain your answer. [3]

19.5 Figure **19.29** shows an electric circuit in which current flows from a 6 V battery through two resistors.

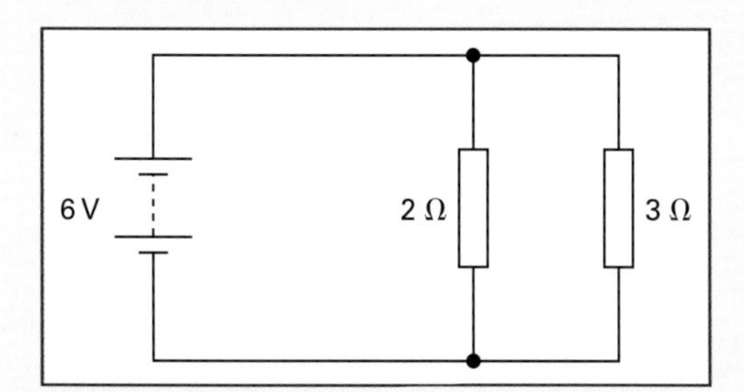

Figure 19.29 For Question **19.5**.

a Are the resistors connected in series or in parallel with each other? [1]

b For each resistor, state the p.d. across it. [2]

c The current flowing from the battery is shared between the resistors. Which resistor will have a bigger share of the current? Explain your answer. [2]

d Calculate the effective resistance of the two resistors, and the current that flows from the battery. [5]

19.6 Figure **19.30** shows a circuit that includes a diode.

a Copy the diagram and label the diode. [1]

b On your diagram, between points A and B, add the symbol for a cell. The cell must be connected in such a way that a current flows through the resistor. [1]

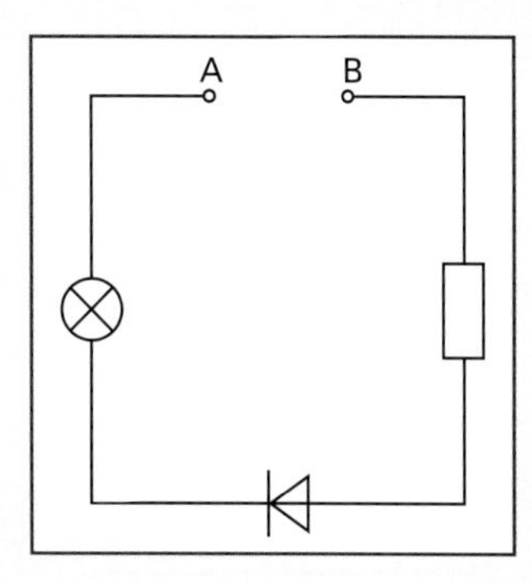

Figure 19.30
For Question **19.6**.

c On your diagram, add a labelled arrow to show the direction in which electrons move through the resistor when the current flows. [1]

19.7 Logic gates are often used in electronic control circuits. The operation of a logic gate can be represented by a truth table.

a What logic gate is represented by the truth table shown in Figure **19.31**? Write a sentence to describe its operation. [2]

b The output of a gate called an 'exclusive OR' gate is ON if just **one** of its two inputs is ON. Otherwise, it is OFF. Draw a truth table to represent this. [4]

input 1	**input 2**	**output**
0	0	0
1	0	0
0	1	0
1	1	1

0 = OFF
1 = ON

Figure 19.31 For Question **19.7a**.

E

c Name the **two** logic gates shown in Figure **19.32**. [2]

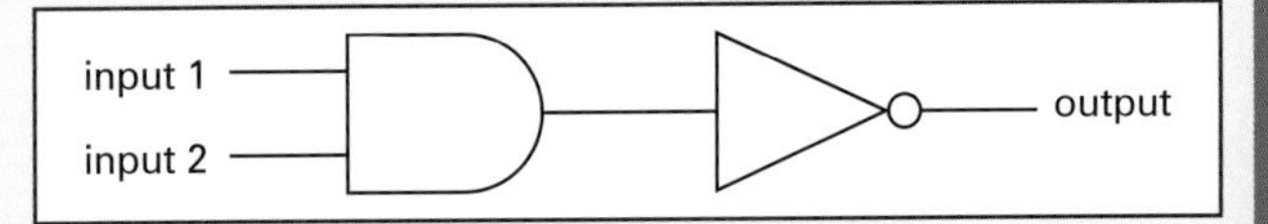

Figure 19.32 For Question **19.7c**.

d What will be the output of the combination of gates shown in Figure **19.32** if both inputs are ON? [1]

19.8 Figure **19.33** shows an electronic control circuit.

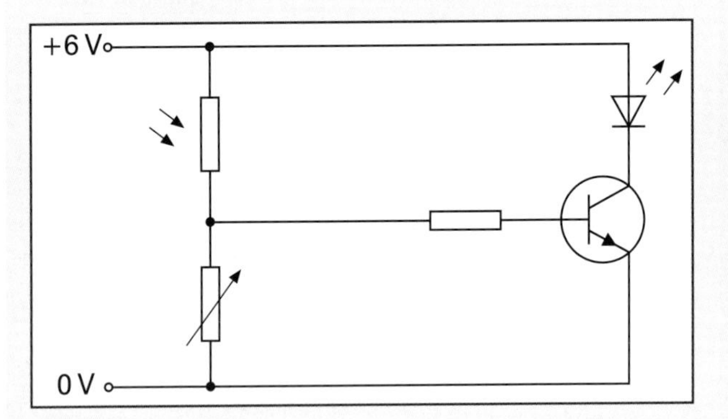

Figure 19.33 For Question **19.8**.

E

a Name the **two** components that form a potential divider. [2]

b Copy the circuit symbol for a transistor and label the three terminals with their appropriate names. [3]

c What is the function of this circuit? [2]

d It is desired to alter the circuit so that it controls a mains-powered lamp rather than an LED. A relay must be incorporated as part of the circuit. Draw a complete circuit diagram to show how the circuit of Figure **19.33** must be modified to achieve this. [3]

20 Electromagnetic forces

Core	Describing uses of electromagnets
Core	Investigating the magnetic force on a current-carrying conductor
E **Extension**	Using Fleming's left-hand rule
Core	Describing the production and detection of cathode rays
E **Extension**	Describing the structure and use of a cathode-ray oscilloscope

Electricity meets magnetism

In Chapter **16**, we saw two ways to produce a magnetic field – using permanent magnets, or by means of an electromagnet, which is a coil of wire through which a current flows. The second of these shows that there is a close connection between electricity and magnetism. This was first discovered almost two centuries ago.

Hans Christian Oersted was a Danish scientist, working in the early years of the 19th century. He noticed that both static electricity and magnetism showed similar patterns – attractive and repulsive forces, two types of charge or pole, a force that gets weaker at a distance, and so on. Most other scientists thought that this was just an interesting coincidence, but Oersted thought that there must be more to it than this. He was sure that he could find a link between electricity and magnetism, and eventually he did so.

It was early in 1820 that Oersted gave a public lecture on electricity. He described a report of a ship that had been struck by lightning. Its compass was affected, so that its north and south poles were reversed. He declared himself certain that this proved that there was a link between electricity and magnetism. Then a sudden thought struck him – there was an experiment he could try there and then to test his idea. On his demonstration bench, he had a wire and a compass (Figure **20.1**). He placed the compass under the wire. When his assistant connected the wire to a

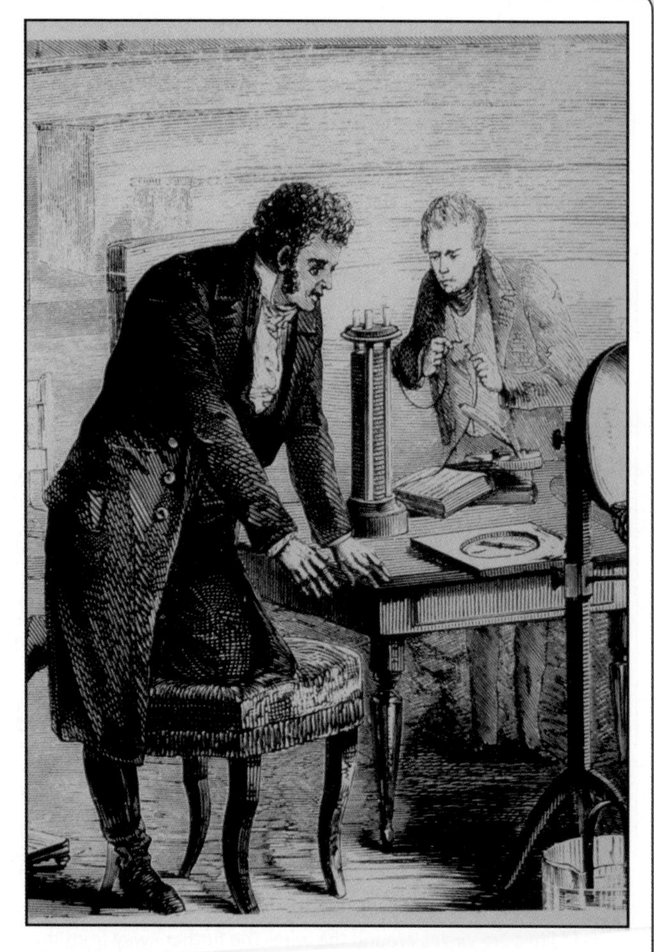

Figure 20.1 During a public lecture, Oersted thought up an experiment that he hoped would show the link between electricity and magnetism. He tested his idea as soon as he had thought of it. Here, you can see Oersted and his assistant, who is holding two wires connected to a tall battery. The compass lies on the table. When a current flowed through the wire, the compass needle moved. It was responding to the magnetic field around the current.

large battery so that a current flowed through it, the compass needle moved. At the time, no-one was very impressed – not even Oersted. However, the more he thought about it, the more he realised that he had observed something of fundamental significance. The current in the wire was producing a magnetic effect, which acted on the compass needle. By moving the compass around near the wire, he discovered that the magnetic effect showed a circular pattern around the current. The study of electromagnetism had begun.

20.1 The magnetic effect of a current

In Chapter **16**, we saw that an electromagnet can be made by passing a current through a coil of wire (a **solenoid**). The flow of current results in a magnetic field around the solenoid. The field is similar to the field around a bar magnet (see Figure **16.9** on page **175**).

If you uncoil a solenoid, you will have a straight wire. With a current flowing through it, it will have a magnetic field around it as shown in Figure **20.2**. The field lines are circles around the current.

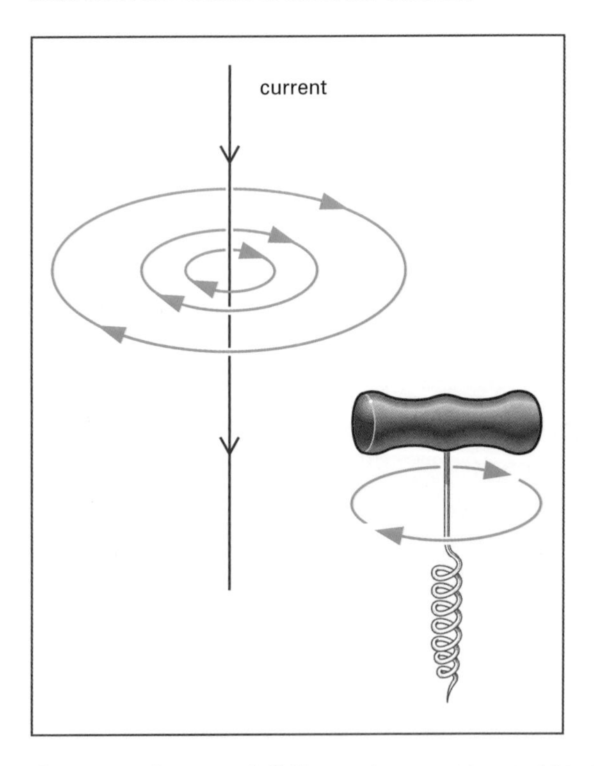

Figure 20.2 The magnetic field around a current in a straight wire. The field lines are circles around the wire. The further away from the wire, the weaker is the field.

Every electric current is surrounded by the magnetic field that it creates. An **electromagnet** is simply a clever way of making use of this, because winding the wire into a coil is a way of concentrating the magnetic field.

The **corkscrew rule** (or pencil sharpener rule) tells you the direction of the field lines. Imagine screwing a corkscrew in the direction of the current. The direction you turn it in is the direction of the magnetic field lines.

Uses of electromagnets

Electromagnets are in use in many different places. An electric motor has at least one – motors are considered in detail shortly. Here are some other applications of electromagnets.

Electric bells

An **electric bell** is a surprisingly clever device. It works using direct current from a battery, but it makes a hammer move repeatedly back and forth to strike the gong and produce the sound, which tells us, for example, that someone is at the door. Figure **20.3** shows the construction of a typical door bell. Notice that the hammer is attached to a springy metal strip, and is normally not in contact with the gong.

- When someone presses on the bell push, the circuit is completed. Current flows from the battery round through the electromagnet coil and the springy strip, and back to the battery via the contact point A.
- The coil is now magnetised and attracts the springy strip. Two things now happen: the hammer strikes the gong, and the circuit breaks at point A.
- The current stops, the coil is no longer magnetised, and the strip springs back to its original position.
- Now the circuit is complete again, and a current flows once more. The coil is magnetised and attracts the iron again, the hammer strikes the gong, and so on.
- This process repeats itself for as long as the bell push is depressed.

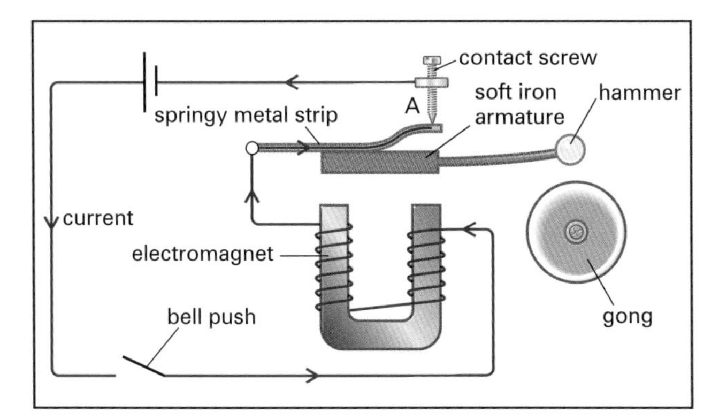

Figure 20.3 The construction of an electric bell. For as long as the bell push is depressed, the hammer springs back and forth, striking the gong. The contact screw at A can be adjusted to ensure that the circuit breaks each time the hammer is attracted by the electromagnet.

Relays

A **relay** is a switch operated by an electromagnet. In Chapter **19**, we saw how a relay can be operated in an electric circuit. One type is shown in Figure **20.4**, together with the circuit symbol.

- When switch A is closed, a small current flows around a circuit through the coil of the electromagnet.
- The electromagnet attracts the iron armature. As the armature tips, it pushes the two contacts at B together, completing the second circuit.

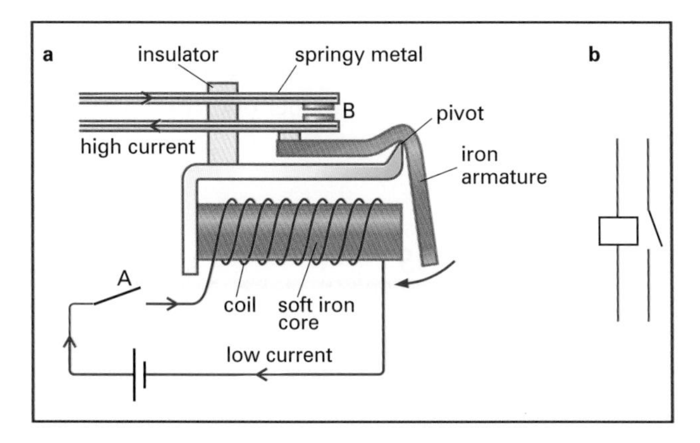

Figure 20.4 **a** A relay, capable of switching a circuit carrying hundreds of amps. **b** The circuit symbol for a relay – the rectangle represents the electromagnet coil.

A relay is used to make a small current switch a larger current on and off. For example, when a driver turns the ignition key to start a car, a small current flows to a relay in the engine compartment. This closes a switch to complete the circuit, which brings a high current to the starter motor from the battery.

 **Activity 20.1** Magnetic movements

Use an electromagnet to make a buzzer.

QUESTIONS

1. A current flows downwards in a wire that passes vertically through a small hole in a table top. Will the magnetic field lines around it go clockwise or anticlockwise (as seen from above)?
2. Look at the magnetic field pattern shown in Figure **20.2**. How can you tell from the pattern that the field gets weaker as you get further from the wire?
3. Look at the diagram of the electric bell (Figure **20.3**).
   a Why is the armature made of iron?
   b Why must soft iron be used?
4. Look at the diagram of the relay (Figure **20.4a**). Why is the coil fitted with a soft iron core?
5. A TV star switches on the Christmas lights in a shopping centre. She closes a switch, which operates a relay. The relay completes the circuit for the lights. Draw a circuit diagram for this, using the symbol shown in Figure **20.4b**. (Remember that there will be two circuits, one to operate the electromagnet coil, the other to power the lights.)

20.2 How electric motors are constructed

The idea of an **electric motor** is this. There is a magnetic field around an electric current. This magnetic field can be attracted or repelled by another magnetic field to produce movement. It is not obvious how to do this so that continuous movement is produced. If you put two magnets together so that they repel, they move apart and stop. Electric motors are cleverly designed to produce movement that continues as long as the current flows.

You may have constructed a model electric motor like the one shown in Figure **20.5**. This is designed to be easy to build and easy to understand. Its essential features are listed below:

- a **coil of wire**, which acts as an electromagnet when a direct current flows through it
- two **magnets**, to provide a steady magnetic field passing through the coil
- a **split-ring commutator**, through which current reaches the coil
- two **brushes**, which are springy wires that press against the two metal sections of the commutator.

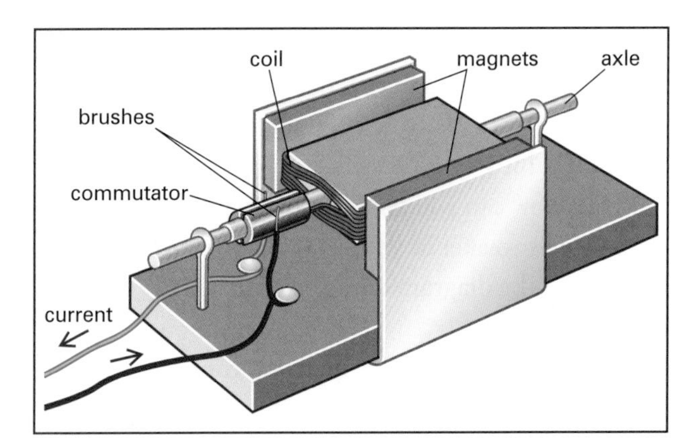

Figure 20.5 This model is used to show the principles of operation of an electric motor.

The important features of the motor are shown in Figure **20.6**. Here is our first explanation of how an electric motor works.

1. A current flows in through the right-hand brush, around the coil, and out through the other brush.
2. When the current is flowing, the coil becomes an **electromagnet**. At the instant shown, the uppermost side of the coil is its north pole and the lowermost side is its south pole (see Figure **16.9**). The north pole of the coil is attracted to the south pole of the permanent magnet on the left, and so the coil starts to turn to the left (anticlockwise).
3. This is where the **commutator** comes in. The coil is attracted round by the two permanent magnets. Its momentum carries it past the vertical position. Now, the brush connections to the two halves of the commutator are reversed. The current flows the opposite way around the coil.
4. We again have a north pole on the uppermost side of the coil, so it turns another 180° anticlockwise.

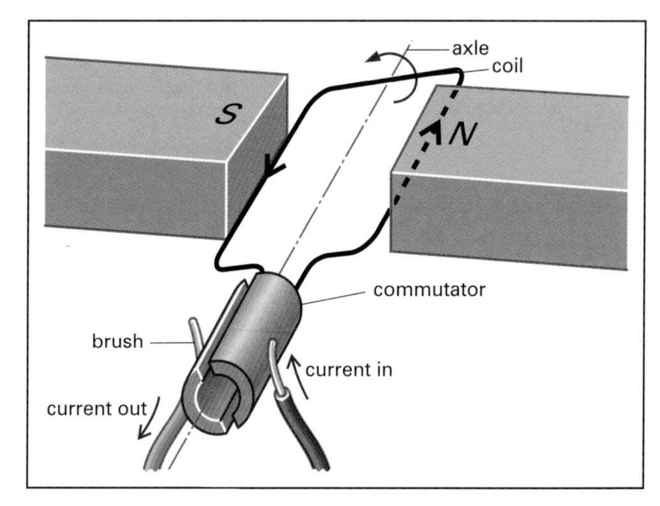

Figure 20.6 A spinning electric motor. The coil is an electromagnet, which is attracted round by the permanent magnets. Every half turn, the commutator reverses the current flowing through the coil, so that it keeps turning in the same direction.

Without the commutator, the coil would simply turn until it was vertical. The commutator cleverly reverses the current through the coil every half turn, so that the coil keeps on turning. If you have made a model like the one shown in Figure **20.5**, you may have noticed electrical sparks flashing around the commutator. These happen as the contact between the brush and one commutator segment is broken, and as it makes contact with the other segment.

For a d.c. motor like this to be of any use, its axle must be connected to something that is to be turned – a wheel, a pulley or a pump, for example. This model motor is not very powerful. The turning effect can be increased by increasing the number of turns of wire on the coil.

QUESTIONS

6 Look at the motor shown in Figure **20.6** and the explanation of how it works. Suppose that the two magnets were turned round so that there was a magnetic north pole on the left. Explain how the coil would move.

7 **a** In a d.c. motor, why must the current to the rotor coil be reversed twice during each rotation?

b What device reverses the current?

E

Making motors more powerful

An electric motor makes use of a coil of wire with a current flowing through it – in other words, an electromagnet. The strength of an electromagnet can be increased by increasing the current flowing through it.

This means that a motor can be made more powerful by increasing the strength of the electromagnet, and there are two ways to do that: by increasing the current or by having more turns of wire on the coil. Alternatively, the permanent magnets can be made stronger.

QUESTION

8 Describe how the turning effect of a d.c. motor will change if the current flowing through the motor coil is increased.

20.3 Force on a current-carrying conductor

An electric motor has a coil with a current flowing around it (an electromagnet) in a magnetic field. It turns because the two magnetic fields interact with each other. However, it is not essential to have a coil to produce movement. The basic requirements are:

- a magnetic field
- a current flowing **across** the magnetic field.

Figure 20.7 shows a way of demonstrating this in the laboratory. The copper rod is free to roll along the two aluminium support rods. The current from the power supply flows along one support rod, through the copper rod, and out through the other support rod. The two magnets provide a vertical magnetic field.

What happens when the current starts to flow? The copper rod rolls horizontally along the support rods. It is pushed by a horizontal force. The force comes about because the magnetic field around the current is repelled by the magnetic field of the permanent magnets. The force can be increased in two ways: by increasing the current, and by using magnets with a stronger magnetic field. This force, which every electric motor makes use of, is known as the **motor effect**.

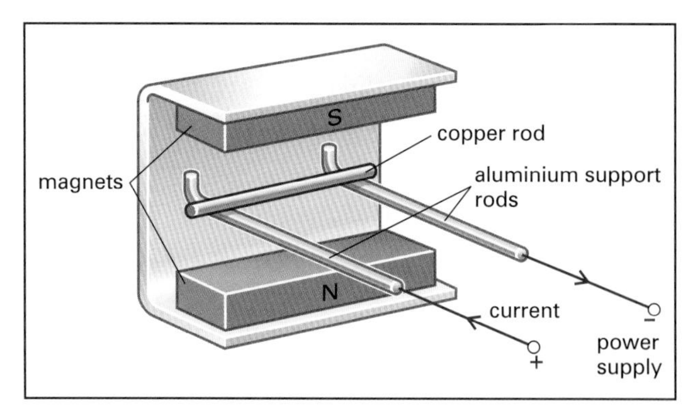

Figure 20.7 Demonstrating the motor effect. There is a magnetic field around the current in the copper rod. This interacts with the field of the magnets, and the result is a horizontal force on the rod. A copper rod is used because it is a non-magnetic material. (A steel rod would be attracted to the magnets.)

By swapping the connections to the power supply, you can reverse the direction of the current in the copper rod. The rod rolls in the opposite direction, showing that the force on it has been reversed. Similarly, if the magnets are reversed so that the magnetic field is in the opposite direction, the force on the copper rod is reversed. So, the force caused by the motor effect is **reversed** if:

- the direction of the current is reversed
- the direction of the magnetic field is reversed.

Activity 20.2 The catapult field

A simple way to show the force on a current-carrying conductor.

QUESTION

9 List **two** ways to reverse the force on a current-carrying conductor in a magnetic field.

Fleming's left-hand rule

In Figure **20.7**, there are three things with direction (three **vector quantities** – see pages **23** and **33**):

- the magnetic field
- the current
- the force.

The magnetic field is vertical. The current and the force are horizontal, and at right angles to each other. Hence we have three things that are all mutually at right angles to each other (Figure **20.8a**). To remember how they are arranged, physicists use **Fleming's left-hand rule** (Figure **20.8b**). It is worth practising holding your thumb and first two fingers at right angles like this. Then learn what each finger represents.

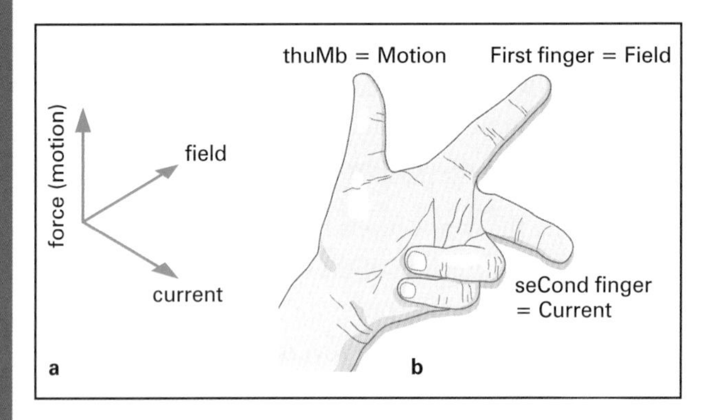

Figure 20.8 **a** Force, field and current are at right angles to each other. **b** Fleming's left-hand rule. Check that it correctly predicts the direction of the force on the current in Figure **20.7**.

We use Fleming's left-hand rule to predict the direction of the force on a current-carrying conductor in a magnetic field. By keeping your thumb and fingers rigidly at right angles to each other, you can show that reversing the direction of the current or field reverses the direction of the force. (Do not try changing the direction of individual fingers. You have to twist your whole hand around at the wrist.)

Electric motors revisited

We can apply Fleming's left-hand rule to an electric motor. Figure **20.9a** shows a simple electric motor with its coil horizontal in a horizontal magnetic field. The coil is rectangular. What forces act on each of its four sides?

- Side **AB**. The current flows from A to B, across the magnetic field. Fleming's left-hand rule shows that a force acts on it, vertically upwards.
- Side **CD**. The current is flowing in the opposite direction to the current in AB, so the force on CD is in the opposite direction, downwards.
- Sides **BC** and **DA**. The current here is parallel to the field. Since it does not cross the field, there is no force on these sides.

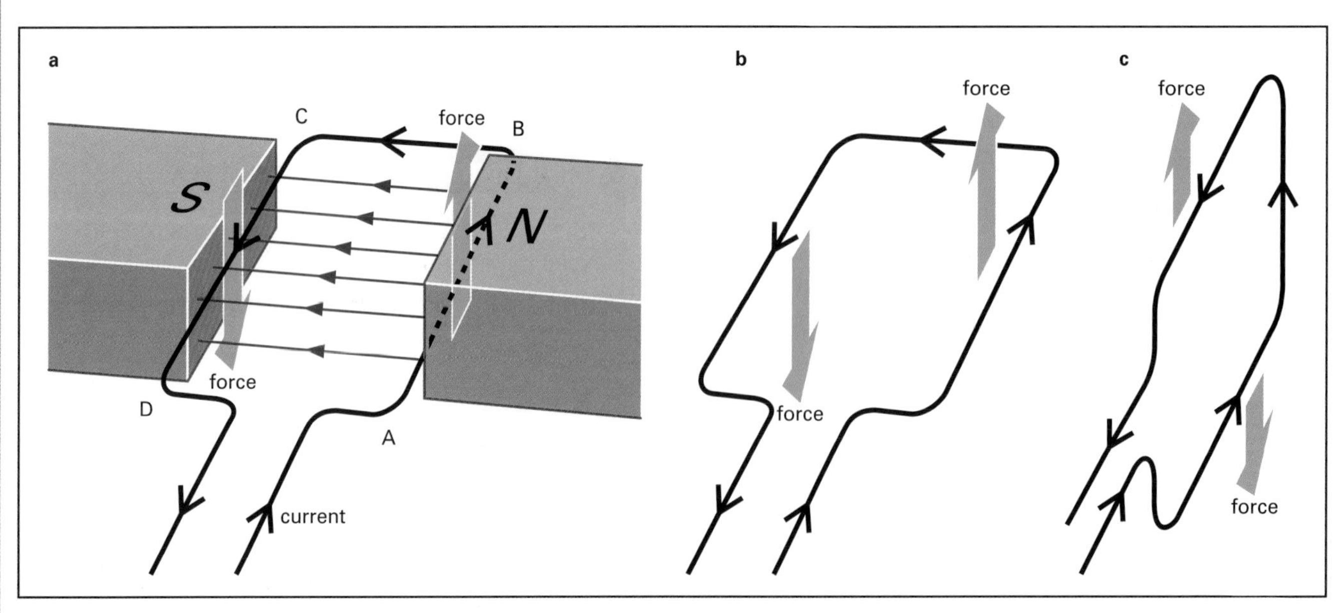

Figure 20.9 **a** A simple electric motor. Only the two longer sides experience a force, since their currents cut across the magnetic field. **b** The two forces provide the turning effect needed to make the coil rotate. **c** When the coil is in the vertical position, the forces have no turning effect.

Figure 20.9b shows a simplified view of the coil. The two forces acting on it are shown. They cause the coil to turn anticlockwise. The two forces provide a turning effect (or torque), which causes the motor to spin. From Figure 20.9c, you can see that the forces will not turn the coil when it is vertical. This is where we have to rely on the coil's momentum to carry it further round.

The diagrams show the coil as if it were a single turn of wire. In practice, the coil might have hundreds of turns of wire, resulting in forces hundreds of times as great. A coil causes the current to flow across the magnetic field many times, and each time it feels a force. A coil is simply a way of multiplying the effect that would be experienced using a single length of wire.

QUESTIONS

10 For Fleming's left-hand rule, write down the three things that are at 90° to each other, and next to each one, write down the finger that represents it.

11 List **two** ways to increase the force on a current-carrying conductor in a magnetic field.

12 What is the force on a current-carrying conductor that is parallel to a magnetic field?

20.4 Cathode rays

In the 1850s, scientists discovered an interesting phenomenon. They were trying to discover whether an electric current could flow through a vacuum, so they set up an evacuated glass tube with metal electrodes at either end. With a high voltage connected to the electrodes, a glowing beam appeared inside the tube. They soon showed that this was coming from the cathode (the negative electrode), and so this radiation came to be known as **cathode rays**. Today, we know that cathode rays are, in fact, rays of fast-moving electrons, particles that have a negative electric charge.

Figure **20.10** shows the principle of a **cathode-ray tube**. The hollow tube is evacuated so that very little gas remains inside it. The cathode is coated with a metal, which, when heated, releases electrons. This process is known as **thermionic emission**. The electrons

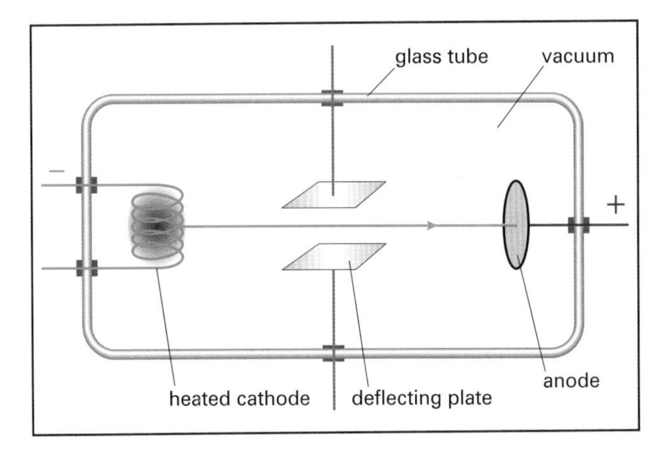

Figure 20.10 The principle of producing cathode rays in a vacuum tube.

are attracted to the anode (because it has a positive charge), and so the electrons travel across the tube.

Cathode rays can be shown to have a negative charge because they can be deflected by an electric field. Two deflecting plates are fitted inside the vacuum tube. One plate is connected to the positive terminal of a supply, and the other is connected to the negative, so that there is an electric field between them. The cathode rays are deflected towards the positive plate and away from the negative.

Cathode rays can be detected in different ways. First, the low-pressure gas through which they are travelling may glow. This is because some of the electrons strike atoms of the gas and give them energy, so that they give out light. Alternatively, the glass of the tube may be coated with a substance that glows when the electrons strike it.

QUESTIONS

13 **a** What is meant by the term **thermionic emission**?

b What particles are released by this process?

14 Two metal plates are mounted one above the other. The upper plate is connected to the positive terminal of a power supply, and the lower plate to the negative terminal. A cathode ray passes between the plates. State how the ray will be deflected and explain your answer.

Using cathode rays

Cathode-ray tubes have two important uses, although they have been largely replaced nowadays by different technologies.

- A **cathode-ray oscilloscope** is used to display traces showing how a voltage varies, for example, the varying voltage produced by a microphone when it detects sound waves.
- A cathode-ray tube is also used in a traditional television set to produce the picture.

For both of these applications, the beam of electrons must be scanned across the screen to produce the image. This can be done using static electricity. Figure **20.11** shows how this works in an oscilloscope. Inside the tube are pairs of plates X_1, X_2 and Y_1, Y_2. If plates X_1 and X_2 are connected to a high voltage, there will be an electric field between them. This will deflect the cathode ray as it passes between the plates. In the case shown in Figure **20.11**, because the plates are vertical, the beam will be deflected horizontally. The electrons will be attracted towards the positive plate and repelled by the negative plate. An electric field between Y_1 and Y_2 will deflect the beam vertically. In this way, the beam can be moved to any point on the screen.

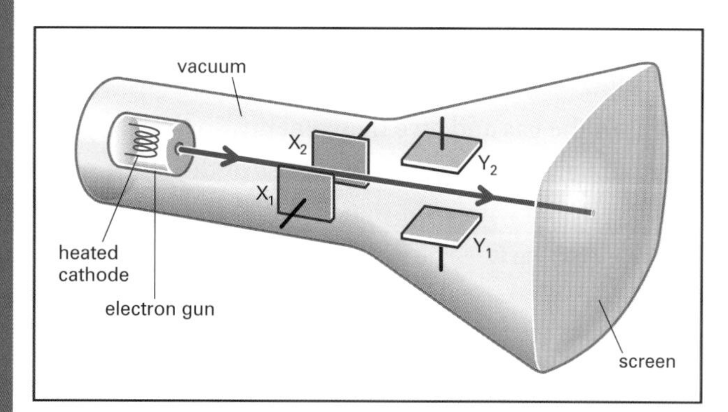

Figure 20.11 The construction of a cathode-ray oscilloscope, showing how cathode rays (beams of electrons) can be produced and then deflected by electric fields.

To display waveforms on the screen of an oscilloscope (as described in Chapter **12**):

- a steadily increasing voltage is applied to the X-plates, so that the spot on the screen moves steadily from left to right
- the varying voltage is applied to the Y-plates, causing the spot to move up and down.

In this way, the varying voltage of the waveform is spread across the screen, giving the traces we discussed in Chapter **12**.

Electron beams and magnetic fields

A magnetic field can also be used to deflect a beam of electrons. This can be demonstrated in the laboratory using a vacuum tube (Figure **20.12**). In this photograph, an electron beam is travelling from left to right in a spherical vacuum tube. Two electromagnet coils (front and back) produce a horizontal magnetic field. The electrons feel an upward force, and this causes the beam to curve.

Figure 20.12 An electron beam in a vacuum tube. Two electromagnet coils provide the magnetic field needed to deflect the beam upwards.

The electrons feel the same force as we saw earlier for a current-carrying conductor in a magnetic field. The direction of the force is given by Fleming's left-hand rule (but recall that the conventional current is in the opposite direction to the electron flow). Check the photograph: the electrons are moving from left to right, so the conventional current is right to left; the magnetic field is pointing towards the front; so the force must be upwards.

In fact, when a current-carrying conductor is placed in a magnetic field, it is the electrons that feel the force. They then transmit it to the conductor. Looking back at Figure 20.7, you can imagine the electrons flowing in the copper rod and being pushed to the right as they cross the magnetic field.

A television tube usually has two sets of electromagnet coils mounted on it (Figure 20.13). The top-and-bottom pair produce a magnetic field that moves the electron beam from side to side. The left-and-right pair deflect the beam up and down. Fleming's left-hand rule should convince you that this is correct. Electromagnet coils are excellent for this job, because the current

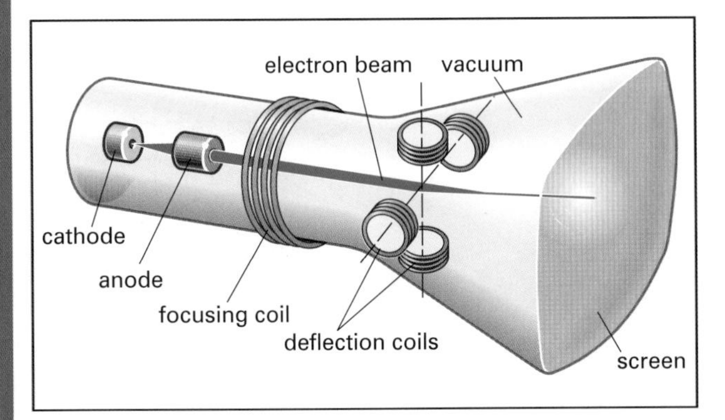

Figure 20.13 Electromagnet coils are mounted on the outside of a traditional television tube, to deflect the electron beam across the screen. One pair of coils moves the beam horizontally, the other vertically. As the beam moves across the screen, it builds up the picture. Usually, each picture is made up of 625 horizontal lines, and 50 or 60 are shown each second.

through them can be changed very rapidly. In this way, the electron beam can be scanned across the screen thousands of times each second to produce the images we see when watching a television programme.

Summary

A current-carrying coil in a magnetic field experiences a turning effect. Use is made of this effect in electric motors.

A force is exerted on any current-carrying conductor that crosses a magnetic field. The direction of the force depends on the direction of the field and the current.

The relative directions of force, field and current are given by Fleming's left-hand rule.

Cathode rays are beams of electrons produced by thermionic emission in a vacuum tube.

Cathode rays are deflected in electric fields.

A beam of charged particles can be thought of as a current, and will experience a force if it crosses a magnetic field. This is used to control the direction of beams of charged particles in television tubes, cathode-ray oscilloscopes and so on.

End-of-chapter questions

20.1 Figure 20.14 shows the construction of an electric bell.

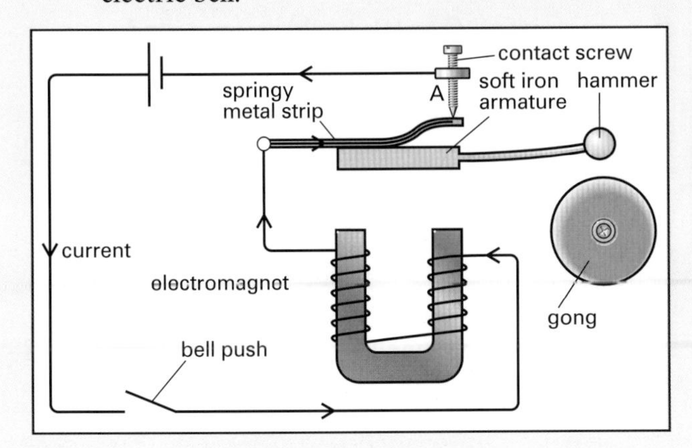

Figure 20.14 For Question **20.1**.

Put the following sentences in the correct order to explain how the bell operates. [7]

- A current flows through the electromagnet.
- At the same time, the circuit is broken at point A.
- Someone presses the bell push.
- The circuit is completed again at A.
- The electromagnet attracts the iron armature.
- The hammer strikes the gong.
- The springy metal pulls the hammer back.

20.2 The motor effect is sometimes demonstrated using the apparatus shown in Figure **20.15**. A current flows through a wire 'swing'. The swing hangs in a magnetic field. When the current is switched on, the swing is pushed out of the field.

a What will happen if the connections to the power supply are reversed? [1]

b What will happen if the magnetic field is reversed? [1]

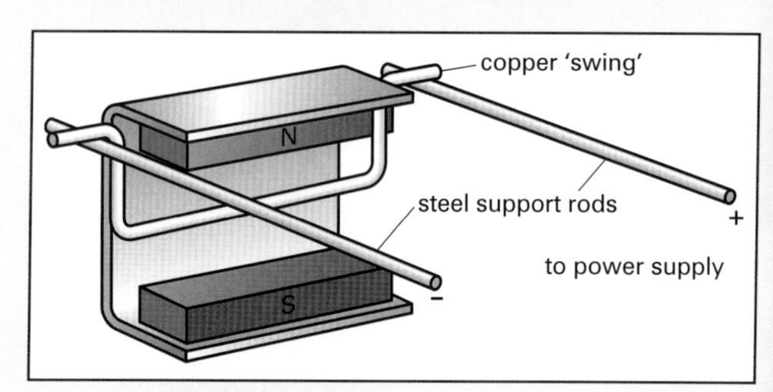

Figure 20.15 For Question **20.2**.

E

20.3 Look again at Figure **20.15**.

a In which direction is the magnetic field between the two magnets? [1]

b In which direction will the swing be pushed? Why? [2]

20.4 Figure **20.16** is a simplified diagram of an electric motor.

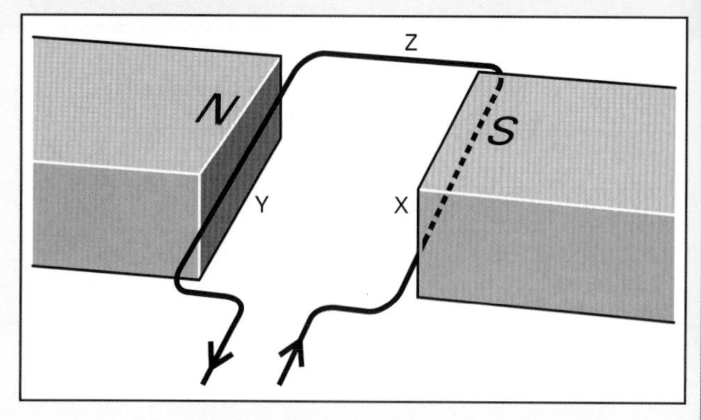

Figure 20.16 For Question **20.4**. The loop of wire is horizontal in a horizontal magnetic field.

E

a In which direction is the force on side X of the wire loop? [1]

b In which direction is the force on side Y of the wire loop? [1]

c Explain how these forces cause the loop to rotate. [2]

d What can you say about the force on side Z? Why? [2]

20.5 Figure **20.17** shows a simple cathode-ray tube. A beam of particles, emitted by the cathode, strikes the centre of the screen.

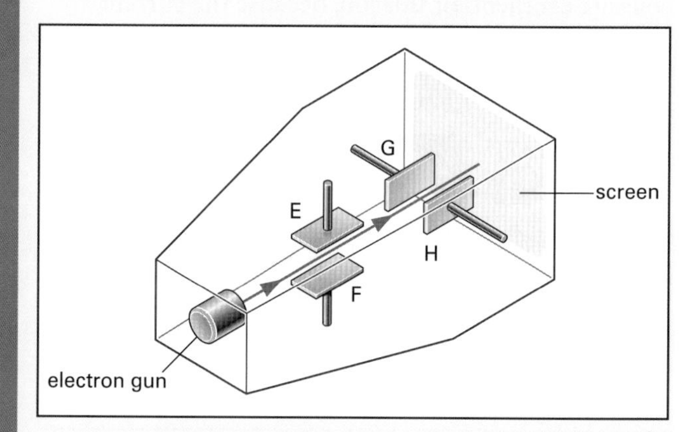

Figure 20.17 For Question **20.5**.

a What are the particles emitted by the cathode? [1]

b Explain why the cathode must be heated. [1]

c What is the principal energy change that occurs when the beam strikes the screen? [2]

d Which of the four deflecting plates (E, F, G or H) should be connected to a positive voltage if the spot on the screen is to be moved upwards? Explain your answer. [2]

e Describe and explain what you would expect to see on the screen if the two deflecting plates G and H were connected to an alternating voltage supply. [2]

21 Electromagnetic induction

Core	Describing how an e.m.f. is induced in a circuit
E **Extension**	Identifying factors affecting the magnitude and direction of an induced e.m.f.
Core	Describing the design of an a.c. generator
Core	Describing the construction of a transformer
Core	Using the transformer equation
E **Extension**	Explaining how transformers work
Extension	Using the power equation for a transformer

Power plant

Modern societies depend greatly on electricity. However, we usually do not have to think about the electricity we use. We plug in a computer or switch on a light – and they work. Often, we have no idea where the electricity we use is generated.

Things can be different in a developing nation. Figure **21.1** shows how electricity is generated and used in the Kenyan village of Tungu-Kabiri, on the slopes of Mount Kenya. This is a micro-hydroelectric scheme. Water is fed by a pipe to a turbine, which causes a generator to spin. This generates electricity at the rate of 14 kW – not a lot, but enough to keep several enterprises working, including a metal workshop, a hairdresser's and a food shop.

Local, environmentally friendly schemes like this can show the way forward for a developing country like Kenya.

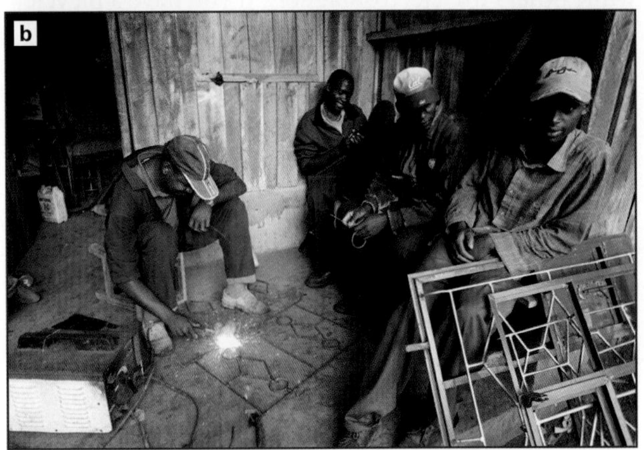

Figure 21.1 The production and use of electricity in Tungu-Kabiri, Kenya. **a** Water from a dam is fed through the yellow pipe to the turbine on the left. A rubber belt transfers the rotation to the generator at the top. The operator is opening the valve to control the flow of water. **b** Welding equipment in use in the workshop in Tungu-Kabiri.

21.1 Generating electricity

A **motor** is a device for transforming electrical energy into mechanical (kinetic) energy. To generate electricity, we need a device that will do the opposite: it must transform mechanical energy into electrical energy. Fortunately, we can simply use an electric motor in reverse. If you connect up an electric motor to a meter and spin its axle, the meter will show that you have generated a voltage (Figure 21.2). Inside the motor, the coil is spinning around in the magnetic field provided by the permanent magnets. The result is that a current flows in the coil, and this is shown by the meter. We say that the current has been induced, and the motor is acting as a **generator**.

Figure 21.2 A motor can act as a generator. Spin the motor and the meter shows that an induced current flows around the circuit.

There are many different designs of generator, just as there are many different designs of electric motor. Some generate direct current, others generate alternating current. Some use permanent magnets, while others use electromagnets. If you have a bicycle, you may have a generator of a different sort – a dynamo, for powering the lights. The power station generators shown in Figure 21.3 generate alternating current at a voltage of about 25 kV. The **turbines** are made to spin by the high-pressure steam from the boiler. The generator is on the same axle as the turbine, so it spins too. A coil inside the generator spins around inside some fixed electromagnets, which provide the magnetic field. A large current is then induced in the rotating coil, and this is the current that the power station supplies to consumers. A fraction of it is used to supply the electromagnets of the generator itself.

All of these generators have three things in common:

- a **magnetic field** (provided by magnets or electromagnets)

Figure 21.3 The turbine and generator in the generating hall of a Canadian nuclear power station. At the back are the turbines, fed by high-pressure steam in pipes. The generator is in the centre.

- a **coil of wire** (fixed or moving)
- **movement** (the coil and magnetic field move relative to one another).

When the coil and the magnetic field move relative to each other, a current flows in the coil if it is part of a complete circuit. This is known as an **induced current**. If the generator is not connected up to a circuit, there will be an **induced e.m.f.** (or **induced voltage**) across its ends, ready to make a current flow around a circuit.

The principles of electromagnetic induction

The process of generating electricity from motion is called **electromagnetic induction**. The science of electromagnetism was largely developed by Michael Faraday (Figure 21.4). He invented the idea of the magnetic field, and drew field lines to represent it. He also invented the first electric motor. Then he extended his studies to show how the motor effect could work in reverse to generate electricity. In this section, we will look at the principles of electromagnetic induction that Faraday discovered.

As we have seen, a coil of wire and a magnet moving relative to each other are needed to induce a voltage across the ends of a wire. This is called the **dynamo effect**. If the coil is part of a complete circuit, the induced e.m.f. will make an induced current flow around the circuit.

Figure 21.4 Michael Faraday delivering a Christmas Lecture at the Royal Institution in London on 27 December 1855. He was a great populariser of science, and his lectures attracted many famous people. The artist, Alexander Blaikley, has included several members of the Royal Family in the audience, as well as famous scientists, including Charles Darwin, although it is unlikely that they were all present at this lecture. The Christmas Lectures started in 1826 and continue to this day. They are presented in the same lecture theatre. You may have seen them on television, as they are broadcast around the world.

In fact, you do not need to use a coil, a single wire is enough to induce an e.m.f., as shown in Figure **21.5a**. The wire is connected to a sensitive meter to show when a current is flowing.

- Move one pole of the magnet downwards past the wire, and a current flows.
- Move the magnet back upwards, and a current flows in the opposite direction.

Alternatively, the magnet can be stationary and the wire can be moved up and down next to it.

You can see similar effects using a magnet and a coil (Figure **21.5b**). Pushing the magnet in to and out of the coil induces a current, which flows back and forth in the coil. Here are two further observations:

- Reverse the magnet to use the opposite pole, and the current flows in the opposite direction.
- Hold the magnet stationary next to the wire or coil, and no current flows. They must move relative to each other, or nothing will happen.

(This provides a good test of how steady your hand is. Hold a strong magnet next to a coil of wire connected to a sensitive meter. If your hand trembles, the meter will show that a current is flowing in the wire.)

An a.c. generator

Faraday's discovery of electromagnetic induction led to the development of the electricity supply industry. In particular, it allowed engineers to design generators that could supply electricity. At first, this was only done on a small scale, but gradually generators got bigger and bigger, until, like the ones shown in Figure **21.3**, they were capable of supplying the electricity demands of thousands of homes.

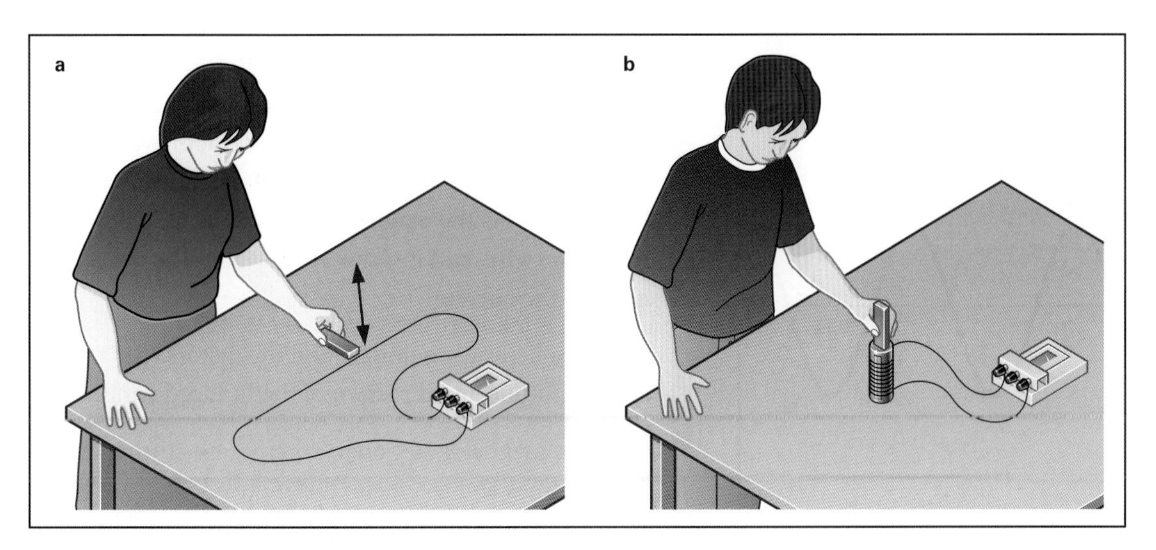

Figure 21.5 **a** Move a magnet up and down next to a stationary wire and an induced current will flow. **b** Similarly, move a magnet in to and out of a coil of wire and an induced current will again flow. Michael Faraday first did experiments like this in 1831.

Figure **21.6** shows a simple **a.c. generator**, which produces alternating current. In principle, this is like a d.c. motor, working in reverse. The axle is made to turn so that the coil spins around in the magnetic field, and a current is induced. The other difference is in the way the coil is connected to the circuit beyond. A d.c. motor uses a split-ring commutator, whereas an a.c. generator uses **slip rings**.

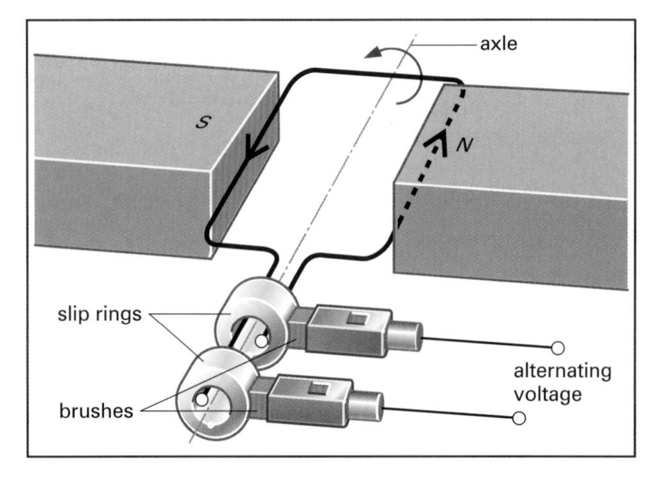

Figure 21.6 A simple a.c. generator works like a motor in reverse. The slip rings and brushes are used to connect the alternating current to the external circuit.

A generator of this type produces **alternating current (a.c.)**. This means that the current is not direct current (d.c.), which always flows in the same direction. Instead, an alternating current flows back and forth. Figure **21.7** shows a graph of this. Half of the time, the current flows in the positive direction. Then it flows in the opposite direction. The **frequency** of an a.c. supply is the number of cycles it produces each second.

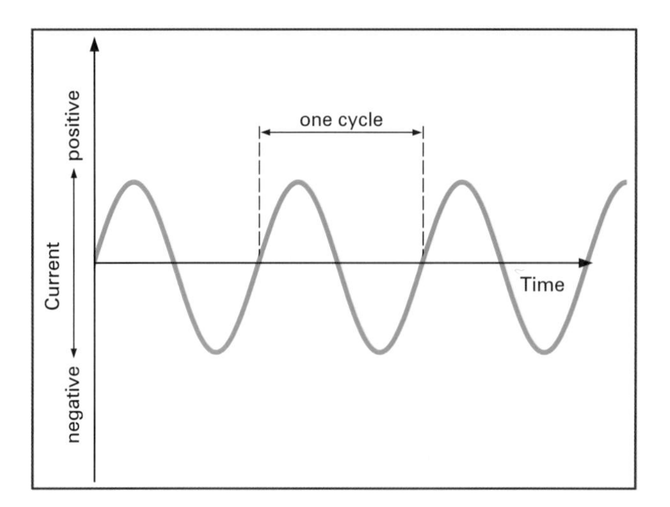

Figure 21.7 A graph to represent an alternating current. For the first half of a cycle, the current flows one way. Then it goes into reverse.

Why does this generator produce alternating current? As the coil rotates, each side of the coil passes first the magnetic north pole and then the south pole. This means that the induced current flows first one way, and then the other. In other words, the current in the coil is alternating.

The current flows out through the slip rings. Each ring is connected to one end of the coil, so the alternating current flows out through the brushes, which press against the rings.

Inducing electricity

Make some observations of electromagnetic induction.

QUESTIONS

1 Draw a diagram to show the energy transformations in:

  a an electric motor

  b a generator.

  Remember that neither is 100% efficient.

2 If you hold a coil of wire next to a magnet, no current will flow. What else is needed to induce a current?

Induction and field lines

We can understand electromagnetic induction using Faraday's idea of magnetic field lines. Picture the field lines coming out of each pole of the magnets shown in Figure 21.5. As the magnet is moved, the field lines are cut by the wire, and it is this cutting of field lines that induces the current.

This idea helps us to understand the factors that affect the magnitude and direction of the induced e.m.f.

- If the magnet is stationary, there is no cutting of field lines and so no e.m.f. is induced.
- If the magnet is further from the wire, the field lines are further apart and so fewer are cut, giving a smaller e.m.f.

- If the magnet is moved quickly, the lines are cut more quickly and a bigger e.m.f. is induced.
- A coil gives a bigger effect than a single wire, because each turn of wire cuts the magnetic field lines and each therefore contributes to the induced e.m.f.

From this, we can see that there are four ways of increasing the voltage generated by an a.c. generator like the one shown in Figure **21.6**:

- turn the coil more rapidly
- use a coil with more turns of wire
- use a coil with a bigger area
- use stronger magnets.

Each of these has the effect of increasing the rate at which magnetic field lines are cut, and so the induced e.m.f. is greater. For the a.c. generator shown in Figure **21.6**, each revolution of the coil generates one cycle of alternating current. Spin the coil 50 times each second and the a.c. generated has a frequency of 50 Hz.

Direction of the induced e.m.f.

How does an induced current 'know' in which direction it must flow? The answer is that the current (like all currents) has a magnetic field around it. This field always pushes back against the field that is inducing the current. So, for the coil shown in Figure **21.5b**, when the magnet's north pole is pushed towards the coil, the current flows so as to produce a north pole at the end of the coil nearest the magnet. These two north poles repel each other. Hence you have to push the magnet towards the coil, and thereby do work. The energy you use in pushing the magnet is transferred to the current. That is where the energy carried by a current comes from. It comes from the work done in making a conductor cut through magnetic field lines.

> An induced current always flows in such a way that its magnetic field opposes the change that causes it.

QUESTIONS

3 The north pole of a magnet is moved towards a coil of wire, as shown in Figure **21.5b**, so that an induced current flows. State **two** ways in which the student could cause an induced current to flow in the opposite direction.

4 State **two** ways in which the current induced in the coil (Figure **21.5b**) could be increased.

5 List the **four** features of a large a.c. generator from a power station (Figure **21.3**) that make it capable of generating a higher voltage than the model a.c. generator shown in Figure **21.6**.

21.2 Power lines and transformers

Power stations may be 100 km or more from the places where the electricity they generate is used. This electricity must be distributed around the country. High-voltage electricity leaves the power station. Its voltage may be as much as one million volts. To avoid danger to people, it is usually carried in cables called **power lines** slung high above the ground between tall pylons. Lines of pylons stride across the countryside, heading for the urban and industrial areas that need the power (Figure **21.8**). This is a country's **national grid**.

Figure 21.8 Electricity is usually generated at a distance from where it is used. If you look on a map, you may be able to trace the power lines that bring electrical power to your neighbourhood.

When the power lines approach the area where the power is to be used, they enter a local distribution centre. Here the voltage is reduced to a less hazardous level, and the power is sent through more cables (overhead or underground) to local substations. In the substation, transformers reduce the voltage to the local supply voltage, typically 230 V. Wherever you live, there is likely to be a substation in the neighbourhood. It may be in a securely locked building, or the electrical equipment may be surrounded by fencing, which carries notices warning of the hazard (Figure **21.9**).

Figure 21.9 An electricity substation has warning signs like this to indicate the extreme hazard of entering the substation.

From the substation, electricity is distributed around the neighbouring houses. In some countries, the power is carried in cables buried underground. Other countries use tall 'poles', which hold the cables above the level of traffic in the street to distribute the power. Overhead power lines and cables can be an eyesore, but the cost of burying cables underground can be ten or a hundred times as great as using poles.

Why use high voltages?

The high voltages used to transmit electrical power around a country are dangerous. That is why the cables that carry the power are supported high above people, traffic and buildings on tall pylons. Sometimes the cables are buried underground, but this is much more expensive, and the cables must be safely insulated. There is a good reason for using high voltages. It means that the current flowing in the cables is relatively low, and this wastes less energy. We can understand this as follows.

When a current flows in a wire or cable, some of the energy it is carrying is lost because of the cable's resistance – the cables get warm. A small current wastes less energy than a high current. Electrical engineers do everything they can to reduce the energy losses in the cables. If they can reduce the current to half its value (by doubling the voltage), the losses will be one-quarter of their previous value. This is because power losses in cables are proportional to the square of the current flowing in the cables:

- double the current gives four times the losses
- three times the current gives nine times the losses.

Transformers

A **transformer** is a device used to increase or decrease the voltage of an electricity supply. They are designed to be as efficient as possible (up to 99.9% efficient). This is because the electricity we use may have passed through as many as 10 transformers before it reaches us from the power station. A loss of 1% of energy in each transformer would represent a total waste of 10% of the energy leaving the power station.

Power stations typically generate electricity at 25 kV. This has to be converted to the grid voltage – say 400 kV – using transformers. For these voltages, we say that the voltage is stepped up by a factor of 16. Figure **21.10** shows the construction of a suitable transformer. Every transformer has three parts:

- a **primary coil** – the incoming voltage V_p is connected across this coil
- a **secondary coil** – this provides the voltage V_s to the external circuit
- an **iron core** – this links the two coils.

Notice that there is **no electrical connection** between the two coils. They are linked together only by the iron core. Notice also that the voltages are both alternating

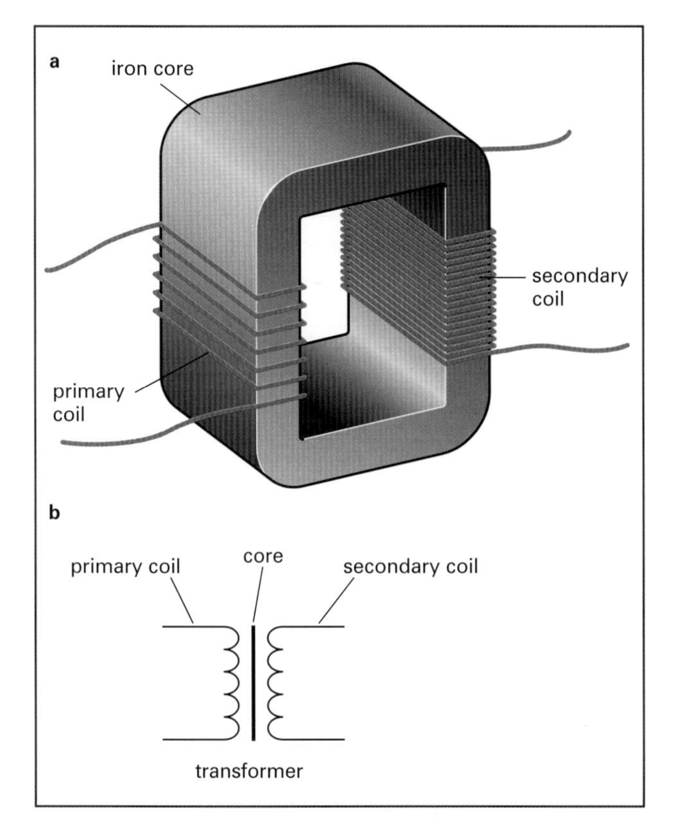

Figure 21.10 **a** The structure of a transformer. This is a step-up transformer because there are more turns on the secondary coil than on the primary. If the connections to it were reversed, it would be a step-down transformer. **b** The circuit symbol for a transformer shows the two coils with the core between them.

voltages – a transformer does not change a.c. to d.c or anything of the sort. It changes the size of an alternating voltage.

To step up the input voltage by a factor of 16, there must be 16 times as many turns on the secondary coil as on the primary coil. Comparing the numbers of turns on the two coils tells us how the voltage will be changed.

- A **step-up transformer** increases the voltage, so there are more turns on the secondary than on the primary.
- A **step-down transformer** reduces the voltage, so there are fewer turns on the secondary than on the primary.

(Note that, if the voltage is stepped up, the current must be stepped down, and vice versa.)

The ratio of the numbers of turns tells us the factor by which the voltage will be changed. Hence we can write an equation, known as the **transformer equation**, relating the two voltages, V_p and V_s, to the numbers of turns on each coil, N_p and N_s:

$$\frac{\text{voltage across primary coil}}{\text{voltage across secondary coil}} = \frac{\text{number of turns on primary}}{\text{number of turns on secondary}}$$

$$\frac{V_p}{V_s} = \frac{N_p}{N_s}$$

It will help you to recall this equation if you remember that the coil with most turns has the higher voltage.

Worked example 1

A transformer is needed to step down the 230 V mains supply to 6 V. If the primary coil has 1000 turns, how many turns must the secondary have?

Step 1: Draw a transformer symbol, and mark on it the information from the question (see Figure 21.11).

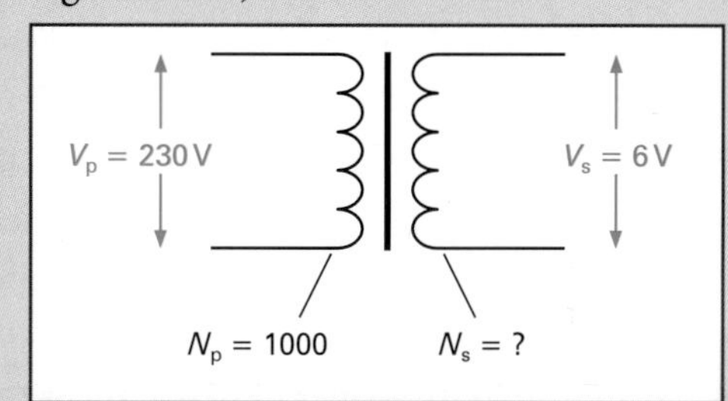

Figure 21.11 See Worked example 1.

Step 2: Write down the transformer equation.

$$\frac{V_p}{V_s} = \frac{N_p}{N_s}$$

Step 3: Substitute values from the question.

$$\frac{230\,\text{V}}{6\,\text{V}} = \frac{1000}{N_s}$$

Step 4: Rearrange and solve for N_s.

$$N_s = \frac{1000 \times 6\,\text{V}}{230\,\text{V}} = 26.1 \text{ turns}$$

So the secondary coil must have 26 turns.

Is the answer to Worked example 1 reasonable? The voltage has to be greatly reduced, so the number of turns on the secondary coil must be much less than 1000. Mental arithmetic shows that the voltage has to be reduced by a factor of about 40 (from 230 V to 6 V), so the number of turns must be reduced by the same factor. So N_s is about 1000/40 = 25. This is an **approximate** answer.

Activity 21.2 The electrical supply system

Find out more about where your electricity comes from.

QUESTIONS

6 Why is electrical power transmitted in the grid at high voltage?

7 Name the **three** essential parts of a transformer?

8 A transformer has 100 turns on the primary coil and 1000 on the secondary. Is it a step-up or a step-down transformer?

9 A portable radio has a built-in transformer so that it can work from the mains instead of batteries. Is this a step-up or step-down transformer?

10 A step-up transformer has 2000 turns on one coil and 5000 on the other. Calculate the ratio N_s/N_p for this transformer.

11 A transformer is designed to provide 20 V from a 240 V supply. If the primary coil has 1200 turns, how many turns must the secondary have?

21.3 How transformers work

Transformers only work with alternating current (a.c.). To understand why this is, we need to look at how a transformer works (Figure **21.12**). It makes use of electromagnetic induction.

- The primary coil has alternating current flowing through it. It is thus an electromagnet, and produces an alternating magnetic field.
- The core transports this alternating field around to the secondary coil.
- Now the secondary coil is a conductor in a changing magnetic field. A current is induced in the coil. (This is another example of electromagnetic induction at work.)

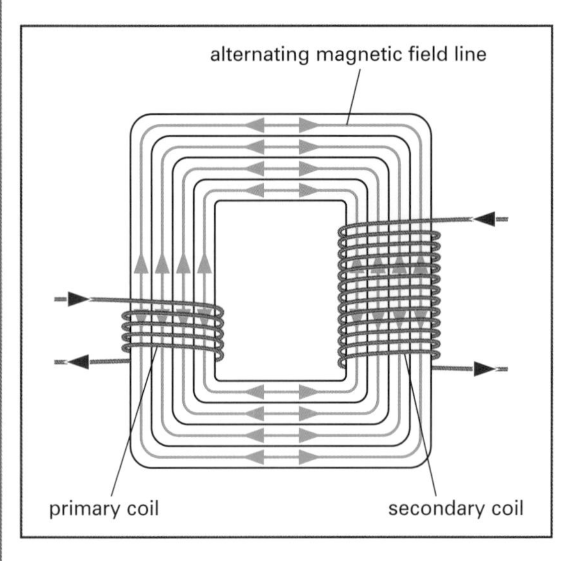

Figure 21.12 The a.c. in the primary coil produces a varying magnetic field in the core. This induces a varying current in the secondary coil. The core of a transformer is often made in sheets (laminated), so that the magnetic field lines follow the sheets around from the primary coil to the secondary.

If the secondary coil has only a few turns, the e.m.f. induced across it is small. If it has a lot of turns, the e.m.f. will be large. Hence, to get a high voltage out, we need a secondary coil with a lot of turns compared to the primary.

If direct current is connected to a transformer, there is no output voltage. This is because the magnetic field produced by the primary coil is unchanging. With an unchanging field passing through the secondary coil, no voltage is induced in it.

Notice from Figure **21.12** that the magnetic field links the primary and secondary coils. The energy being brought by the current in the primary coil is transferred to the secondary by the magnetic field. This means that the core must be very good at transferring magnetic energy. A **soft magnetic material** must be used – usually an alloy of iron with a small amount of silicon. (Recall that soft magnetic materials are ones that can be magnetised and demagnetised easily.) Even in a well-designed transformer, some energy is lost because of the resistance of the wires, and because the core 'resists the flow' of the changing magnetic field.

QUESTIONS

12 **a** What is the function of the core of a transformer?

b Why must the core be made of a soft magnetic material?

13 Explain why a transformer will not work with direct current.

Calculating current

To transmit a certain power P, we can use a small current I if we transmit the power at high voltage V. This follows from the equation for electrical power (see Chapter **18**):

electrical power, $P = IV$

Worked example **2** shows how this works.

Worked example 2

Suppose that a power station generates 500 MW of power. What current will flow from the power station if it transmits this power at 50 kV? What current will flow if it transmits it at 1 MV?

Step 1: Rearranging $P = IV$, we have the equation we need to use.

$$I = \frac{P}{V}$$

Step 2: Substituting values for the first case (P = 500 MW = 500×10^6 W, V = 50 kV = 50×10^3 V) gives the current as

$$I = \frac{500 \times 10^6 \text{ W}}{50 \times 10^3 \text{ V}} = 10\,000 \text{ A}$$

Step 3: Now consider the second case, when the power is transmitted at 1 MV (10^6 V), which is the operating voltage of some national grids. The current is now given by

$$I = \frac{500 \times 10^6 \text{ W}}{10^6 \text{ V}} = 500 \text{ A}$$

Energy saving

Now you should be able to understand why electricity is transmitted around the country at high voltages. The higher the voltage, the smaller the current in the cables and the smaller the energy losses. You can see this from Worked example **2**. Increasing the voltage by a factor of 20 reduces the current by a factor of 20. This means that the power lost in the cables is greatly reduced (in fact, it is reduced by a factor of 20^2, which is 400), and so thinner cables can safely be used.

The current flowing in the cables is a flow of coulombs of charge. At high voltage, we have fewer coulombs flowing, but each coulomb carries more energy with it.

Thinking about power

If a transformer is 100% efficient, no power is lost in its coils or core. This is a reasonable approximation, because well-designed transformers waste only about 0.1% of the power transferred through them. This allows us to write an equation relating the primary and secondary voltages, V_p and V_s, to the primary and secondary currents, I_p and I_s, flowing in the primary and secondary coils, using $P = IV$:

power into primary coil = power out of secondary coil

$$I_p \times V_p = I_s \times V_s$$

Worked example **3** shows how to use this equation.

Worked example 3

The primary coil of a transformer is connected to a 12 V alternating supply, and carries a current of 5 A. If the output voltage is 240 V, what current flows in the secondary circuit? Assume that the transformer is 100% efficient.

Step 1: Draw a transformer symbol and mark on it the information from the question (see Figure **21.13**).

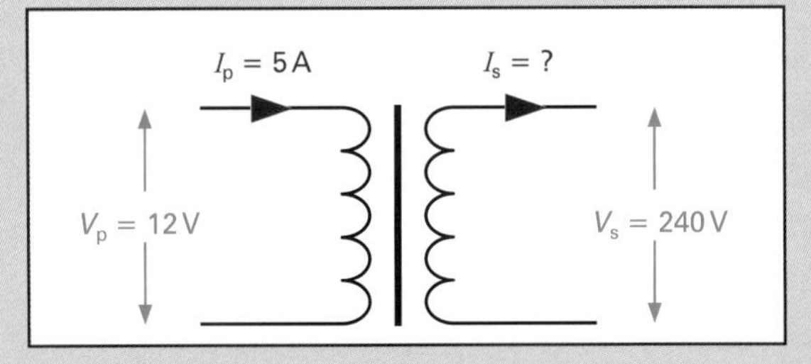

Figure 21.13 See Worked example **3**.

Step 2: Think about what a reasonable answer might be. The voltage is being stepped up by a factor of 20 (from 12 V to 240 V). So the current will be stepped down by the same factor. You can probably see that the secondary current will be one-twentieth of the primary current, that is 5/20 = 1/4 = 0.25 A. (This is the correct answer, but we will press on with the formal calculation.)

Step 3: Write down the transformer power equation.

$$I_p \times V_p = I_s \times V_s$$

Step 4: Substitute values from the question.

$$5\,\text{A} \times 12\,\text{V} = I_s \times 240\,\text{V}$$

Step 5: Rearrange and solve for I_s.

$$I_s = \frac{5\text{A} \times 12\text{V}}{240\text{ V}} = 0.25\text{ A}$$

Hence the current supplied by the secondary coil is 0.25 A. So, in stepping up the voltage, the transformer has stepped down the current. If both had been stepped up, we would be getting something for nothing – which in physics is impossible!

QUESTIONS

14 In a step-up transformer, is the current in the secondary coil greater than or less than the current in the primary coil?

15 **a** A power distribution system transmits 200 MW of power at a current of 500 A. At what voltage is the power distributed? Give your answer in kV.

b It is proposed to double the distribution voltage. What current will now flow in the cables?

c If power losses in the existing system are 6 MW, what will they be if the higher-voltage system is adopted?

16 A transformer is used to reduce a 230 V mains supply to 6 V, to power a radio.

a If the primary coil has 6000 turns, how many turns must the secondary have?

b If, in normal use, a current of 0.04 A flows in the primary coil, what current flows in the secondary?

c What assumption must be made to solve part **b**?

Summary

When a conductor is moved so that it cuts across a magnetic field, an e.m.f. is induced between its ends. If the conductor is part of a complete circuit, an induced current will flow. This is electromagnetic induction.

To generate electricity, a magnet or electromagnet is rotated inside a coil of wire. An induced e.m.f. appears across the ends of the coil.

The direction of an induced e.m.f. is such as to oppose the change causing it.

Electrical power is transmitted at high voltages so that less energy is wasted in the cables.

At higher voltages, the current is relatively low, so that resistive losses in the cables are low, and thinner cables can be used.

A transformer changes the voltage of an alternating supply.

A transformer consists of a primary coil and a secondary coil, linked by an iron core.

The changing magnetic field produced by the primary coil induces an alternating current in the secondary coil.

A step-up transformer increases the voltage of the supply.

The transformer equation relates the voltages and the numbers of turns:

$$\frac{V_p}{V_s} = \frac{N_p}{N_s}$$

For a 100% efficient transformer (in which no power is wasted):

power into primary coil = power out of secondary coil

$$I_p \times V_p = I_s \times V_s$$

End-of-chapter questions

21.1 A student holds a bent piece of wire in a horizontal magnetic field, as shown in Figure **21.14**. She moves the wire downwards through the field, and then upwards.

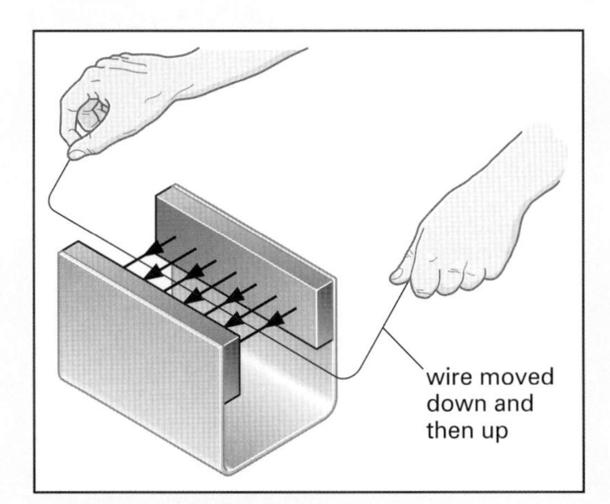

Figure 21.14 For Question **21.1**.

a Explain why an e.m.f. is induced between the ends of the wire. [1]

b How will the e.m.f. differ between moving the wire downwards and moving it upwards? [1]

c Suggest how she could move the wire to induce a bigger e.m.f. across its ends. [1]

d She now moves the wire horizontally from side to side in the magnetic field. Will an e.m.f. be induced? Give a reason to support your answer. [2]

21.2 Figure **21.15** shows a simple a.c. generator.

a Name the parts labelled X, Y and Z. [3]

b Describe the essential difference between alternating current and direct current. Include a diagram to support your answer. [3]

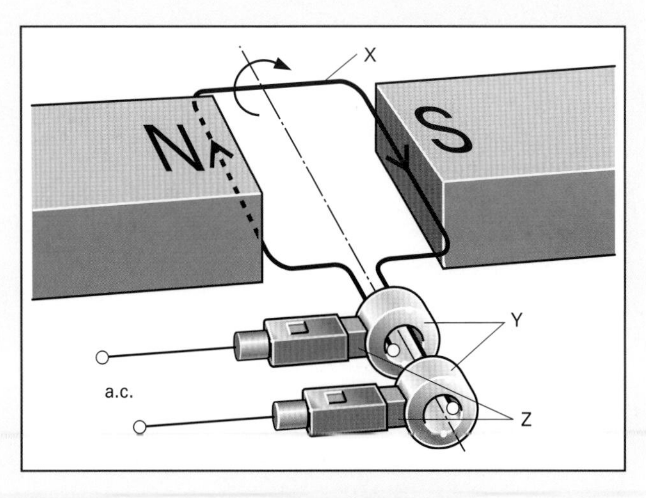

Figure 21.15 For Questions **21.2** and **21.4**.

21.3 Electrical power is often transmitted over long distances in high-voltage power lines (cables). Transformers are used to increase the voltage provided by the power station, and to reduce the voltage for the final user.

- **a** Explain why electricity is transmitted at high voltages like this. [1]
- **b** A transformer has 10 turns of wire on its primary coil, and 200 turns on its secondary coil. If the p.d. across the primary coil is 3 V a.c., what will the e.m.f. across the secondary be? [3]
- **c** How could the same transformer be used as a step-down transformer? [1]

E **21.4** Look again at the generator shown in Figure **21.15** on page **237**.

- **a** Suggest **two** ways in which the coil could be altered to induce a bigger e.m.f. [2]
- **b** Suggest **two** other ways in which the e.m.f. could be increased. [2]

21.5 A transformer is used to transform a 230 V mains supply to 12 V for a computer games console.

- **a** The primary coil has 5 000 turns. How many turns should there be on the secondary coil? [3]
- **b** In normal use, a current of 0.40 A flows in the secondary coil. What current flows in the primary coil? Assume that there are no power losses in the transformer. [3]

Block 5

Atomic physics

A century ago, many physicists were still reluctant to believe that matter is made of atoms. How could you believe in particles that were invisible? Today, it is generally accepted that atoms are made of protons, neutrons and electrons, and that protons and neutrons are themselves made up of even smaller particles called quarks.

There are many other particles, too, which we have come to understand (although we cannot see them). There are particles associated with the fundamental forces of nature. The photograph shows a small section of the Large Hadron Collider (LHC) at the CERN laboratory near Geneva. Its task is to look for the Higgs boson, a fundamental particle that is thought to give matter its mass.

When the atomic nucleus was discovered, the apparatus used was small enough to fit on a laboratory bench. As physicists have searched for even smaller particles, the instruments they have used (like the LHC) have become bigger and bigger.

Another surprising fact is that, in learning more about the underlying structure of matter, we have learned more about the origin and development of the Universe itself. The LHC is described as 'taking us back to within a fraction of a second of the Big Bang'.

A view of part of the 27 km long tunnel of the Large Hadron Collider at the European Centre for Nuclear Research (CERN). The LHC is so big that technicians travel around on bicycles.

22 The nuclear atom

Core Describing the structure of atoms

E **Extension** Describing evidence for the nuclear model of the atom

Core Describing the composition of the nucleus

Core Representing nuclides in the form $^{A}_{Z}X$

E **Extension** Using the term 'isotope'

Matter and atoms

You probably have the idea in your head that 'All matter is made of atoms', and that is more or less true. Most of the matter around you – buildings, the air, your body, this book – is made of tiny atoms, far too small to be seen individually. You have probably also seen an image of an atom like the one shown on the coin in Figure **22.1**. This is a Greek 10-drachma coin (an old coin that was used before Greece changed to the Euro). On the other side, it shows Democritus, the Greek philosopher who is usually credited with first suggesting the idea that matter was made of tiny, indivisible particles called atoms.

If you look at the image of the atom, you will see that it shows a tiny nucleus at the centre, with three electrons travelling around it along circular paths. (The paths are shown as ellipses because of the perspective view.) This image of an atom is like a tiny solar system, and it is not how Democritus would have pictured an atom. He believed that atoms were the smallest building blocks of matter, so they could not be divided into anything smaller. The word '*a-tom*' means 'not divisible'.

The mini-solar-system picture of the atom was developed in the early years of the 20th century, and it is still quite a useful model. Today, most scientists would picture an atom rather differently. This chapter looks the structure of atoms and the particles they are made of.

Figure 22.1 One side of this old Greek 10-drachma coin shows Democritus, a philosopher who lived almost 2500 years ago. According to his atomic theory, matter is made up of vast numbers of tiny particles, which come together in different combinations to make the things we see around us. The reverse side of the coin shows a modern image of a single atom. Democritus would not have imagined that an atom could be subdivided into a nucleus and electrons.

22.1 Atomic structure

At one time, physics textbooks would have said that atoms are very tiny, too tiny ever to be seen. Certainly, a single atom is too small to be seen using a conventional light microscope. But technology has made great advances. Now there is more than one kind of microscope that can be used to show individual atoms. Figure **22.2** shows a photograph made using a scanning tunnelling microscope. The picture shows silicon atoms on the surface of a crystal of silicon (the material that transistors and computer chips are made from). The diamond shape shows a group of 12 atoms. The whole crystal is made up of vast numbers of groups of atoms like this.

In 1910, Ernest Rutherford and his colleagues discovered that every atom has a tiny central nucleus. This gave rise to the 'solar system' model of the atom shown in Figure **22.3**. In this model, the negatively charged electrons orbit the positively charged nucleus. The electrons are attracted to the nucleus (because of its opposite charge), but their speed prevents them from falling into it.

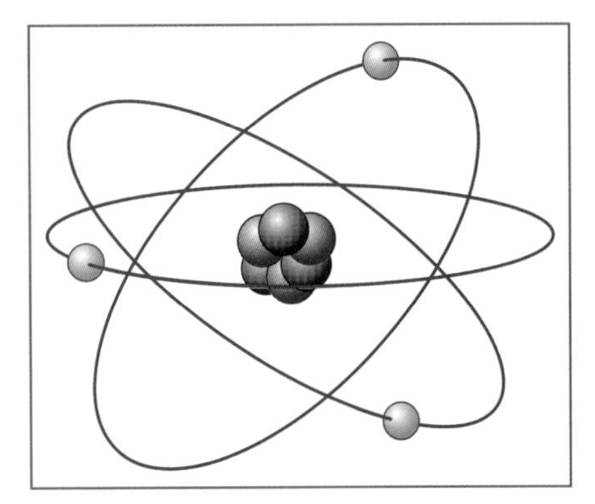

Figure 22.3 The nuclear model of the atom – three electrons are orbiting a nucleus made up of three protons and three neutrons.

Discovering the nucleus

Electrons were discovered in 1896 by the English physicist, J. J. Thomson. He realised that electrons were much smaller than atoms, at least one thousand times lighter than a hydrogen atom. (Now we can be more accurate and say that the mass of an electron is about 1/1836 of the mass of a hydrogen atom.) He guessed, correctly, that electrons were part of atoms. He even suggested that atoms were made up entirely of electrons, spinning in such a way that they stuck together. This was not a very successful model.

Other scientists argued that, since electrons had negative charge, there must be other particles in an atom with an equal amount of positive charge, so that an atom has no overall charge – it is neutral. Since electrons have very little mass, the positive charge must also account for most of the mass of the atom. Figure **22.4** shows a model that illustrates this. The atom is formed from a sphere of positively charged matter with tiny, negatively charged electrons embedded in it. This is the famous 'plum pudding model', where the electrons are the negatively charged plums in the positively charged pudding. You can see that this is a different model from the 'solar system' model we described earlier (Figure **22.3**).

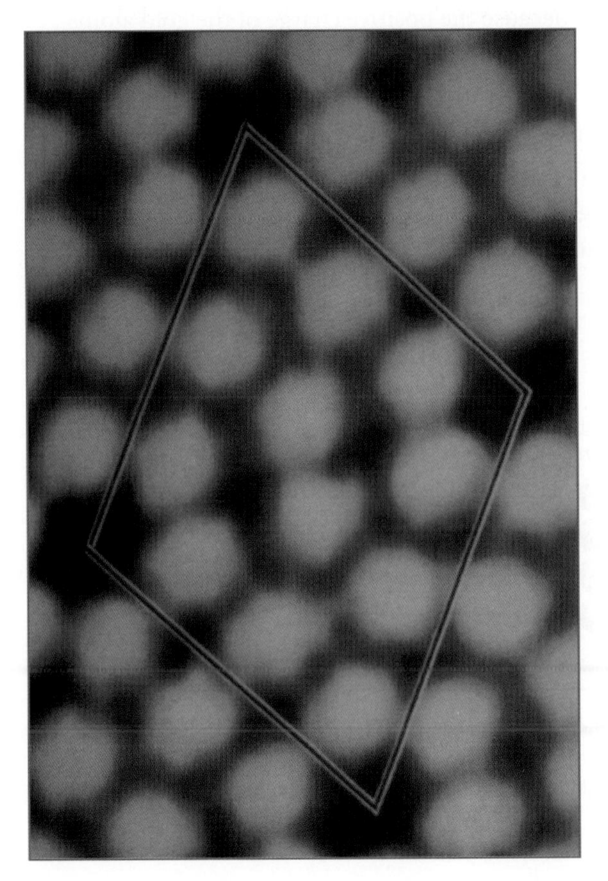

Figure 22.2 Individual silicon atoms (bright spots, artificially coloured by a computer) on the surface of a silicon crystal, observed using a scanning tunnelling microscope. The diamond shape (which has been drawn on the image) indicates the basic repeating pattern that makes up the crystal structure of silicon. In this photograph, the silicon atoms are magnified 100 million times. (A good light microscope can only magnify by about 1000 times.) Roughly speaking, 4 000 000 000 atoms would fit into a length of 1 m.

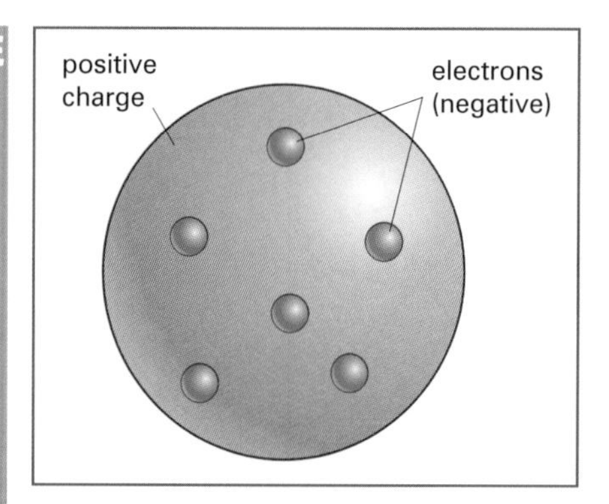

Figure 22.4 The plum pudding model of an atom. Electrons (negatively charged) form the plums stuck in a spherical pudding (positively charged).

So why do we no longer think that atoms are like plum puddings? The answer comes from some work done by the New Zealander Ernest Rutherford and his colleagues, about ten years after Thomson's discovery of the electron.

Radioactivity had been discovered at about the same time as the electron. Rutherford understood that the radiation coming from radioactive substances – alpha, beta and gamma radiation – was a result of changes happening in individual atoms. (You will find more about radioactivity and the different types of radiation in Chapter **23**.) Tiny particles or rays were being spat out from inside atoms, and Rutherford thought that he could use this radiation to investigate other atoms. Rutherford decided to use alpha radiation to probe the atoms in a sample of gold.

Alpha radiation consists of tiny, fast-moving positively charged particles. Rutherford's colleagues Geiger and Marsden set up an experiment (Figure **22.5**) in which alpha radiation was directed at a thin gold foil. They expected the alpha particles to be deflected as they passed through the foil, because their paths would be affected by the positive and negative charges of the atoms. (You might be surprised to think of anything passing through something as solid as gold. Rutherford pictured the alpha particles as tiny bullets, fired through a wall of plum puddings.)

Geiger and Marsden found that most of the alpha particles passed straight through the gold foil, scarcely deflected. However, a very few bounced back towards

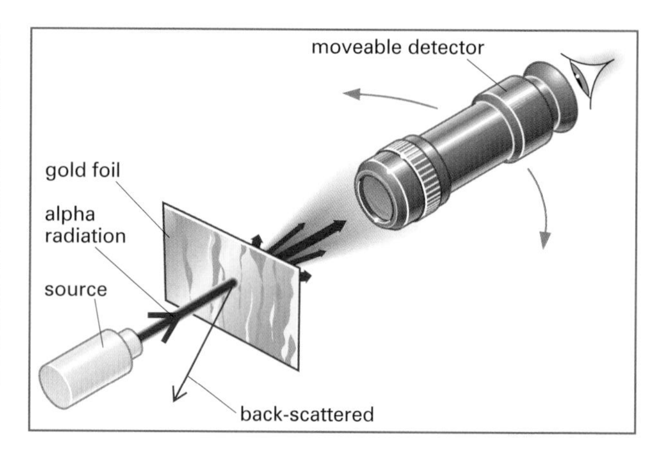

Figure 22.5 The experiment to show alpha particle scattering by a gold foil, also now known as Rutherford scattering. Alpha particles from the source on the left strike the gold foil. Most pass straight through, but a few – about one in 8000 – are scattered back towards the source.

the source of the radiation. It was as if there was something very hard in the gold foil – like a ball-bearing buried inside the plum pudding. What was going on?

Rutherford realised that the answer was to do with static electricity. Alpha particles are positively charged. If they are repelled back from the gold foil, it must be by another positive charge. If only a few were repelled, it was because the positive charge of the gold atoms was concentrated in a tiny space within each atom. Most alpha particles passed straight through because they never went near this concentration of charge (see Figure **22.6**). This speck of concentrated positive charge, at the heart of every atom, is what we now call the atom's **nucleus**.

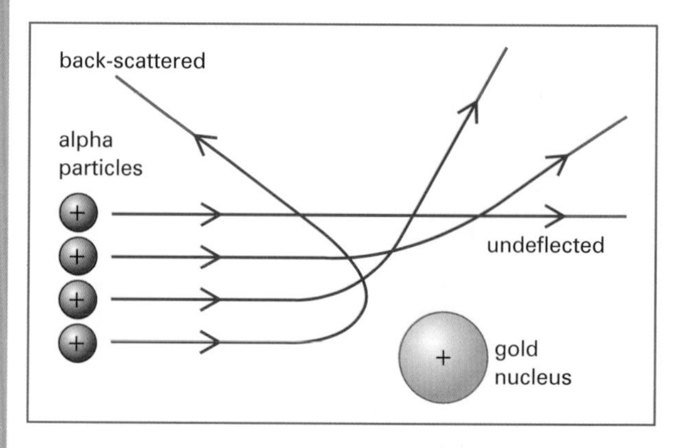

Figure 22.6 Most alpha particles pass straight through the gold foil, because they do not pass close to the atomic nucleus. The closer they get to the nucleus, the more they are deflected or scattered. Only those which score a 'direct hit' are reflected straight back.

In later years, Rutherford often spoke of the surprising results of the alpha scattering experiment. He said:

> 'It was quite the most incredible event that ever happened to me in my life. It was as if you fired a fifteen-inch artillery shell at a piece of tissue paper and it came back and hit you.'

A sense of scale

Rutherford was able to analyse the results from Geiger and Marsden's experiment to work out just how big the nucleus of a gold atom was. An atom is small (about 10^{-10} m across) but its nucleus is very much smaller (about 10^{-15} m in diameter). Around the nucleus travel the electrons. They are even tinier than the nucleus. And the rest of the atom is simply empty space.

It is hard to imagine these relative sizes. Try picturing a glass marble about 1 cm in diameter, placed at the centre of a football pitch, to represent the nucleus of an atom. Then the electrons are like tiny grains of dust, orbiting the nucleus at different distances, right out to the edge of the football ground.

It is even harder to imagine, when you stub your toe on a rock, that the atoms of the rock (and your toe) are almost entirely empty space!

A successful model

Rutherford's picture of the atom rapidly gained acceptance among scientists. It gave a clear explanation of the alpha particle scattering experiment, and further tests with other metals confirmed Rutherford's ideas. Thomson had the idea that the atom was made of many electrons spinning in such a way that they stuck together. This was a rather unclear model, and it was swept away by Rutherford's simpler picture.

Rutherford's model also allowed scientists to think about other questions. Chemists wanted to know how atoms bonded together. Physicists wanted to understand why some atoms are unstable and emit radiation. How were X-rays produced? These were all questions to which we now have good answers, and Rutherford's discovery of the atomic nucleus did a lot to help answer them.

Today, practising scientists have rather different ideas about atoms. They want to calculate many different quantities, and so models of the atom are much more mathematical. Quantum theory, developed not long after Rutherford's work, made the atom seem like a much fuzzier thing, not a collection of little spheres orbiting around each other. However, the important thing about a model is that it should help us to understand things better, and help us to make new predictions, and Rutherford's model of the nuclear atom has certainly done that.

QUESTIONS

1 Explain why, in Geiger and Marsden's experiment, some alpha particles were 'back-scattered' when they came near to the nucleus of a gold atom.

2 Explain why only a very few alpha particles were back-scattered.

3 Think about the plum pudding model of the atom.
  - **a** What are the plums?
  - **b** What is the pudding?

4 In the 'solar system' model of the atom, what force holds the electrons in their orbits around the nucleus?

22.2 Protons, neutrons and electrons

Nowadays, we know that the atomic nucleus is made up of two types of particle, **protons** and **neutrons**. The protons carry the positive charge of the nucleus, while the neutrons are neutral. Negatively charged **electrons** orbit the positively charged nucleus. Protons and neutrons have similar masses, and they account for most of the mass of the atom (because electrons are so light). Together, protons and neutrons are known as **nucleons**.

Table **22.1** summarises information about the masses and charges of the three sub-atomic particles. The columns headed 'Relative charge' and 'Relative mass' give the charge and mass of each particle compared to that of a proton. It is much easier to remember these values, rather than the actual values in coulombs (C) and kilograms (kg).

Particle	Position	Charge / C	Relative charge	Mass / kg	Relative mass
proton	in nucleus	$+1.6 \times 10^{-19}$	+1	1.67×10^{-27}	1
neutron	in nucleus	0	0	1.67×10^{-27}	1
electron	orbiting nucleus	-1.6×10^{-19}	−1	9.11×10^{-31}	$\frac{1}{1836}$ (approx. 0)

Table 22.1 Charges and masses of the three sub-atomic particles.

Atoms and elements

Once the particles that make up atoms were identified, it was much easier to understand the **Periodic Table** of the elements (Figure 22.7). This shows the elements in order, starting with the lightest (hydrogen, then helium) and working up to the heaviest. In fact, it is not the masses of the atoms that determine the order in which they appear, but the **number of protons** in the nucleus of each atom. Every atom of hydrogen has one proton in its nucleus, so hydrogen is element number 1. Every helium atom has two protons, so helium is element number 2, and so on.

Each element has its own symbol, consisting of one or two letters, such as H for hydrogen, and He for helium. Sometimes, the symbol for an atom may be written with two numbers in front of it, one above the other, such as:

$${}^{4}_{2}\text{He}$$

This represents an atom of helium. The bottom number tells us that there are 2 protons in the nucleus of an atom of helium, and the top number tells us that there is a total of 4 nucleons in the nucleus of an atom of helium. (From this, it is simple to work out that there must be 2 neutrons in the nucleus.)

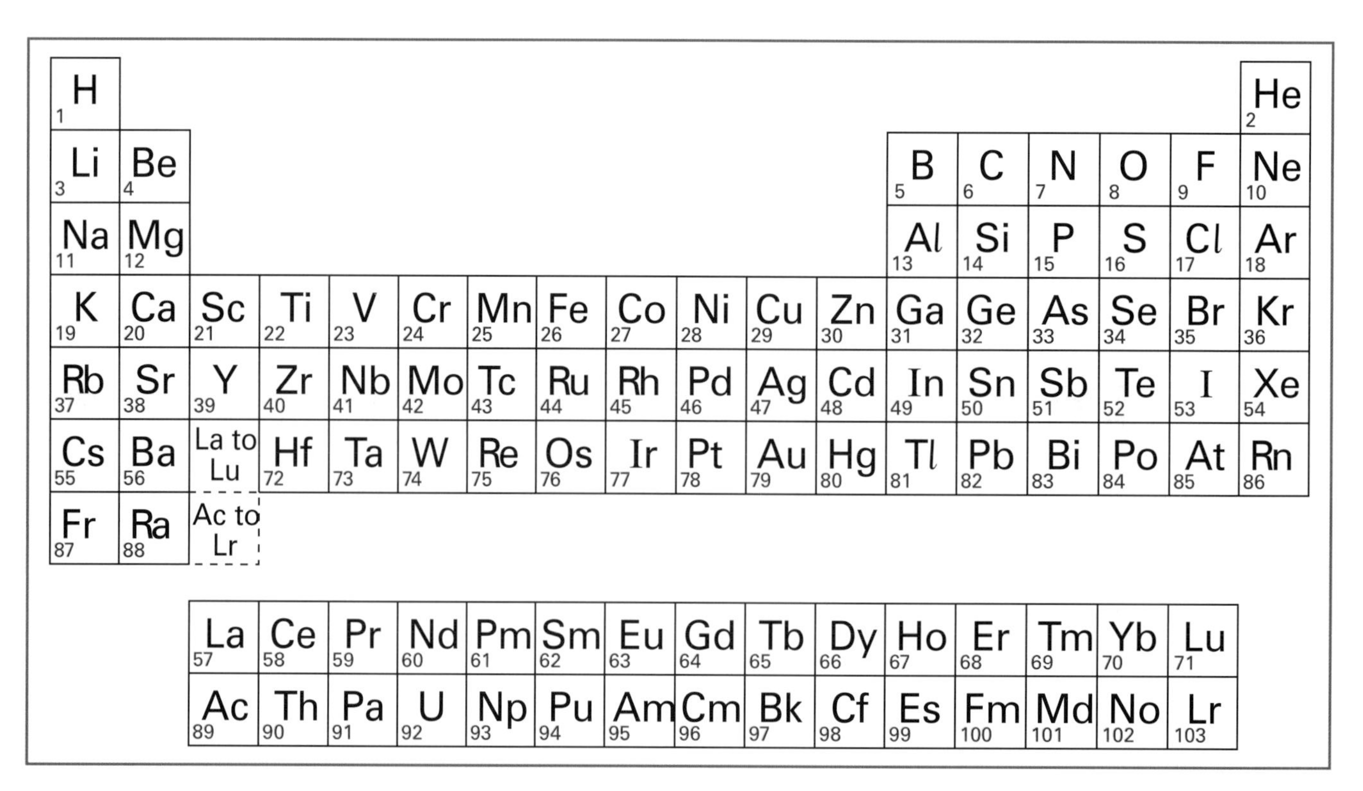

Figure 22.7 The Periodic Table of the elements is a way of organising what we know about the different elements, based on their atomic structures. The elements are arranged in order according to the number of protons in the nucleus (their proton number *Z*).

We can write the general symbol for an element X with its **proton number** Z, which is the number of protons in the nucleus, and **nucleon number** A, which is the number of nucleons (protons and neutrons) in the nucleus, as follows.

The nucleus of an atom of element X is written as

$${}^{A}_{Z}\mathrm{X}$$

where Z is the proton number and A is the nucleon number.

A neutral atom of element X will also have Z electrons orbiting the nucleus. There are just over a hundred different elements X, all of which have different possible combinations of Z and A, each giving a different type of nucleus. Each type is called a **nuclide**. (Sometimes physicists refer to nuclides as **nuclear species**, as if all the different species make up a 'zoo' of nuclei.)

QUESTIONS

5 a Which particles make up the nucleus of an atom?
 b Which particles orbit around the nucleus?

6 An atom of a particular isotope of oxygen is written as ${}^{17}_{8}\mathrm{O}$.
 a What is its nucleon number?
 b What is its proton number?

7 An atom of a particular isotope of lead (symbol Pb) contains 82 protons and 128 neutrons. Write down the full symbol for this atom.

8 How many protons, neutrons and electrons are there in a neutral atom silver atom, with the symbol ${}^{107}_{47}\mathrm{Ag}$?

9 How many times greater is the mass of a proton than the mass of an electron?

Activity 22.1 The atomic zoo

Solve some problems involving nuclides.

Elements and isotopes

It is the proton number Z that tells us which element an atom belongs to. For example, a small atom with just 2 protons in its nucleus ($Z = 2$) is a helium atom. A much bigger atom with 92 protons in its nucleus is a uranium atom, because uranium is element 92.

From Z and A you can work out a third number, the **neutron number** N, which is the number of neutrons in the nucleus.

$$\text{proton number} + \text{neutron number} = \text{nucleon number}$$

$$Z + N = A$$

The atoms of all elements exist in more than one form. For example, Table **22.2** shows three types of hydrogen atom. Each has just one proton in its nucleus, but they have different numbers of neutrons (0, 1 and 2). They are described as different isotopes of hydrogen.

Symbol for isotope	Proton number Z	Neutron number N	Nucleon number A
${}^{1}_{1}\mathrm{H}$	1	0	1
${}^{2}_{1}\mathrm{H}$	1	1	2
${}^{3}_{1}\mathrm{H}$	1	2	3
Symbol for isotope	**Proton number Z**	**Neutron number N**	**Nucleon number A**
${}^{235}_{92}\mathrm{U}$	92	143	235
${}^{238}_{92}\mathrm{U}$	92	146	238

Table 22.2 Three isotopes of hydrogen, and two isotopes of uranium.

The different **isotopes** of an element all have the same chemical properties, but those with a greater number of neutrons are heavier.

> The isotopes of an element have the same number of protons but different numbers of neutrons in their nuclei.

Figure **22.8** shows atoms of two isotopes of helium, ${}^{4}_{2}He$ (the commonest isotope) and ${}^{3}_{2}He$ (a lighter and much rarer isotope). Each has two protons in the nucleus and two electrons orbiting it, but the lighter isotope ${}^{3}_{2}He$ has only one neutron.

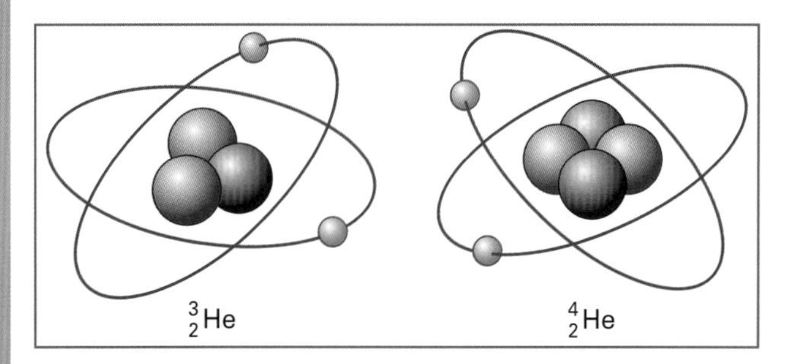

Figure 22.8 These drawings represent two isotopes of helium.

Isotopes at work

For most chemical elements, at least one isotope is stable; however, other isotopes may be unstable. By this we mean that they undergo radioactive decay, emitting radiation as they change from one element to another. In Chapter **23**, you will learn about how this can be put to practical use.

QUESTIONS

10 a What is the same for the atoms of two different isotopes of an element?

b What is different for them?

11 Table **22.3** lists the proton and nucleon numbers of six different nuclides.

a Copy the table and fill in the empty spaces.

b Which **three** nuclides are isotopes of one element?

c Which **two** nuclides are isotopes of another element?

d Use the Periodic Table of the elements (Figure **22.7**) to identify the three elements in Table **22.3**.

Isotope	Proton number, *Z*	Neutron number, *N*	Nucleon number, *A*
I-1	6	6	
I-2		6	13
I-3	7		14
I-4		8	14
I-5		6	11
I-6		7	13

Table 22.3 For Question **11**.

Summary

Negatively charged electrons orbit around the positively charged nucleus of an atom, held in their orbits by the electrostatic attraction between them.

In Rutherford's alpha particle scattering experiment, fast-moving alpha particles were deflected as they passed through a gold foil. Some were back-scattered, which showed that the mass and positive charge of the atom were concentrated in a tiny space at the centre of the atom.

The nucleus of an atom is made up of protons and neutrons.

The nucleus of an atom of element X can be represented as ${}^{A}_{Z}X$, where Z is the proton number and A is the nucleon number.

Atoms of isotopes of an element have the same number of protons in their nuclei, but different numbers of neutrons.

End-of-chapter questions

22.1 Diamond is a form of carbon. It is made up (almost entirely) of carbon-12 atoms. The symbol for the nucleus of a carbon-12 atom is $^{12}_{6}C$.

- **a** How many protons are there in a carbon-12 atom? [1]
- **b** How many neutrons are there in a carbon-12 atom? [1]
- **c** How many electrons are there in a neutral carbon-12 atom? [1]

22.2 A particular atom of gold (chemical symbol Au) contains 79 protons and 118 neutrons.

- **a** How many nucleons are there in the nucleus of this atom? [2]
- **b** Write the symbol for this nuclide in the form $^{A}_{Z}X$. [2]

E

22.3 A particular isotope of potassium is represented by the symbol $^{39}_{19}K$.

- **a** What is the proton number of this isotope? [1]
- **b** What is the nucleon number of this isotope? [1]
- **c** A second isotope of potassium has one more neutron in its nucleus. Write down its symbol in the form $^{A}_{Z}X$. [2]

E

22.4 Ernest Rutherford devised an experiment in which alpha particles were directed at a thin gold foil. The results of this experiment showed that every atom has a nucleus, and the 'plum pudding' model of the atom had to be discarded.

- **a** Consider these three particles: alpha particle, gold nucleus, and electron.
  - **i** What type of charge (positive or negative) does each have? [3]
  - **ii** List the three particles in order, from smallest to largest. [3]
- **b** Describe the 'plum pudding' model of the atom. [2]
- **c** Draw a diagram to show how an alpha particle could be scattered backwards by a gold atom, towards the source from which it came. [2]
- **d** Explain why most alpha particles passed straight through the gold foil. [2]

23 Radioactivity

Core Describing background and artificial radiation
Core Detecting radiation
Core Describing the nature of alpha (α), beta (β) and gamma (γ) radiation
Core Interpreting nuclear equations to represent decay
E **Extension** Describing how radiation behaves in electric and magnetic fields
Core Describing the ionising and penetrating behaviour of radiation
Core Carrying out calculations involving radioactive half-life
E **Extension** Using radioactive substances

Making sense of radioactivity

Radioactivity is a serious topic. You can tell that because people make lots of jokes about it. If you go on a school visit to a nuclear power station, you will come back with two heads. If you have radiation treatment in hospital, you will glow in the dark. As with many jokes, there is a small element of truth here, and a great deal of fear of the unknown.

When radioactivity was first discovered, people became very excited by it. Some doctors claimed that it had great health-promoting effects. They sold radioactive water, and added radioactive substances to chocolate, bread and toothpaste (see Figure 23.1). There were radioactive cures for baldness, and contraceptive cream containing radium. This attitude still lingers on today, with some Alpine spas offering residents the chance to breathe radioactive air in old mine tunnels.

Our use of radioactive substances, particularly by the most technologically advanced countries, has had some very damaging effects, which stick in people's imaginations. The dropping of atomic bombs on the Japanese cities of Hiroshima and Nagasaki at the end of the Second World War is one example (Figure 23.2). The positive side of our use of radioactive materials has been less obvious, but today millions of people who would once have died of cancer are now alive thanks to radiation therapy (Figure 23.3).

Radioactive materials produce radiation. We have eyes to see light, and we can detect infrared radiation with our skin. But we have no organ for detecting the radiation from radioactive materials that is all around us. We make little use of radioactive materials in our everyday lives, so they remain unfamiliar to us. We learn about them from a teacher who handles

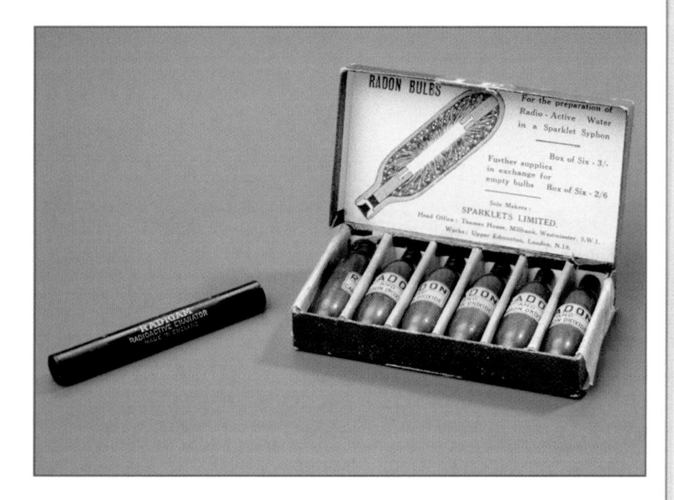

Figure 23.1 In the 1930s, you could buy bulbs of radioactive radon gas to dissolve in your drinking water. An American called Ethan Byers drank a bottle a day for five years – he died of cancer of the jaw.

Figure 23.2 An atomic bomb being tested. Only two atomic bombs have been used in warfare, but many others have been tested. Such tests have released large quantities of radioactive materials into the environment. Their main impact has been on the power struggles between different countries and blocs around the world.

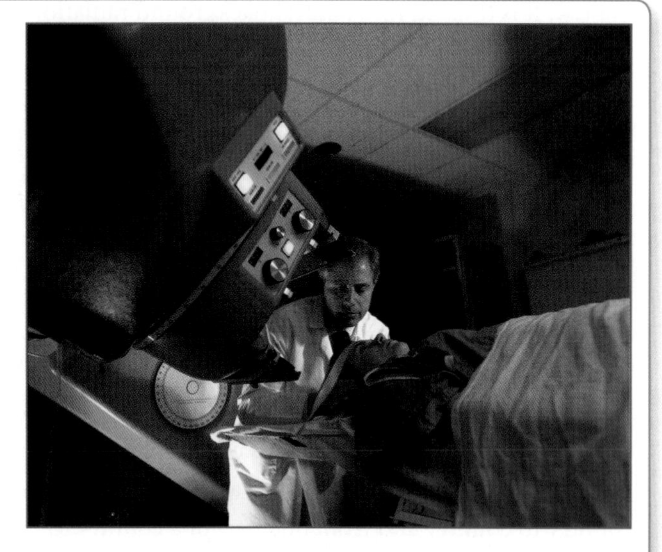

Figure 23.3 Radiation can cause cancer, but it can also be used in its cure. This patient is being exposed to gamma rays from a radioactive source. The rays are directed at the patient's tumour in order to destroy the cancerous cells.

radioactive samples with great care. It is not surprising that we are cautious, if not downright scared, when the topic of radioactivity is raised.

In this chapter, we will look at radioactive substances and the radiation they produce, and discuss how they can be used safely.

23.1 Radioactivity all around

We need to distinguish between two things: **radioactive substances** and the **radiation** that they give out. Many naturally occurring substances are radioactive. Usually these are not very concentrated, so that they do not cause a problem. There are two ways in which radioactive substances can cause us problems:

- If a radioactive substance gets inside us, its radiation can harm us. We say that we have been **contaminated**.
- If the radiation they produce hits our bodies, we say that we have received a dose of radiation. We have been **irradiated**.

In fact, we are exposed to low levels of radiation all the time – this is known as **background radiation**. In addition, we may be exposed to radiation from artificial sources, such as the radiation we receive if we have a medical X-ray.

Figure 23.4 shows the different sources that contribute to the average dose of radiation received by people in

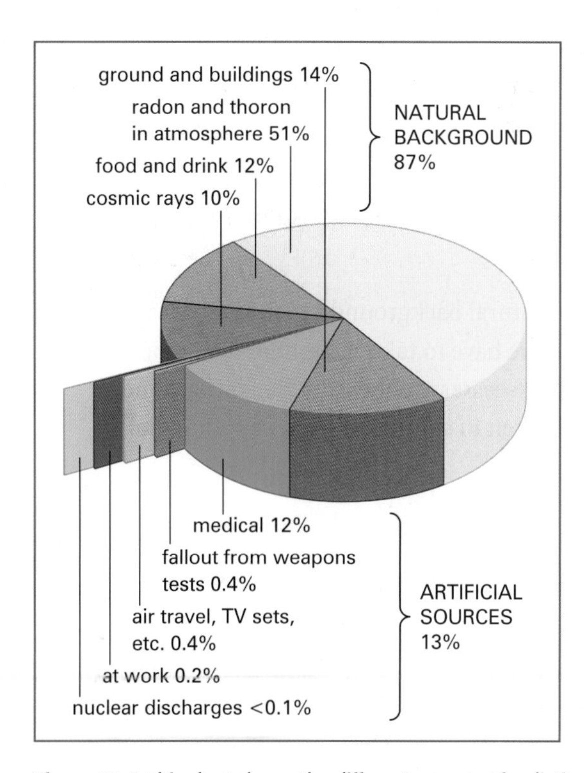

Figure 23.4 This chart shows the different sources of radiation and how they contribute to the average dose of radiation received each year by someone living in the UK. The main division is between natural background radiation and radiation from artificial sources.

the UK. It is divided into natural background radiation (about 87%) and radiation from artificial sources (about 13%). We will look at these different sources in turn.

Sources of background radiation

The air is radioactive. It contains a radioactive gas called radon, which seeps up to the Earth's surface from radioactive uranium rocks underground. Because we breathe in air all the time, we are exposed to radiation from this substance. This contributes about half of our annual exposure. (This varies widely from country to country and from one part of a country to another, depending on how much uranium there is in the underlying rocks.)

The ground contains radioactive substances. We use materials from the ground to build our houses, so we are exposed to radiation from these.

Our food and drink is also slightly radioactive. Living things grow by taking in materials from the air and the ground, so they are bound to be radioactive. Inside our bodies, our food then exposes us to radiation.

Finally, radiation reaches us from space in the form of cosmic rays. Some of this radiation comes from the Sun, some from further out in space. Most cosmic rays are stopped by the Earth's atmosphere. If you live up a mountain, you will be exposed to more radiation from this source.

Because natural background radiation is around us all the time, we have to take account of it in experiments. It may be necessary to measure the background level and then to subtract it from experimental measurements.

Sources of artificial radiation

Most radiation from artificial sources comes from medical sources. This includes the use of X-rays and gamma rays for seeing inside the body, and the use of radiation for destroying cancer cells. There is always a danger that exposure to such radiation may trigger cancer. Medical physicists are always working to reduce the levels of radiation used in medical procedures. Overall, many more lives are saved than lost through this beneficial use of radiation.

Today, most nuclear weapons testing is done underground. In the past, bombs were detonated on land (see Figure **23.2**) or in the air, and this contributed much more to the radiation dose received by people around the world.

If you fly in an aircraft, you are high in the atmosphere. You are exposed to more cosmic rays. This is not a serious problem for the occasional flier, but airline crews have to keep a check on their exposure.

Many people, such as medical radiographers and staff in a nuclear power station, work with radiation. Overall, this does not add much to the national average dose, but for individuals it can increase their dose by up to 10%.

Finally, small amounts of radioactive substances escape from the nuclear industry, which processes uranium for use as the fuel in nuclear power stations, and handles the highly radioactive spent fuel after it has been used.

Detecting radiation

Radioactivity was discovered by a French physicist, Henri Becquerel, in 1896. He had been investigating some phosphorescent rocks – rocks that glow for a while after they have been left under a bright light. His method was to leave a rock on his window sill in the light. Then he put it in a dark drawer on a piece of photographic film to record the light it gave out. He suspected that rocks containing uranium might be good for this. But he discovered something even more dramatic: the photographic film was blackened even when the rock had not been exposed to light. He realised that some kind of invisible radiation was coming from the uranium. What was more, the longer he left it, the darker the photographic film became. Uranium gives out radiation all the time, without any obvious supply of energy.

Becquerel had discovered a way of revealing the presence of invisible radiation, using photographic film. This method is still used today. One of his first photographs of radiation is shown in Figure **23.5**.

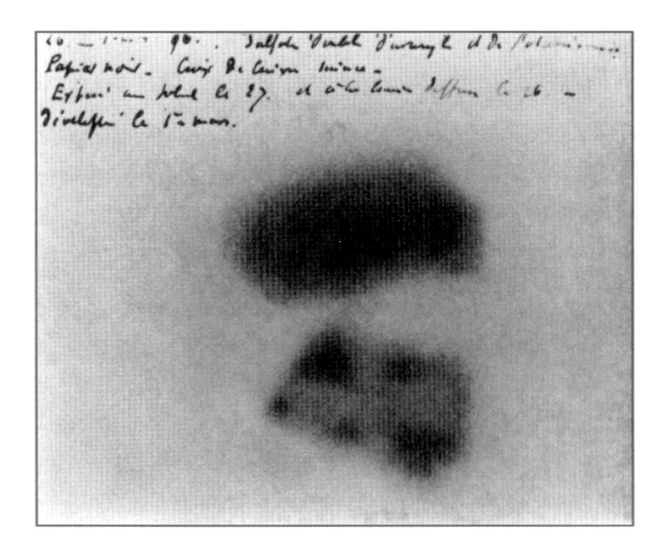

Figure 23.5 One of Henri Becquerel's first photographic records of the radiation produced by uranium. The two black blobs are the outlines of two crystals containing uranium. To show that the radiation would pass through metal, he placed a copper cross between one of the crystals and the photographic film. You can see the 'shadow' of the cross on the photograph. The writing is Becquerel's; the last line says 'développé le 1er mars' – developed on 1st March (1896).

It takes a while to expose and develop a photographic film. For a quicker measurement of radiation, we can use a Geiger counter. The detector is a Geiger–Muller tube, which is held close to a suspected source of radiation (Figure 23.6). The radiation enters the tube, which produces an electrical pulse every time it detects any radiation. The electronic counter (in the man's left hand) adds up these pulses. It can give a click or beep for each pulse. In the photograph, the Geiger counter is being used to check the radiation levels of moss gathered from a mountainside in France. Regular checks are made on samples of air, soil, vegetation and water for 20 km around nuclear power stations. Other analytical equipment can also be seen on the table.

Figure 23.6 Using a Geiger counter to monitor radiation levels.

The randomness of radioactive decay

If you listen to the clicks or beeps of a Geiger counter, you may notice that it is impossible to predict when the next sound will come. This is because radioactive decay is a **random process**. If you study a sample of a radioactive material, you cannot predict when the next atom will decay. Atoms decay randomly over time.

Similarly, it is impossible to point at an individual atom and say that it will be the next one to decay. If an atom on the left of the sample has just decayed, we cannot predict that an atom on the right of the sample will be the next to decay.

To sum up this randomness, we say that radioactive decay occurs randomly over space and time.

Activity 23.1 Observing radioactivity

Watch some demonstrations that illustrate the properties of radiation

QUESTIONS

1. What is the biggest contributor to background radiation?
2. Why are people who live high above sea level likely to be exposed to higher levels of background radiation?
3. What fraction of our annual average dose of radiation is from artificial sources?
4. List **three** sources of exposure to artificial radiation.
5. Name **two** methods of detecting radiation from radioactive materials.

23.2 The microscopic picture

To understand the nature of radioactivity, we need to picture what is going on at a microscopic level, on the level of atoms and nuclei. Two questions we need to answer are: Why are some atoms radioactive while others are not? What is the nature of the radiation they produce?

Radiation is emitted by the nucleus of an atom (Figure 23.7). We say that the nucleus is unstable. An unstable nucleus emits radiation in an attempt to become more stable. This is known as **radioactive decay**. Fortunately, most of the atoms around us have stable nuclei. When the Earth formed, about 4 500 million years ago, there were many more radioactive atoms around. However, as those millions of years have passed, most have decayed to become stable. In the distant past, the level of background radiation was much higher than it is today.

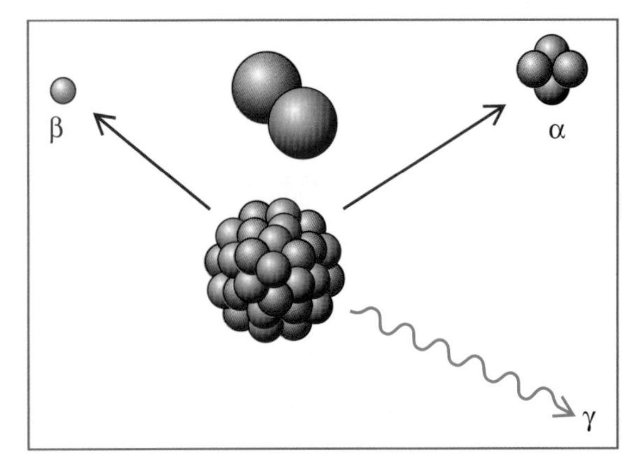

Figure 23.7 Radiation comes from the nucleus of a radioactive atom.

Three types of radiation

There are three types of radiation emitted by radioactive substances (Table 23.1). These are named after the first three letters of the Greek alphabet, alpha (α), beta (β) and gamma (γ). Alpha and beta are particles; gamma is a form of electromagnetic radiation (see Chapter 15).

- An **alpha particle (α-particle)** is made up of two protons and two neutrons. (This is the same as the nucleus of a helium atom, ^4_2He.) Because it contains protons, it is positively charged.
- A **beta particle (β-particle)** is an electron. It is not one of the electrons that orbit the nucleus – it comes from inside the nucleus. It is negatively charged, and its mass is much less than that of an alpha particle.
- A **gamma ray (γ-ray)** is a form of electromagnetic radiation. We can think of it as a wave with a very short wavelength (similar to an X-ray, but even more energetic). Alternatively, we can picture it as a 'photon', a particle of electromagnetic energy.

An atom of a radioactive substance emits either an alpha particle or a beta particle. In addition, it may emit some energy in the form of a gamma ray. The gamma ray is usually emitted at the same time as the alpha or beta, but it may be emitted some time later.

Alpha particles have a much greater mass than beta particles, so they travel more slowly. Gamma rays travel at the speed of light.

Name	Symbol	Made of	Mass	Charge	Speed / m/s
alpha	α or ^4_2He	2 protons + 2 neutrons	approx. (mass of proton) × 4	+2	$\sim 3 \times 10^7$
beta	β or $^{0}_{-1}e$	an electron	approx. (mass of proton) / 1 840	−1	$\sim 2.9 \times 10^8$
gamma	γ	photon of electromagnetic radiation	0	0	3×10^8

Table 23.1 Three types of radiation produced by naturally occurring radioactive substances. To these we should add neutrons and positively charged beta radiation, produced by some artificial radioactive substances.

QUESTIONS

6 Name **four** types of ionising radiation.

7 a Which radiation from a radioactive substance is positively charged?
  b Which radiation from a radioactive substance is negatively charged?

8 What type of sub-atomic particle is a β-particle?

9 Which type of radiation is a form of electromagnetic radiation?

10 a Which radiation travels fastest, alpha, beta or gamma?
  b Which radiation travels most slowly?

Radioactive decay equations

When an atom of a radioactive substance decays, it becomes an atom of another element. This is because, in alpha and beta decay, the number of protons in the nucleus changes. We can represent any radioactive decay by an equation using the notation explained in Chapter **22** (pages **244–245**).

Here is an example of an equation for **alpha decay**:

$$^{241}_{94}\text{Am} \rightarrow {}^{237}_{92}\text{U} + {}^{4}_{2}\text{He} + \text{energy}$$

This represents the decay of americium-241, the isotope used in smoke detectors. It emits an alpha particle (represented as a helium nucleus) and becomes an isotope of uranium. Notice that the numbers in this equation must balance, because we cannot lose mass or charge. So

nucleon numbers: $241 \rightarrow 237 + 4$
proton numbers: $94 \rightarrow 92 + 2$

Here is an example of an equation for **beta decay**:

$$^{14}_{6}\text{C} \rightarrow {}^{14}_{7}\text{N} + {}^{0}_{-1}\text{e} + \text{energy}$$

This is the decay that is used in radiocarbon dating. A carbon-14 nucleus decays to become a nitrogen-14 nucleus. (The beta particle, an electron, is represented by ${}^{0}_{-1}\text{e}$.) If we could see inside the nucleus, we would see that a single neutron has decayed to become a proton. So

$$^{1}_{0}\text{n} \rightarrow {}^{1}_{1}\text{p} + {}^{0}_{-1}\text{e}$$

For each of these two beta decay equations, you should be able to see that the nucleon numbers and proton numbers are balanced. We say that, in radioactive decay, nucleon number and proton number are **conserved**.

QUESTION

11 The equation below represents the decay of a polonium nucleus to form a lead nucleus. An alpha (α-) particle is emitted.

$$^{210}_{84}\text{Po} \rightarrow {}^{206}_{82}\text{Pb} + \ldots\ldots\ldots\ldots + \text{energy}$$

  a Copy and complete the equation.
  b Show that proton numbers are equal on each side of the equation.
  c Show that nucleon numbers are equal on each side of the equation.

E

Deflecting radiation

How can we tell the difference between these three types of radiation? One method is to see how they behave in electric and magnetic fields.

Because they have opposite charges, alpha and beta particles are deflected in opposite directions when they pass through an electric field (Figure **23.8a**). Alpha particles are attracted towards a negatively charged plate, while beta particles are attracted towards a positively charged plate. Gamma rays are not deflected because they are uncharged.

Alpha and beta particles are charged, so, when they move, they constitute an electric current. Because of their opposite signs, the forces on them in a magnetic field are in opposite directions (Figure **23.8b**). This is an example of the motor effect (Chapter **20**). The direction in which the particles are deflected can be predicted using Fleming's left-hand rule. As in an electric field, gamma rays are not deflected because they are uncharged.

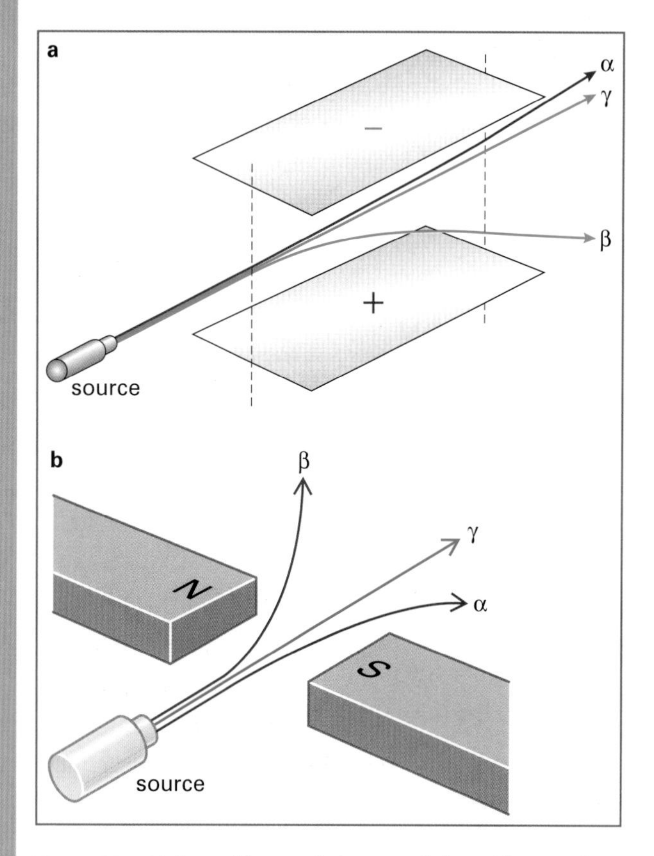

Figure 23.8 Alpha and beta radiations are deflected in opposite directions: **a** in an electric field, and **b** in a magnetic field.

Energy released

Radioactive substances release energy when they decay. Before they decay, this energy is stored in the nucleus of the atom. When it is released, it is in two forms:

- An alpha or beta particle is fast-moving. The nucleus that has emitted it recoils. Both particles have **kinetic energy**.
- A gamma ray transfers energy as **electromagnetic radiation**.

Penetrating power

When physicists were trying to understand the nature of radioactivity, they noticed that radiation can pass through solid materials. (In Figure 23.5, we saw how Becquerel showed that some of the radiation from uranium could pass through copper.) Different types of radiation can penetrate different thicknesses of materials.

- Alpha particles are the most easily absorbed. They can travel about 5 cm in air before they are absorbed. They are absorbed by a thin sheet of paper.
- Beta particles can travel fairly easily through air or paper. But they are absorbed by a few millimetres of metal.
- Gamma radiation is the most penetrating. It takes several centimetres of a dense metal like lead, or several metres of concrete, to absorb most of the gamma radiation.

These ideas about **penetrating power** are represented in Figure 23.9.

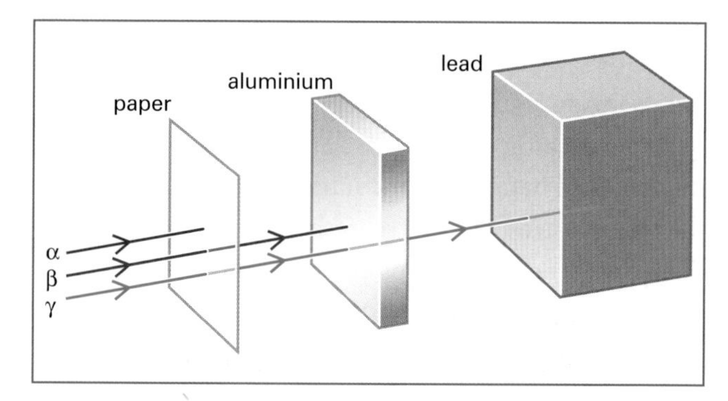

Figure 23.9 The penetrating power of radiation is greatest for gamma radiation and least for alpha radiation. This is related to their ability to ionise the materials they are passing through.

Ionisation

When radiation passes through air, it may interact with air molecules, knocking electrons from them, so that the air molecules become charged. We say that the air molecules have become ionised. The relative ionising effects are as follows:

- alpha particles are the most ionising
- gamma radiation is the least ionising.

Because the radiation from radioactive substances causes **ionisation** of the materials that absorb it, it is often known as **ionising radiation**.

Consider an alpha particle passing through the air. An alpha particle is the slowest moving of all the three radiations and has the largest charge. As the alpha particle collides with an air molecule, it may knock an electron from the air molecule, so that it becomes charged. The alpha particle loses a little of its energy. It must ionise thousands of molecules before it loses all of its energy and comes to a halt. Nonetheless, alpha radiation is the most strongly ionising radiation.

A beta particle can similarly ionise air molecules. However, it is less ionising for two reasons: its charge is less than that of an alpha particle, and it is moving faster, so that it is more likely to travel straight past an air molecule without interacting with it. This is why beta radiation can travel further through air without being absorbed.

Gamma radiation is uncharged and it moves fastest of all, so it is the least readily absorbed in air, and therefore is the least ionising. Lead is a good absorber because it is dense (its atoms are packed closely together), and its nuclei are relatively large, so they present an easy target for the gamma rays.

You should be able to see the pattern linking ionising power and absorption:

- Alpha radiation is the most strongly ionising, so it is the most easily absorbed and the least penetrating.
- Gamma radiation is the least strongly ionising, so it is the least easily absorbed and the most penetrating.

X-rays also cause ionisation in the materials they pass through, and so they are also classed as ionising radiation. X-rays are very similar to gamma rays. But X-rays usually have less energy (longer wavelength) than gamma rays, and they are produced by X-ray machines, stars, and so on, rather than by radioactive substances.

When something has been exposed to radiation, we say that it has been **irradiated**. Although it absorbs the radiation, it does not itself become radioactive. Things only become radioactive if they absorb a radioactive substance. So you do not become radioactive if you absorb cosmic rays (which you do all the time). But you do become radioactive if you consume a radioactive substance – coffee, for example, contains measurable amounts of radioactive potassium.

QUESTIONS

12 Why are gamma rays undeflected in a magnetic field?

13 a Which type of radiation from a radioactive source is the most highly ionising?

b What does this tell you about how easily it is absorbed?

Safe handling

Knowing about the radiation produced by radioactive materials tells us how to handle them as safely as possible.

Radioactive sources should be stored in a container that will absorb as much as possible of the radiation coming from them. Lead is a good material for this as it is a strong absorber of all three types of radiation.

Figure **23.10** shows a storage box used for keeping radioactive sources in a school laboratory. Each source is kept in its own lead-lined compartment, and the whole box should be stored in a metal cabinet with a hazard warning sign.

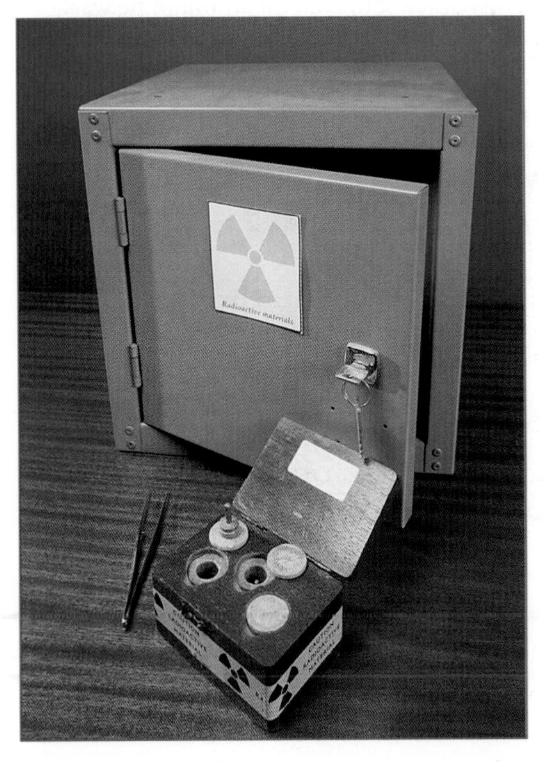

Figure 23.10 A storage box for laboratory radioactive sources, and the metal cupboard in which it is stored when not in use.

When sources are not in their protective container, they should be handled carefully. To avoid contamination, tongs can be used so that the user does not come into direct contact with the source. During any experiment, the user should stand at a safe distance from the source.

Activity 23.2 Safety first

Observe and explain how radioactive materials can be handled safely.

23.3 Radioactive decay

Henri Becquerel discovered the radioactivity of uranium. What surprised him was that uranium appears to be able to emit radiation endlessly, without ever running out of energy. This would go against the principle of conservation of energy. What he did not realise was that the uranium he used was undergoing very gradual **radioactive decay**. The problem was that uranium decays very slowly, so that, even if Becquerel had carried on with his experiments for a thousand years, he would not have noticed any decrease in the activity of his samples. In fact, the uranium he was working with had been decaying gradually ever since the Earth formed, over 4500 million years ago.

All radioactive substances decay with the same pattern, as shown in Figure **23.11a**. The graph shows that the amount of a radioactive substance decreases rapidly at first, and then more and more slowly. In fact, because the graph tails off more and more slowly, we cannot say when the last atoms will decay. Different radioactive substances decay at different rates, some much faster than others, as shown in Figure **23.11b**.

Because we cannot say when the substance will have entirely decayed, we have to think of another way of describing the rate of decay. As shown on the graph in Figure **23.11a**, we identify the **half-life** of the substance.

The half-life of a radioactive substance is the average time taken for half of the atoms in a sample to decay.

Uranium decays slowly because it has a very long half-life. The radioactive samples used in schools usually have half-lives of a few years, so that they have to be replaced once in a while. Some radioactive substances have half-lives that are less than a microsecond. No sooner are they formed than they decay into something else.

Explaining half-life

After one half-life, half of the atoms in a radioactive sample have decayed. However, this does not mean that all of the atoms will have decayed after two half-lives. From the graph of Figure **23.11a**, you can see that one-quarter will still remain after two half-lives. Why is this?

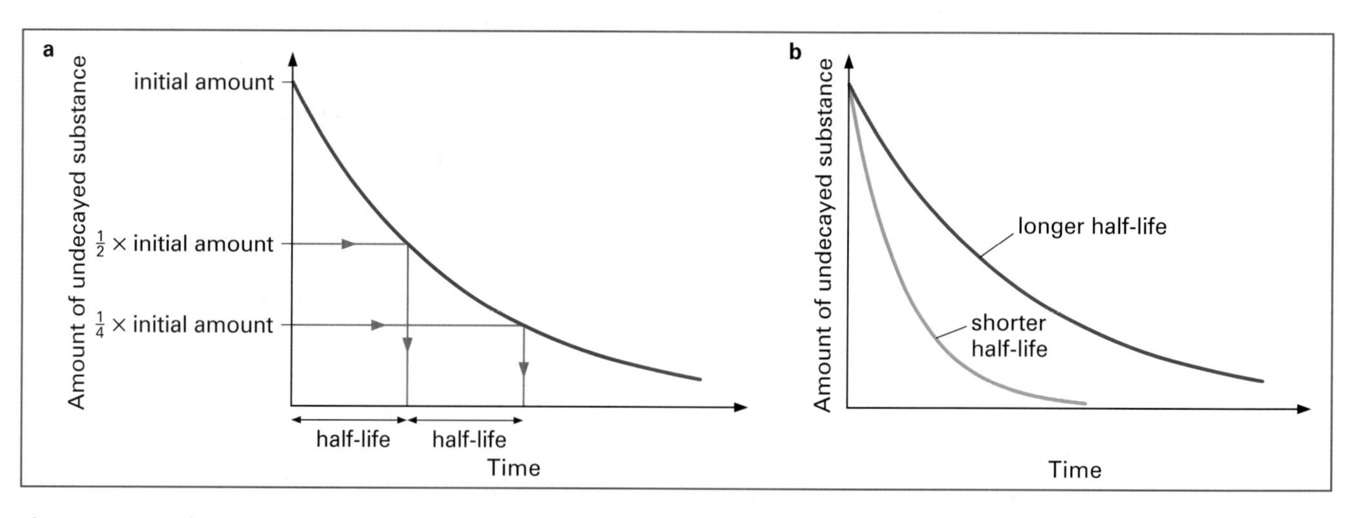

Figure 23.11 **a** A decay graph for a radioactive substance. A curve of this shape is known as an exponential decay graph. **b** A steep graph shows that a substance has a short half-life.

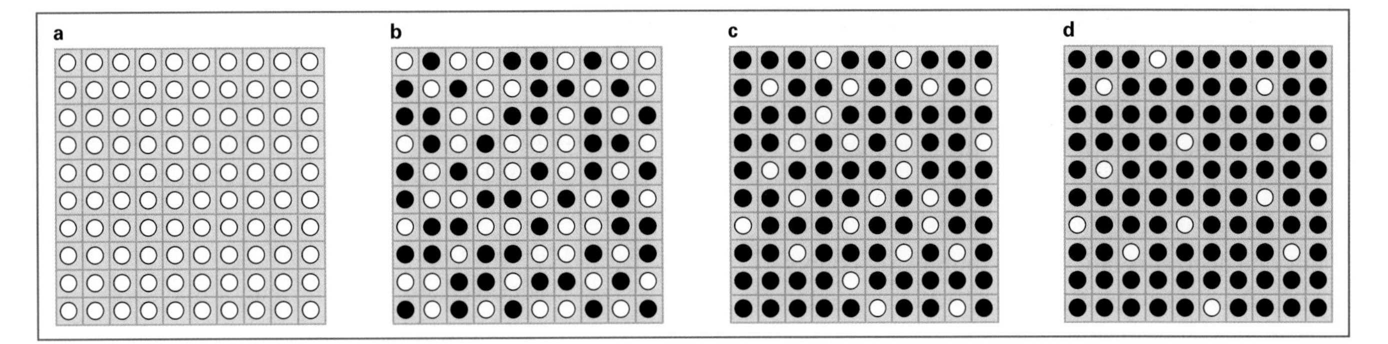

Figure 23.12 The pattern of radioactive decay comes about because the decay of individual atoms is random. Half of the atoms decay during each half-life, but we have no way of predicting which individual atoms decay.

Figure **23.12** shows one way of thinking about what is going on. We picture a sample of 100 undecayed atoms of a radioactive substance (white circles in Figure **23.12a**). They decay randomly – each one has a 50 : 50 chance of decaying in the course of one half-life.

a At the start, there are 100 undecayed atoms.

b After one half-life, a random selection of 50 atoms has decayed.

c During the next half-life, a random selection of half of the remaining 50 atoms decays, leaving 25 undecayed.

d During the third half-life, half of the remaining atoms decay, leaving 12 or 13. (Of course, you cannot have half an atom.)

So the number of undecayed atoms goes 100–50–25–12–... and so on. It is because radioactive atoms decay in a random fashion that we get this pattern of decay. Notice that, just because one atom has not decayed in the first half-life does not mean that it is more likely to decay in the next half-life. It has no way of remembering its past.

Usually, we cannot measure the numbers of atoms in a sample. Instead, we measure the **count rate** using a Geiger counter or some other detector. We might also determine the **activity** of a sample. This is the number of atoms that decay each second, and is measured in **becquerels (Bq)**. An activity of 1 Bq is one decay per second. The count rate and activity both decrease following the same pattern as the number of undecayed atoms.

Worked example 1

A sample of radioactive element X has an activity of 240 Bq. If the half-life of X is 3 years, what will its activity be after 12 years?

Step 1: Calculate the number of half-lives in 12 years.

$$\frac{12\text{ years}}{3\text{ years}} = 4\text{ half-lives}$$

Hence we want to know the activity of the sample after 4 half-lives.

Step 2: Calculate the activity after 1, 2, 3 and 4 half-lives (divide by 2 each time).

initial activity = 240 Bq
activity after 1 half-life = 120 Bq
activity after 2 half-lives = 60 Bq
activity after 3 half-lives = 30 Bq
activity after 4 half-lives = 15 Bq

So the activity of the sample has fallen to 15 Bq after 12 years.

(Another way to do this is as follows. We have found that 12 years is 4 half-lives, so we need to divide the initial activity by 2^4, which is 16, giving

$$\frac{240\text{ Bq}}{16} = 15\text{ Bq}$$

So the activity is 15 Bq after 12 years, as before.)

Measuring a half-life

Figure **23.13** shows how the half-life of a particular substance, protactinium-234, is measured in the lab. After the bottle has been shaken, the upper layer of liquid contains protactinium, which emits beta radiation as it decays. Because its half-life is 70 s, the count rate decreases quickly. The number of counts in successive intervals of 10 s is recorded. A graph is plotted of number of counts in each interval against time, as in Figure **23.14**. The half-life can then be deduced from the decay graph, as shown.

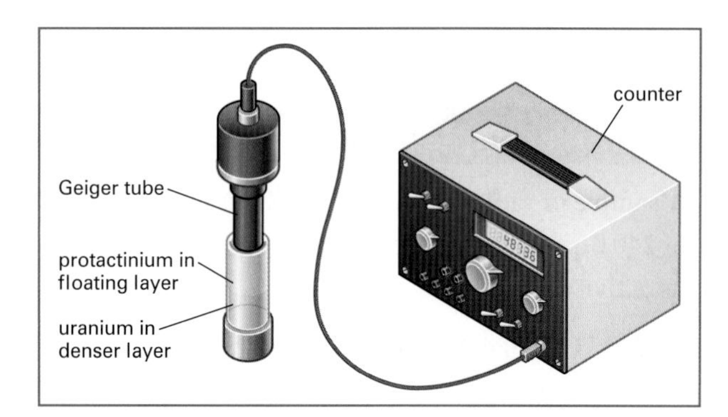

Figure 23.13 A practical arrangement for measuring the half-life of the radioactive decay of protactinium-234.

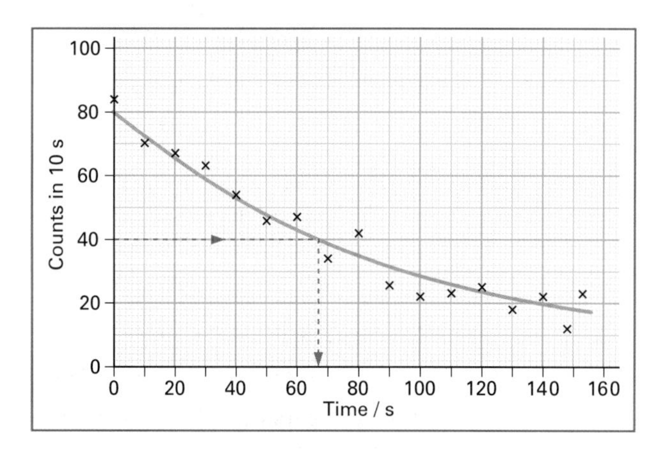

Figure 23.14 The count rate for the radioactive decay of protactinium-234 decreases rapidly. The points show some experimental scatter, so a smooth curve is drawn. From this, the half-life can be deduced. Here the initial count rate is 80. Half of this is 40. Reading across from this value to the curve and then down to the time axis gives the half-life as 67 s.

Activity 23.3 Radioactive decay

Solve some problems involving radioactive decay and half-life.

QUESTIONS

14 The half-life of a radioactive substance is the time taken for half of the atoms in a sample to decay. What word is missing from this definition of half-life?

15 A sample of a radioactive substance contains 200 undecayed atoms. How many will remain undecayed after 3 half-lives?

16 The half-life of radioisotope X is 10 days. A sample gives an initial count rate of 440 counts per second. What will be the count rate after 30 days?

17 Radioisotope Y has a half-life of 2000 years. How long will it take the activity of a sample of Y to decrease to one-eighth of its initial value?

23.4 Using radioisotopes

Any element comes in several forms or isotopes (see pages **245–6**). Some may be stable, but others are unstable – in other words, they are radioactive. For example, carbon has two stable isotopes ${}^{12}_{6}C$ and ${}^{13}_{6}C$), but ${}^{14}_{6}C$ is an unstable isotope. Unstable (radioactive) isotopes are known as **radioisotopes**.

Effects of radioisotopes on cells

Safe handling of radioisotopes requires an understanding of how radiation affects cells. There are three ways in which radiation can damage living cells.

- An intense dose of radiation causes a lot of ionisation in a cell, which can kill the cell. This is what happens when someone suffers radiation burns. The cells affected simply die, as if they had been burned. If the sufferer is lucky and receives suitable treatment, the tissue may regrow.
- If the DNA in the cell nucleus is damaged, the mechanisms that control the cell may break down. The cell may divide uncontrollably, and a tumour forms. This is how radiation can cause cancer.
- If the affected cell is a gamete (a sperm or egg cell), the damaged DNA of its genes may be passed on to future generations. This is how radiation can produce genetic mutations. Occasionally, a mutation can be beneficial to the offspring, but

more usually it is harmful. A fertilised egg cell may not develop at all, or the baby may have some form of genetic disorder.

We are least likely to be harmed by alpha radiation coming from a source outside our bodies. This is because the radiation is entirely absorbed by the layer of dead skin cells on the outside of our bodies (and by our clothes). However, if an alpha source gets inside us, it can be very damaging, because its radiation is highly ionising. That is why radon and thoron gases are so dangerous. We breathe them into our lungs, where they irradiate us from the inside. The result may be lung cancer.

Today, we know more about radiation and the safe handling of radioactive materials than ever before. Knowing how to reduce the hazards of radiation means that we can learn to live safely with it and put it to many worthwhile purposes.

Radioisotopes at work

Now we will look at some of the many uses to which radioisotopes have been put. We will look at these uses in four separate groups:

- uses related to their different penetrating powers
- uses related to the damage their radiation causes to living cells
- uses related to the fact that we can detect tiny quantities of radioactive substances
- uses related to radioactive decay and half-life.

Uses related to penetrating power

Smoke detectors

These are often found in domestic kitchens, and in public buildings such as offices and hotels. If you open a smoke detector to replace the battery, you may see a yellow and black radiation hazard warning sign (Figure **23.15a**). The radioactive material used is americium-241, a source of alpha radiation. Figure **23.15b** shows how it works.

- Radiation from the source falls on a detector. Since alpha radiation is charged, a small current flows in the detector. The output from the processing circuit is OFF, so the alarm is silent.
- When smoke enters the gap between the source and the detector, it absorbs the alpha radiation. Now no current flows in the detector, and the processing circuit switches ON, sounding the alarm.

In this application, a source of alpha radiation is chosen because alpha radiation is easily absorbed by the smoke particles.

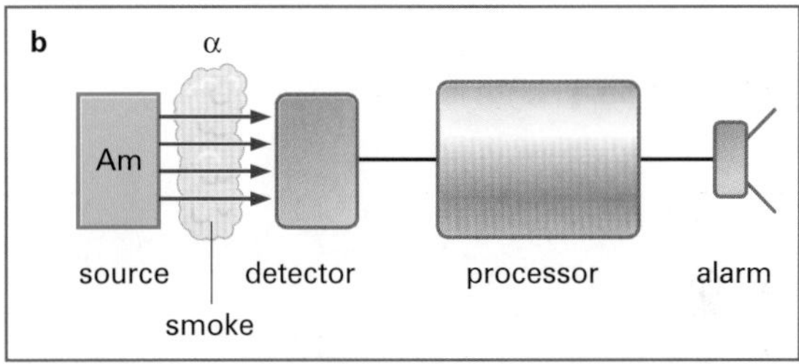

Figure 23.15 a The inside of a smoke detector. The source of radiation is a small amount of americium-241. **b** Block diagram of a smoke detector. The alarm sounds when smoke absorbs the alpha radiation.

Thickness measurements

In industry, beta radiation is often used in measuring thickness. Manufacturers of paper need to be sure that their product is of a uniform thickness. To do this, beta radiation is directed through the paper as it comes off the production line. A detector measures the amount of radiation getting through. If the paper is too thick, the radiation level will be low and an automatic control system adjusts the thickness. The same technique is used in the manufacture of plastic sheeting.

Beta radiation is used in this application because alpha radiation would be entirely absorbed by the paper or plastic. Gamma radiation would hardly be affected, because it is the most penetrating.

Medical diagnosis

The diagnosis of some diseases may be carried out using a source of gamma radiation. The patient is injected with a radioactive chemical that targets the problem area (it may accumulate in bone, for example). Then a camera detects the radiation coming from the chemical and gives an image of the tissue under investigation.

Fault detection

Fault detection in manufactured goods sometimes makes use of gamma rays. Figure **23.16** shows an example, where engineers are looking for any faults in some pipework. A photographic film is strapped to the outside of the pipe, and the radioactive source is placed on the inside. When the film is developed, it looks like an X-ray picture, and shows any faults in the welding.

Figure 23.16 Checking for faults in a metal pipe. The engineers are checking that no radiation is escaping from the pipe. The gamma source is stored in the black box in the foreground, but can be pushed through the long tube along the pipe to reach the part to be checked.

Uses related to cell damage

Radiation therapy

The patient shown earlier in Figure **23.3** is receiving radiation treatment as part of a cure for cancer. A source of gamma rays (or X-rays) is directed at the tumour that is to be destroyed. The source moves around the patient, always aiming at the tumour. In this way, other tissues receive only a small dose of radiation. Radiation therapy is often combined with chemotherapy, using chemical drugs to target and kill the cancerous cells.

Food irradiation

This is a way of preserving food. Food often decays because of the action of microbes. These can be killed using intense gamma rays. Because these organisms are single-celled, any cell damage kills the entire organism. Different countries permit different foods to be irradiated. The sterile food that results has been used on space missions (where long life is important) and for some hospital patients whose resistance to infection by microbes may be low.

Sterilisation

Sterilisation of medical products works in the same way as food irradiation. Syringes, scalpels and other instruments are sealed in plastic bags and then exposed to gamma radiation. Any microbes present are killed so that, when the packaging is opened, the item can be guaranteed to be sterile. The same technique is used to sterilise sanitary towels and tampons.

Uses related to detectability

Radioactive tracing

Every time you hear a Geiger counter click, it has detected the radioactive decay of a single atom. This means that we can use radiation to detect tiny quantities of substances, far smaller than can be detected by chemical means. Such techniques are often known as **radioactive tracing**.

Engineers may want to trace underground water flow, for example. They may be constructing a new waste dump, and they need to be sure that poisonous water from the dump will not flow into the local water supply. Under high pressure, they inject water containing a radioactive chemical into a hole in the ground (Figure **23.17**). Then they monitor how it moves through underground cracks using gamma detectors at ground level.

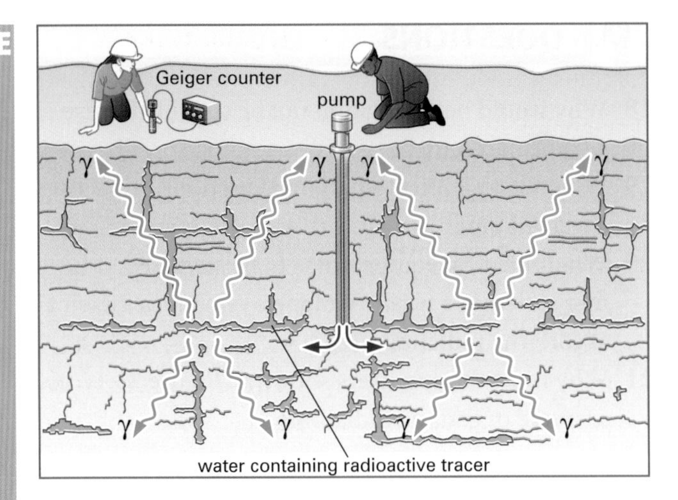

Figure 23.17 Detecting the movement of underground water. Engineers need to know how water will move underground. This can also affect the stability of buildings on the site. Water containing a source of gamma radiation is pumped underground and its passage through cracks is monitored at ground level.

Radioactive labelling and genetic fingerprinting

Biochemists use radioactively labelled chemicals to monitor chemical reactions. The chemicals bond to particular parts of the molecules of interest, so that they can be tracked throughout a complicated sequence of reactions. The same technique is used to show up the pattern of a genetic fingerprint (Figure **23.18**).

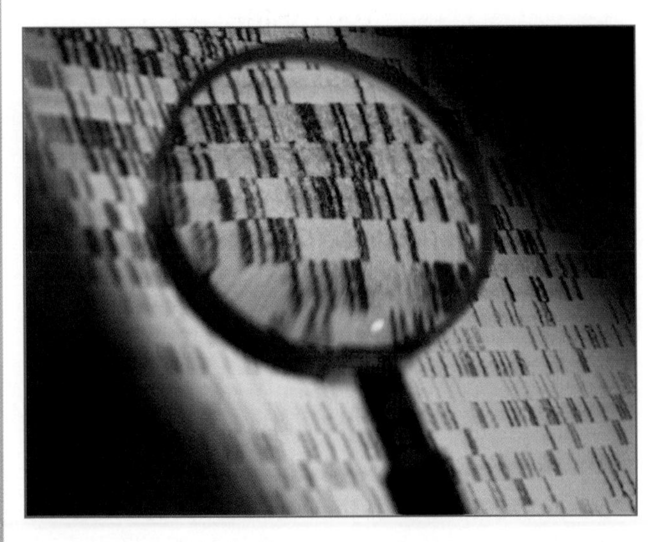

Figure 23.18 A DNA (genetic) fingerprint appears as a series of bands. Each band comes from a fragment of DNA labelled with a radioactive chemical. The bands (and thus particular DNA fragments) show up on a photographic film.

Half-life and dating

Here is another application of radioisotopes. Because radioactive substances decay at a rate we can determine, we can use them to discover how old objects and materials are. The best-known example of this is **radiocarbon dating**.

All living things contain carbon. Plants get this from atmospheric carbon dioxide, which they use in photosynthesis. Plant-eating animals get it from the plants they eat to build their bodies. Meat-eating animals get it from their prey. Most carbon is carbon-12 ($^{12}_{6}C$), which is not radioactive. A tiny fraction is radioactive carbon-14 ($^{14}_{6}C$), with a half-life of 5370 years. (It emits beta radiation.)

The idea of radiocarbon dating is this. When a living organism dies, the carbon-14 in its body decays. As time passes, the amount remaining decreases. If we can measure the amount remaining, we can work out when the organism was alive.

There are two ways to measure the amount of carbon-14 present in an object:

- by measuring the activity of the sample using a detector such as a Geiger counter
- by counting the number of carbon-14 atoms using a mass spectrometer.

The Turin shroud (Figure **23.19**) was famously dated in 1988 using a mass spectrometer. (This is a large machine that uses magnetic fields to separate atoms according to their mass and charge.) The shroud was dated to 1325 ± 33 AD, showing that it did not date from biblical times.

Problems can arise with radiocarbon dating. It may be that the amount of carbon-14 present in the atmosphere was different in the past. Certainly, nuclear weapons testing added extra carbon-14 to the atmosphere during the 1950s and 1960s. This means that living objects that died then have an excess of carbon-14, making them appear younger than they really are.

Geologists use a radioactive dating technique to find the age of some rocks. Many rocks contain a

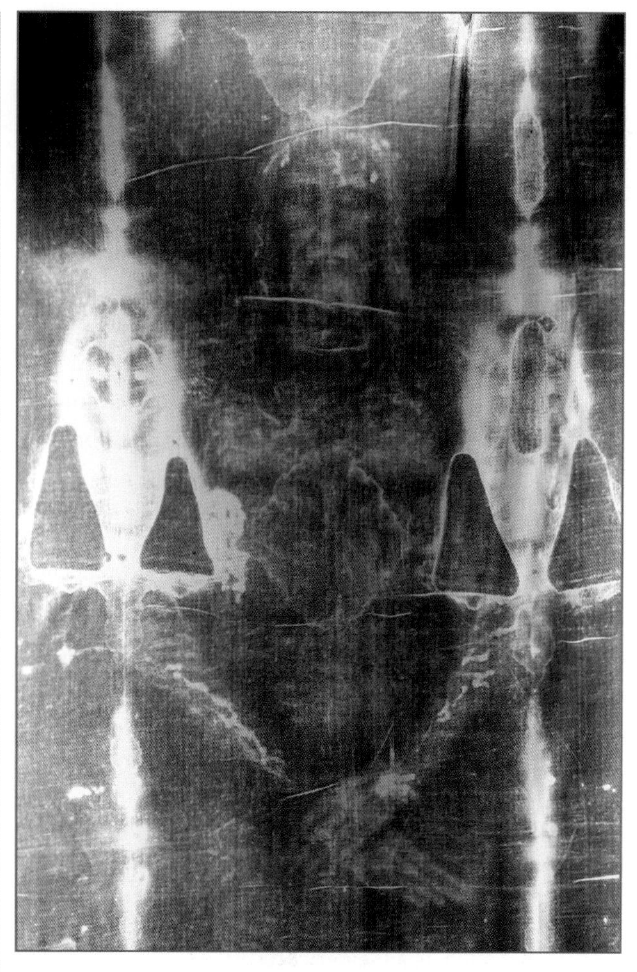

Figure 23.19 The Turin shroud was dated by radiocarbon dating. It was found to date from the 14th century, which matched the dates of the earliest historical records of its existence.

radioactive isotope, potassium-40 ($^{40}_{19}K$), which decays by beta emission to a stable isotope of argon ($^{40}_{18}Ar$). Argon is a gas, and it is trapped in the rock as the potassium decays. Here is how the dating system works.

The rocks of interest form from molten material (for example, in a volcano). There is no argon in the molten rock because it can bubble out. After the rock solidifies, the amount of trapped argon gradually increases as the potassium decays. Geologists take a sample and measure the relative amounts of argon and potassium. The greater the proportion of argon, the older the rock must be.

QUESTIONS

18 Why would beta radiation not be suitable for use in a smoke detector?

19 Why must gamma radiation be used for inspecting a welded pipe?

20 When medical equipment is to be sterilised, it is first sealed in a plastic wrapper. Why does this not absorb the radiation used?

21 Why must the engineers shown in Figure **23.17** use a source of gamma radiation?

Summary

We are constantly exposed to background radiation from a variety of sources. We are also exposed to radiation from artificial sources.

Radiation can be detected using photographic film, Geiger counters and other detectors.

Naturally occurring radioactive substances produce alpha (α) and beta (β) particles, and gamma (γ) rays (a form of electromagnetic radiation).

The radioactive decay of a radioisotope may be represented by a balanced equation. In radioactive decay, nucleon number and proton number are both conserved.

The radiations from radioactive substances causes ionisation in the materials they pass through, and so they are known as ionising radiations.

The half-life of a radioactive substance is the average time taken for half of the atoms in a sample to decay.

Radioactive substances have many uses. Suitable substances are selected according to the penetrating power of their radiation, the effect of their radiation on cells, their detectability (as tracers) and their half-life.

End-of-chapter questions

Radiation	Symbol	Type of particle or electromagnetic radiation	Mass	Charge
alpha	......	......	......	......
beta	......	......	......	......
gamma	......	......	......	......

Table 23.2 For Question **23.1**.

23.1 Alpha, beta and gamma radiations are three types of radiation produced by radioactive substances. Copy and complete Table **23.2** to show the nature of these radiations. [12]

23.2 A school has two radioactive sources for use in physics experiments. One is a source of α radiation, and the other is a source of β radiation. They have lost their labels, and the teacher wants to check which is which. Use your knowledge of the different penetrating powers of these radiations to suggest how this might be done. [4]

23.3 In an experiment to determine the half-life of a radioisotope, the graph shown in Figure **23.20** was obtained.

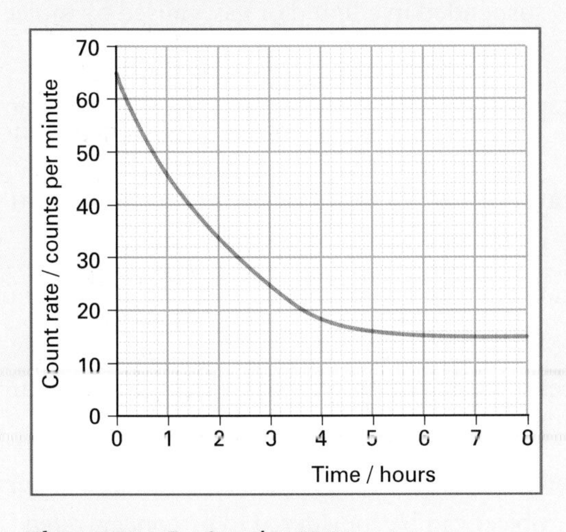

Figure 23.20 For Question **23.3**.

a From the graph, what was the background count rate? [2]

b What was the initial count rate from the sample (that is, disregarding the background count rate)? [2]

c From the graph, deduce the half-life of the radioisotope. Draw a sketch graph to illustrate your method. [3]

23.4 Alpha, beta and gamma radiations are produced by radioactive substances. They are sometimes described as **ionising radiations.**

a Explain what is meant by the term **ionisation.** [2]

b Name another type of ionising radiation. [1]

E

23.5 The radiation produced by radioactive substances has many uses.

a Describe a use of γ radiation that makes use of its ability to damage living tissues. [3]

b Describe a use of β radiation that makes use of the fact that it is absorbed by a few millimetres of solid matter. [3]

Glossary

a.c. generator a device, such as a dynamo, used to generate alternating current (a.c.)

acceleration the rate of change of an object's velocity

acceleration due to gravity the acceleration of an object falling freely under gravity

acceleration of free fall *see* acceleration due to gravity

activity the rate at which nuclei decay in a sample of a radioactive substance

air resistance the frictional force on an object moving through air

alpha decay the decay of a radioactive nucleus by emission of an α-particle

alpha particle (α-particle) a particle of two protons and two neutrons emitted by an atomic nucleus during radioactive decay

alternating current (a.c.) electric current that flows first one way, then the other, in a circuit

ammeter a meter for measuring electric current

amp, ampere (A) the SI unit of electric current

amplitude the greatest height of a wave above its undisturbed level

angle of incidence the angle between an incident ray and the normal to the surface at the point where it meets a surface

angle of reflection the angle between a reflected ray and the normal to the surface at the point where it reflects from a surface

angle of refraction the angle between a refracted ray and the normal to the surface at the point where it passes from one material to another

average speed speed calculated from total distance travelled divided by total time taken

axis the line passing through the centre of a lens, perpendicular to its surface

background radiation the radiation from the environment to which we are exposed all the time

barometer an instrument used to measure atmospheric pressure

battery two or more electrical cells connected together in series; the word may also be used to mean a single cell

becquerel (Bq) the SI unit of activity; 1 Bq = one decay per second

beta decay the decay of a radioactive nucleus by emission of a beta particle

beta particle (β-particle) a particle (an electron) emitted by an atomic nucleus during radioactive decay

biomass fuel a material, recently living, used as a fuel

boiling point the temperature at which a liquid changes to a gas (at constant pressure)

Boyle's law the law that relates the pressure and volume of a fixed mass of gas (pV = constant at constant temperature)

Brownian motion the motion of small particles suspended in a liquid or gas, caused by molecular bombardment

capacitor a device used for storing energy in an electric circuit

cathode ray a ray of electrons travelling from cathode to anode in a vacuum tube

cathode-ray oscilloscope a device used to show the pattern of a varying voltage

cell a device that provides a voltage in a circuit by means of a chemical reaction

centre of mass the point at which the mass of an object can be considered to be concentrated

charge *see* electrostatic charge

chemical energy energy stored in chemical substances and which can be released in a chemical reaction

circuit breaker a safety device that automatically switches off a circuit when the current becomes too high

commutator a device used to allow current to flow to and from the coil of a d.c. motor or generator

compression a region of a sound wave where the particles are pushed close together

conduction the transfer of heat energy or electrical energy through a material without the material itself moving

conductor a substance that transmits heat or allows an electric current to pass through it

contaminated when an object has acquired some unwanted radioactive substance

convection the transfer of heat energy through a material by movement of the material itself

converging lens a lens that causes rays of light parallel to the axis to converge at the principal focus

corkscrew rule the rule used to determine the direction of the magnetic field around an electric current

coulomb (C) the SI unit of electric charge; 1 C = 1 A s

count rate the number of decaying radioactive atoms detected each second (or minute, or hour)

crest the highest point of a wave

critical angle the minimum angle of incidence at which total internal reflection occurs

current the rate at which electric charge flows in a circuit

demagnetisation destroying the magnetisation of a piece of material

density the ratio of mass to volume for a substance

diffraction when a wave spreads out as it travels through a gap or past the edge of an object

diode an electrical component that allows electric current to flow in one direction only

direct current (d.c.) electric current that flows in the same direction all the time

dispersion the separation of different wavelengths of light because they are refracted through different angles

diverging lens a lens that causes rays of light parallel to the axis to diverge from the principal focus

doing work transferring energy by means of a force

drag the frictional force when an object moves through a fluid (a liquid or a gas)

dynamo effect electricity is generated when a coil moves near a magnet

earthed when the case of an electrical appliance is connected to the earth wire (for safety)

efficiency the fraction of energy that is converted into a useful form

electric field a region of space in which an electric charge will feel a force

electrical energy energy transferred by an electric current

electrical resistance *see* resistance

electromagnet a coil of wire that, when a current flows in it, becomes a magnet

electromagnetic radiation energy travelling in the form of waves

electromagnetic spectrum the family of radiations similar to light

electron a negatively charged particle, smaller than an atom

electron charge the electric charge of a single electron; -1.6×10^{-19} C

electrostatic charge a property of an object that causes it to attract or repel other objects with charge

e.m.f. (electro-motive force) the voltage across the terminals of a source of electrical energy (for example, a cell or power supply)

energy the capacity to do work

equilibrium when no net force and no net moment act on a body

evaporation when a liquid changes to a gas at a temperature below its boiling point

extension the increase in length of a spring when a load is attached

Fleming's left-hand rule a rule that gives the relationship between the directions of force, field and current when a current flows across a magnetic field

focal length the distance from the centre of a lens to its principal focus

focal point *see* principal focus

fossil fuel a material, formed from long-dead material, used as a fuel

frequency the number of vibrations or waves per second

friction the force that acts when two surfaces rub over one another

fuse a device used to prevent excessive currents flowing in a circuit

gamma ray (γ-ray) electromagnetic radiation emitted by an atomic nucleus during radioactive decay

geothermal energy the energy stored in hot rocks underground

gravitational potential energy (g.p.e.) the energy of an object raised up against the force of gravity

gravity the force that exists between any two objects with mass

half-life the average time taken for half the atoms in a sample of a radioactive material to decay

hard a material that, once magnetised, is difficult to demagnetise

Hooke's law the extension of an object is proportional to the load producing it, provided that the limit of proportionality is not exceeded

image what we see when we view an object by means of reflected or refracted rays

incident ray a ray of light striking a surface

induction a method of giving an object an electric charge without making contact with another charged object

infrared radiation electromagnetic radiation whose wavelength is greater than that of visible light; sometimes known as heat radiation

infrasound sound waves whose frequency is so low that they cannot be heard

insulator a substance that transmits heat very poorly or does not conduct electricity

internal energy the energy of an object; the total kinetic and potential energies of its particles

interrupt card a piece of card that breaks the light beam of a light gate

ionisation when a particle (atom or molecule) becomes electrically charged by losing or gaining electrons

ionising radiation radiation, for example from radioactive substances, that causes ionisation

irradiated when an object has been exposed to radiation

isotope isotopes of an element have the same proton number but different nucleon numbers

joule (J) the SI unit of work or energy

kinetic energy (k.e.) the energy of a moving object

kinetic model of matter a model in which matter consists of molecules in motion

lamina a flat object of uniform thickness

laser a device for producing a narrow beam of light of a single colour or wavelength

latent heat the energy needed to melt or boil a material

law of reflection the law relating the angle of incidence of a light ray to the angle of reflection ($i = r$)

light-dependent resistor (LDR) a device whose resistance decreases when light shines on it

light-emitting diode (LED) a type of diode that emits light when a current flows through it

light energy energy emitted in the form of visible radiation

light gate a device for recording the passage of a moving object when it breaks a light beam

limit of proportionality the point beyond which the extension of an object is no longer proportional to the load producing it

load a force that causes a spring to extend

logic gate an electronic component whose output voltage depends on the input voltage(s)

longitudinal wave a wave in which the vibration is forward and back, along the direction in which the wave is travelling

magnetic field the region of space around a magnet or electric current in which a magnet will feel a force

magnetisation causing a piece of material to be magnetised; a material is magnetised when it produces a magnetic field around itself

manometer a device used to measure the pressure difference between two points

mass the property of an object that causes it to have a gravitational attraction for other objects, and that causes it to resist changes in its motion

melting point the temperature at which a solid melts to become a liquid

model a way of representing a system in order to understand its functioning; usually mathematical

moment the turning effect of a force about a point, given by force × perpendicular distance from point

monochromatic describes a ray of light (or other electromagnetic radiation) of a single wavelength

national grid the system of power lines, pylons and transformers used to carry electricity around a country

negative charge one type of electric charge

neutral having no overall positive or negative electric charge

neutron an electrically neutral particle found in the atomic nucleus

neutron number (*N*) the number of neutrons in the nucleus of an atom

newton (N) the SI unit of force

non-renewable energy resource which, once used, is gone forever

normal the line drawn at right angles to a surface at the point where a ray strikes the surface

nuclear energy energy stored in the nucleus of an atom

nuclear fission the process by which energy is released by the splitting of a large heavy nucleus into two or more lighter nuclei

nuclear fusion the process by which energy is released by the joining together of two small light nuclei to form a new heavier nucleus

nucleon a particle found in the atomic nucleus: a proton or a neutron

nucleon number (*A*) the number of protons and neutrons in an atomic nucleus

nuclide a 'species' of nucleus having particular values of proton number and nucleon number

ohm (Ω) the SI unit of electrical resistance; $1\,\Omega = 1\,V/A$

pascal (Pa) the SI unit of pressure; $1\,Pa = 1\,N/m^2$

p.d. (potential difference) another name for the voltage between two points

penetrating power how far radiation can penetrate into different materials

period the time for one complete oscillation of a pendulum, one complete vibration or the passage of one complete wave

photocell *see* solar cell

pitch how high or low a note sounds

pivot the fixed point about which a lever turns

positive charge one type of electric charge

potential divider a part of a circuit consisting of two resistors connected in series

power the rate at which energy is transferred or work is done

power lines cables used to carry electricity from power stations to consumers

pressure the force acting per unit area at right angles to a surface

principal focus the point at which rays of light parallel to the axis converge after passing through a converging lens

principle of conservation of energy the total energy of interacting objects is constant provided no external force acts

proton a positively charged particle found in the atomic nucleus

proton charge the electric charge of a single proton; $+1.6 \times 10^{-19}$ C

proton number (Z) the number of protons in an atomic nucleus

radiation energy spreading out from a source carried by particles or waves

radioactive decay the decay of a radioactive substance when its atomic nuclei emit radiation

radioactive substance a substance that decays by emitting radiation from its atomic nuclei

radioactive tracing a technique that uses a radioactive substance to trace the flow of liquid or gas, or to find the position of cancerous tissue in the body

radiocarbon dating a technique that uses the known rate of decay of radioactive carbon-14 to find the approximate age of an object made from dead organic material

radioisotope a radioactive isotope of an element

random process a process that happens at a random rate rather than at a steady rate; in radioactive decay, it is impossible to predict which atom will be the next to decay, or when a given atom will decay

rarefaction a region of a sound wave where the particles are further apart

ray diagram a diagram showing the paths of typical rays of light

real image an image that can be formed on screen

rectifier an electric circuit in which one or more diodes are used to convert alternating current to direct current

reflected ray a ray of light that has been reflected after striking a surface

reflection the change in direction of a ray of light when it strikes a surface without passing through it

refracted ray a ray of light that has changed direction on passing from one material to another

refraction the bending of a ray of light on passing from one material to another

refractive index the property of a material that determines the extent to which it causes rays of light to be refracted

relay an electromagnetically operated switch

renewable energy resource which, when used, will be replenished naturally

residual-current device (RCD) a device used to protect the user in case of an electrical fault

resistance a measure of the difficulty of making an electric current flow through a device or a component in a circuit

resistor a component in an electric circuit whose resistance reduces the current flowing

resultant force the single force that has the same effect on a body as two or more forces

ripple a small, uniform wave on the surface of water

scalar quantity a quantity that has only magnitude

slip rings a device used to allow current to flow to and from the coil of an a.c. motor or generator

Snell's law the law that relates the angles of incidence and refraction: refractive index $= \frac{\sin i}{\sin r}$

soft describes a material that, once magnetised, can easily be demagnetised

solar cell an electrical device that transfers the energy of sunlight directly to electricity, by producing a voltage when light falls on it

solenoid a coil of wire that becomes magnetised when a current flows through it

sound energy energy being transferred in the form of sound waves

sound wave a wave that carries sound from place to place

specific heat capacity (s.h.c.) a measure of how much thermal (heat) energy a material can hold

specific latent heat the energy required to melt or boil 1 kg of a substance

spectrum waves, or colours of light, separated out in order according to their wavelengths

speed the distance travelled by an object in unit time

speed of light the speed at which light travels (usually in a vacuum: 3.0×10^8 m/s)

static electricity electric charge held by a charged insulator

strain energy energy of an object due to its having been stretched or compressed

temperature a measure of how hot or cold something is

terminal velocity the greatest speed reached by an object when moving through a fluid

thermal (heat) energy energy being transferred from a hotter place to a colder place because of the temperature difference between them

thermal equilibrium describes the state of two objects (or an object and its surroundings) that are at the same temperature so that there is no heat flow between them

thermal expansion the expansion of a material when its temperature rises

thermionic emission the process by which cathode rays (electrons) are released from the heated cathode of a cathode-ray tube

thermistor a resistor whose resistance changes a lot over a small temperature range

thermocouple an electrical device made of two different metals, used as an electrical thermometer

total internal reflection (TIR) when a ray of light strikes the inner surface of a solid material and 100% of the light reflects back inside it

transducer any device that converts energy from one form to another

transformer a device used to change the voltage of an a.c. electricity supply

transistor an electrical component that can act as an electronic switch

transverse wave a wave in which the vibration is at right angles to the direction in which the wave is travelling

trip switch a device used to protect an electric circuit in case of an electrical fault

trough the lowest point on a wave

truth table a way of summarising the operation of a combination of logic gates

turbine a device that is caused to turn by moving air, steam or water, often used to generate electricity

ultrasound sound waves whose frequency is so high that they cannot be heard

ultraviolet radiation electromagnetic radiation whose frequency is higher than that of visible light

upper limit of hearing the highest frequency of sound that a person can just hear

variable resistor a resistor whose resistance can be changed, for example by turning a knob

vector quantity a quantity that has both magnitude and direction

vector triangle a method for finding the vector sum of two vector quantities

velocity the speed of an object in a stated direction

virtual image an image that cannot be formed on a screen; formed when rays of light appear to be spreading out from a point

volt (V) the SI unit of voltage (p.d. or e.m.f.); 1 V = 1 J/C

voltage the 'push' of a battery or power supply in a circuit

voltmeter a meter for measuring the p.d. (voltage) between two points

wave speed the speed at which a wave travels

wavefront a line joining adjacent points on a wave that are all in step with each other

wavelength the distance between adjacent crests (or troughs) of a wave

weight the downward force of gravity that acts on an object because of its mass

work done the amount of energy transferred when one body exerts a force on another; the energy transferred by a force when it moves

Index